CALCIUM AND CELLULAR FUNCTION

BIOLOGICAL COUNCIL

The Co-ordinating Committee for Symposia
on Drug Action

A Symposium on

CALCIUM AND CELLULAR FUNCTION

Edited by

A. W. Cuthbert

Lecturer in Pharmacology
Cambridge University

Palgrave Macmillan

First published 1970

Published by
MACMILLAN AND CO LTD

London and Basingstoke, Great Britain
Bombay Calcutta Madras Hong Kong and Singapore
Macmillan South Africa (Publishers)
The Macmillan Company of Australia
The Macmillan Company of Canada
St Martin's Press New York

SBN 333 11176 1

ISBN 978-1-349-00907-7   ISBN 978-1-349-00905-3 (eBook)
DOI 10.1007/978-1-349-00905-3

FOREWORD

At first sight it might seem surprising that an inorganic ion should be chosen as the subject for an inter-society symposium on the mechanism of drug action. Calcium, however, has a unique position in its relation with excitable tissues as was made very clear in Sidney Ringer's classical studies. The papers in this symposium illustrate how calcium, because of its importance in an extraordinary range of biological activities, well merits classification as a drug, and indeed as a drug with very complex actions. The interpretation of complicated biological systems is not easy. It is difficult to ask the critical question, and this is especially so in pharmacology where complexity exists in the drug as well as in the responsive systems. In the case of calcium one is confronted with the problem of why calcium rather than its closely related ions should have such interesting properties. What is it that sets calcium apart in this way? The striking advances in inorganic chemistry in recent years should make it possible to answer this question.

Much of the present intense interest in calcium stems from the observation of Heilbrunn and Wierczinski that a small amount of calcium injected into a muscle fibre caused it to contract. This in turn has led to the discovery that the level of ionic calcium in cell cytoplasm is extremely low and is held low by an efficient calcium pump and segregation mechanism. The full implications of the low intracellular calcium as far as cellular functions are concerned is still not wholly clear. However, the low calcium concentration relative to that in the extracellular fluid and in the sarcoplasmic reticulum may represent a poised system with important controlling functions in addition to effects on the contractile machinery. This appears to be fertile locus for drug actions.

Despite the general interest in calcium, it appears that this symposium is the first of its kind to be held, and this volume is the most comprehensive collection of essays on the biology of this ion that has so far appeared. We hope that it will be useful and will stimulate further work.

A. S. V. BURGEN
Chairman of the Organizing Committee

EDITOR'S ACKNOWLEDGEMENTS

The symposium was organized by a Committee consisting of A. S. V. Burgen (Chairman), A. W. Cuthbert (Secretary), P. F. Baker, H. O. Schild, and L. Wolpert.

The following societies helped to sponsor the symposium: Biochemical Society, British Pharmacological Society, British Society for Cell Biology, Chemical Society, Nutrition Society, Pharmaceutical Society of Great Britain, Physiological Society, and Society for Drug Research.

The meeting was held on March 24–25, 1969, in the Mechanical Engineering Department, Imperial College, London. It was attended by some 370 members mainly from societies sponsoring the symposium. A considerable number of the equivalent European and North American societies were represented at the symposium.

Acknowledgement is made to the Wellcome Trust for their generosity in providing a grant which enabled us to invite speakers from many distant parts of the world.

The editor is grateful to the publishers of journals for permission to reproduce some of their figures, and to Macmillan & Co. Ltd. for the courteous and efficient way in which they have dealt with this publication.

A. W. CUTHBERT

Cambridge
September, 1969

CONTENTS

Session I: Chemistry of the Alkaline Earth Metals

Chairman: A. S. V. BURGEN

Page

The simple chemistry of calcium and its relevance to biological systems   R. D. GILLARD   3

Calcium chelation and buffers   P. C. CALDWELL   10

Binding of calcium to phospholipids
R. M. C. DAWSON and H. HAUSER   17

Aequorin-calcium luminescence and its application to muscle physiology   C. C. ASHLEY and E. B. RIDGWAY   42

Discussion BAKER, GOODFORD, HAGUE, COWAN, BIRKS   54

Session II: Calcium and Cell Surfaces

Chairman: L. WOLPERT

Divalent cations and cell adhesion
D. GINGELL, D. R. GARROD and J. F. PALMER   59

Calcium and cellular function after freezing and thawing
J. FARRANT   65

Discussion JONES, GILLARD   72

Session III: Calcium Permeability and Transport

Chairman: D. H. SMYTH

Activation and inhibition of the sarcoplasmic calcium transport
W. HASSELBACH, M. MAKINOSE and W. FIEHN   75

Transmembrane calcium movements in resealed human red cells
H. J. SCHATZMANN   85

Sodium–calcium exchange across the nerve cell membrane

P. F. BAKER   96

Active transport of calcium by intestine: studies with a calcium activity electrode

D. SCHACHTER, S. KOWARSKI and PHYLLIS REID   108

Discussion REUTER, LOEWENSTEIN, GOODFORD, BIRKS, WOODIN, HODGKIN, CALDWELL, NORMAN, BREWER, RASMUSSEN, ROBINSON   124

Session IV: Calcium and Cellular Function

Chairmen: N. HALES and P. F. BAKER

Role of calcium ions in neuromuscular transmission

R. RAHAMIMOFF   131

Involvement of calcium in the secretion of catecholamines

P. BANKS   148

Calcium and hormone release   E. K. MATTHEWS   163

Site of protein secretion and calcium accumulation in the polymorphonuclear leucocyte treated with leucocidin

A. M. WOODIN and ANTOINETTE A. WEINEKE   183

Hormones, cell calcium and cyclic AMP

H. RASMUSSEN and N. NAGATA   198

The S-S polypeptide receptor as a metal receptor   H. O. SCHILD   214

Interaction of local anaesthetics and calcium with erythrocyte membranes   J. C. METCALFE   219

Caffeine, calcium and the activation of contraction

H. C. LÜTTGAU   241

Calcium and the action potential in smooth muscle

EDITH BÜLBRING and T. TOMITA   249

Kinetic aspects of calcium current in ventricular myocardial fibres

H. REUTER   261

Cardioactive steroids with special reference to calcium

M. REITER   270

Discussion MONGAR, JONES, BAKER, VRBOVA, CALDWELL, RÜDEL, HODGKIN, BOUMAN   280

LIST OF PARTICIPANTS

(This list was correct at the time of the Symposium, March 1969)

C. ASHLEY	Department of Zoology, University of Bristol
P. F. BAKER	Physiological Laboratory, University of Cambridge
P. BANKS	Department of Biochemistry, University of Sheffield
R. I. BIRKS	Department of Physiology, McGill University, Montreal
L. N. BOUMAN	Department of Physiology, University of Amsterdam
J. E. BREWER	Department of Zoology, University of Southampton
E. BÜLBRING	Department of Pharmacology, University of Oxford
A. S. V. BURGEN	Department of Pharmacology, University of Cambridge
P. C. CALDWELL	Department of Zoology, University of Bristol
J. B. CHAPPELL	Department of Biochemistry, University of Bristol
S. L. COWAN	Exposure Trials Station, Ministry of Defence (Navy Department), Portsmouth
R. M. C. DAWSON	Biochemistry Department, A.R.C. Institute of Animal Physiology, Babraham, Cambridge
J. FARRANT	Clinical Research Centre Laboratories, National Institute for Medical Research, Mill Hill, London
D. GARROD	Department of Biology as Applied to Medicine, Middlesex Hospital Medical School, London
R. D. GILLARD	University Chemical Laboratory, University of Kent, Canterbury
D. GINGELL	Department of Biology as Applied to Medicine, Middlesex Hospital Medical School, London
P. J. GOODFORD	Biophysics and Chemistry Department, The Wellcome Foundation, Ltd., London
D. N. HAGUE	University Chemical Laboratory, University of Kent, Canterbury
N. HALES	Department of Biochemistry, University of Cambridge
W. HASSELBACH	Max-Planck Institut für Medizinische Forschung, Heidelberg
H. HAUSER	Unilever Research Laboratories, Welwyn Garden City, Hertfordshire
A. L. HODGKIN	Physiological Laboratory, University of Cambridge

P. C. T. JONES   Department of Zoology, University College of Wales, Aberystwyth

S. KOWARSKI   Department of Physiology, Columbia University, New York

W. R. LOEWENSTEIN   Department of Physiology, Columbia University, New York

H. C. LÜTTGAU   Institut für Zellphysiologie, Ruhr Universität Bochum

E. K. MATTHEWS   Department of Pharmacology, University of Cambridge

J. C. METCALFE   M.R.C. Molecular Pharmacology Unit, Department of Pharmacology, University of Cambridge

J. L. MONGAR   Department of Pharmacology, University College London

N. NAGATA   Department of Biochemistry, University of Pennsylvania, Philadelphia

A. W. NORMAN   Department of Biochemistry, University of California

J. F. PALMER   Department of Physiology, Middlesex Hospital Medical School, London

R. RAHAMIMOFF   Department of Physiology, Hebrew University, Hadassah Medical School, Jerusalem

H. RASMUSSEN   Department of Biochemistry, University of Pennsylvania, Philadelphia

P. REID   Department of Physiology, Columbia University, New York

M. REITER   Pharmakologisches Institut der Universität, Munich

H. REUTER   Department of Pharmacology, University of Bern

E. RIDGWAY   Department of Zoology, University of Bristol

C. J. ROBINSON   National Institute for Medical Research, Mill Hill, London

R. RÜDEL   Department of Physiology, University College London

O. SCHACHTER   Department of Physiology, Columbia University, New York

H. J. SCHATZMANN   Veterinär-Pharmakologisches Institut, Universität Bern

H. O. SCHILD   Department of Pharmacology, University College London

D. H. SMYTH   Department of Physiology, University of Sheffield

T. TOMITA   Department of Physiology, Kyushu University, Hakozaki, Fukuoka City, Fukuoka Prefecture, Japan

G. VRBOVA   Department of Physiology, University of Birmingham

A. A. WEINEKE   Dunn School of Pathology, University of Oxford
L. WOLPERT   Department of Biology as Applied to Medicine, Middlesex Hospital Medical School, London
A. M. WOODIN   Dunn School of Pathology, University of Oxford

The papers published here are in the form and units in which they were presented at the symposium. As many of the authors did not use SI units, the SI equivalent of the units are given below.

Physical quantity	Name of unit	Abbreviation	SI equivalent
Length	ångström	Å	10^{-10} m $= 10^{-1}$ nm
Length	micron	μm	10^{-6} m
Length	inch	in	$2 \cdot 54 \times 10^{-2}$ m
Energy	erg	erg	10^{-7} J
Force	dyne	dyn	10^{-5} N
Magnetic flux density	gauss	G	10^{-4} T
Dynamic viscosity	poise	P	10^{-1} kg m^{-1} s^{-1}

Session I

Chemistry of the Alkaline Earth Metals

Chairman:
A. S. V. BURGEN,
Department of Pharmacology, University of Cambridge

THE SIMPLE CHEMISTRY OF CALCIUM AND ITS RELEVANCE TO BIOLOGICAL SYSTEMS

R. D. GILLARD
University of Kent at Canterbury

In this paper I consider some aspects of the chemistry of calcium which are relevant to biological systems. Calcium is essential in so many biochemical operations that it is difficult to decide what general questions require answering.

The aim of this paper is to try, in very simple terms, to set up and comment on partial answers to the following questions:

(i) Why calcium at all?

(ii) Why calcium rather than, say, magnesium or strontium [or even copper(II)]?

(iii) What is the function of the calcium ion?

(iv) What explanation can be offered for specific biological facts, such as the antagonism occasionally observed between calcium and magnesium?

I should like to set up some general statements about the chemistry of calcium and see if these will help in answering these questions.

Occurrence

The widespread occurrence of calcium as an essential factor in cellular function is mirrored in its common observation in biological artefacts. For example, many urinary calculi (Lonsdale, 1968) contain forms of insoluble calcium oxalate—Weddellite, $CaC_2O_4 . 2H_2O$, or Whewellite, $CaC_2O_4 . H_2O$—and we know that absorption of dietary calcium from the intestinal tract is prevented by certain ingested organic acids (such as oxalic or phytic) which give insoluble calcium salts. Similarly, the small rectangular crystals of the oothecae of the praying mantis are calcium citrate, $6H_2O$ (Parker & Rudall, 1955). Such common requirement or observation of calcium is not altogether surprising when we consider the composition of the Earth itself, which may be divided into the lithosphere and the (presumably later (Oparin, 1969)) hydrosphere, as in Table 1.

TABLE 1. *Composition of the Earth* (Wells, 1962)

Lithosphere		Hydrosphere	
Element	% wt	Element	% wt
O	46·6	O	85·9
Si	27·7	H	10·8
Al	8·13	Cl	1·93
Fe	5·0	Na	1·07
Mg	2·1	Mg	0·13
Ca	3·6	S	0·09
Na	2·85	Ca	0·04
K	2·60	K	0·04
Ti	0·63	Others less	
Others less		than 0·01	

We might perhaps reasonably conclude, in partial answer to the first question, that calcium is extensively used as a component of cells because it is extensively available. It is interesting to note the almost equal occurrence of the other alkali and alkaline earth metals, sodium, potassium, and magnesium. In the context of the ready availability of calcium, it is relevant that the most ancient probable evidence (derived from Oparin, 1969) of life is the calcareous secretions called 'algal limestones' found in Southern Rhodesia, apparently $2·7 \times 10^9$ years old.

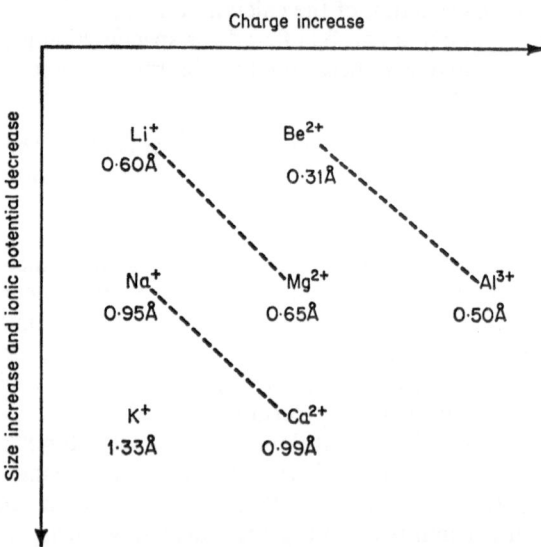

FIG. 1. Calcium and its nearest neighbours in the periodic table, using the ionic formulae most appropriate for aqueous chemistry.

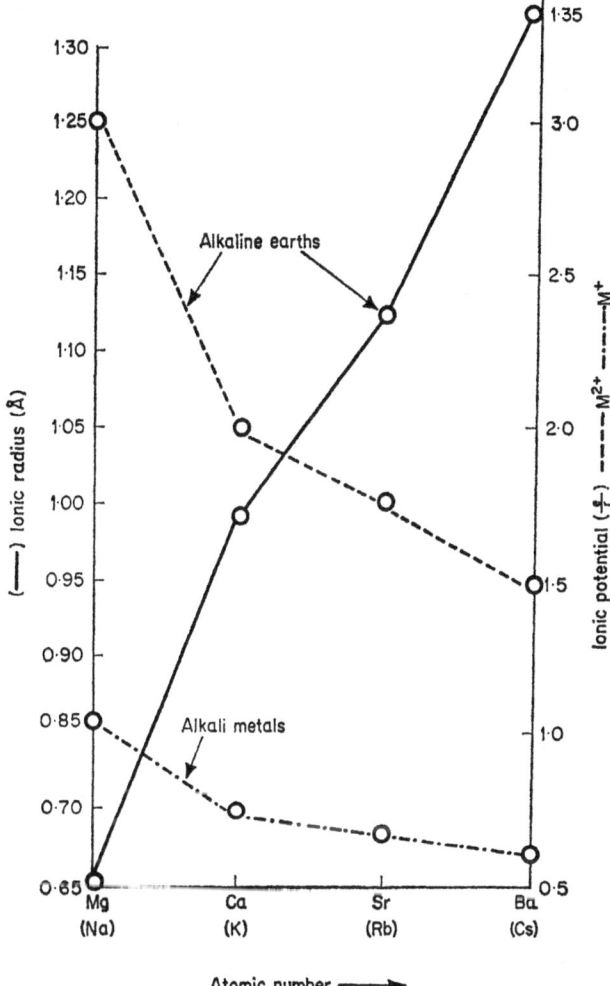

FIG. 2. Ionic radii and ionic potentials of alkali metals and alkaline earths.

General properties

The relevant part of the periodic table of the elements is shown in Fig. 1, with the ionic formulae most appropriate for aqueous chemistry. Ionic potential, a very useful property in comparative inorganic chemistry, is defined as e/r, where e is the charge and r the radius of the ion. The ionic potential indicates how strong are the ionic bonds formed by the ion. The ionic radii (Wells, 1962) are indicated in Figs. 1 and 2. In addition, ionic potentials of the alkaline earth ions are plotted, with some other information, in Fig. 2. The diagonal lines of Fig. 1, the so-called diagonal relationship of inorganic chemistry, remind us that, in terms of size, we have the pairs Li^+ and Mg^{2+}, but Na^+ and Ca^{2+}. Bearing in mind the common observation that isomorphous replacement may occur when the radius of two ions is equal within 10%, we expect (and often find) that lithium may isomorphously replace magnesium, and sodium may isomorphously replace calcium (in both cases, with the necessary charge adjustment elsewhere in the system). It is interesting in this connection that Mg^{2+} ($r = 0.65$ Å) in silicate minerals can (Greenwood, 1968) isomorphously substitute Fe^{2+} ($r = 0.76$ Å) or Fe^{3+} ($r = 0.64$ Å) but *not* Ca^{2+} (0.99 Å).

Effects of ionic size

(a) *Bond strengths.* We have already seen that, for the pair Mg^{2+} and Ca^{2+}, the increase in ionic radius gives a marked decrease in ionic potential.

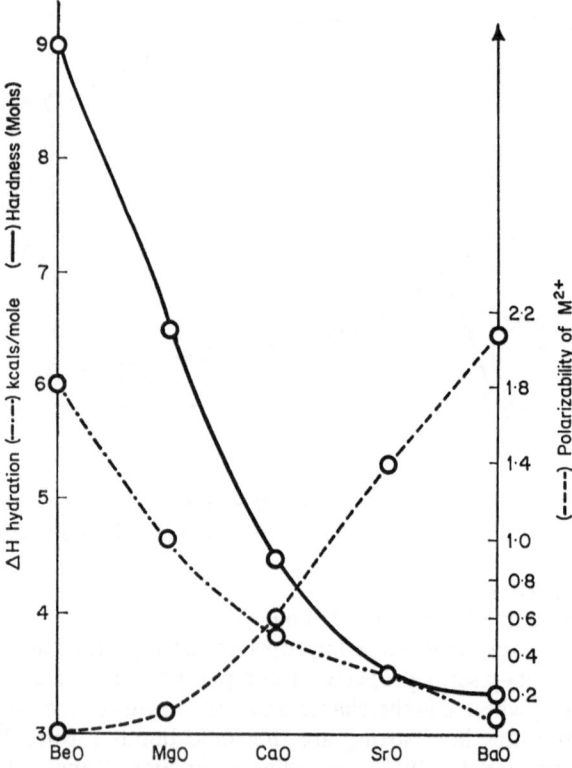

FIG. 3. The hardnesses (on Mohs' scale), i.e., the resistance to distortion and the polarizability, i.e., ease of distortion, of the alkaline earth oxides.

When we consider that, for a purely ionic bond, $M^{n+}X^{m-}$, the coulombic binding energy is given by

$$E = \frac{nm}{r_m r_x} = \text{(ionic potential of } M^{n+}\text{)(Ionic potential of } X^{m-}\text{)},$$

then clearly, for a given X^-, magnesium will form a stronger ionic bond. As an example of this, I have plotted the hardnesses (on Mohs' scale) of the alkaline earth oxides (essentially the resistance to distortion) in Fig. 3. For related reasons, magnesium carbonate is less stable on heating than is calcium carbonate; the equilibrium $MCO_3 \rightarrow MO + CO_2$ has a $p(CO_2)$ equal to 1 atmosphere for $M = Mg^{2+}$ at 540°C, but for $M = Ca^{2+}$ at 900°C. The increasing size of the alkaline earth metal ions also effects an increasing tendency toward partly covalent bonding. This is reflected in the polarizability (or ease of distortion) data of Fig. 3 (based on the value for $F^{(-)} = 1.0$).

TABLE 2. *Effect of ionic size on the co-ordination number*

Compound	Radius ratio $(r_{M^{2+}}/r_{F^-})$	Structure[a]	Co-ordination numbers A : B
BeF_2	0·32	Cristobalite	4 : 2
MgF_2	0·60	Rutile	6 : 3
CaF_2	0·87	Fluorite	8 : 4

[a] This column gives the name of the prototype structure.

A. This is the co-ordination number of the metal ion (e.g., each beryllium ion in BeF_2 is surrounded by four F^- ions).

B. The co-ordination number of the F^- ions in the particular fluoride.

(b) *Co-ordination number.* The co-ordination number of an ion is the number of nearest neighbouring atoms. For example, in the famous cubic structure of sodium chloride, both sodium and chloride ions each have six nearest neighbours, and a co-ordination number of six. Clearly, on geometrical grounds, the larger an ion, the more groups of the opposite electronegativity can surround it. This offers a strong distinction between calcium and magnesium. In calcium compounds of known structure we commonly find co-ordination numbers of eight for oxygen, but for similar magnesium compounds we find only six. For example, in the silicate mineral diopside (Warner & Bragg, 1928), $CaMg(SiO_3)_2$, both magnesium and calcium ions are in an oxygen matrix, but each magnesium is surrounded by six oxygen atoms, whereas each calcium, being larger, is surrounded by eight oxygen atoms. This difference in co-ordination numbers is commonly found in silicate minerals and elsewhere. For a given compound, MX_2, the details of the packing of X around M in the crystal is determined by the ratio of the radii of M and X. Table 2 gives an example of this, for the fluorides of the alkaline earths; again calcium has a higher co-ordination number (8) than magnesium (6).

Stabilities

The behaviour of metal ions in their complexes may be classified on the basis of their relative affinities for light atoms and for heavy atoms. In the classification due to Chatt, Ahrland, & Davies (1957), we have:

$$\text{Class } a: \quad K_M\text{--O} > K_M\text{--S}$$
$$K_M\text{--N} > K_M\text{--P}$$
$$K_M\text{--F}^{(-)} > K_M\text{--Cl}^{(-)}$$

$$\text{Class } b: \quad K_M\text{--O} < K_M\text{--S}$$
$$K_M\text{--N} < K_M\text{--P}$$
$$K_M\text{--F}^{(-)} < K_M\text{--Cl}^{(-)}$$

Clearly, the alkaline earth metals are firmly in class (a), in which the bonding might loosely be described as largely ionic.

To illustrate this preference of calcium for ionic bonding in complexes, consider the well known equilibria (represented simply as (1)) involving cobalt(II) in water in the presence of chloride ion.

$$[\text{CoCl}_4]^{2-} + 6\text{H}_2\text{O} \underset{\text{Ca}^{2+}}{\overset{\text{Zn}^{2+}}{\rightleftharpoons}} [\text{Co(H}_2\text{O)}_6]^{2+} + 4\text{Cl}^- \qquad (1)$$
$$\text{blue} \qquad\qquad\qquad\qquad\qquad \text{pink}$$

When calcium ions (with high affinity for water) are added, the solutions become more blue. The addition of zinc salts (zinc being strongly class (b)) makes the solutions more pink, as the zinc binds the chloride ions.

The high affinity of calcium for oxygen is reflected in the enthalpies of hydration plotted in Fig. 3. It seems likely that the most common function of calcium in cells will be as a binder of oxygen ligands.

A large number of stability sequences for alkaline earth metals are known (partly as a consequence of the extensive work during the 1950s aimed at preferential removal of $(\text{Sr}^{90})^{2+}$ from bone-forming media containing calcium). Attention is given here to three such sequences:

(a) For $\text{CH}_3\text{COO}^{(-)}$: $\log K\ \text{Mg}^{2+} > \text{Ca}^{2+} > \text{Sr}^{2+} > \text{Ba}^{2+}$.
(b) For tartrate: $\log K\ \text{Ca}^{2+} > \text{Sr}^{2+} > \text{Ba}^{2+} > \text{Mg}^{2+}$.
(c) In sugar chemistry, both calcium and strontium form stable complexes with saccharic acid, whereas Mg^{2+} does not.

It seems at least possible that the germ of calcium specificity is contained in such series, and that polyhydroxy ligands favour calcium over magnesium (perhaps for steric reasons).

One further significant feature of the simple chemistry of calcium centres is the rate at which they undergo substitutions. For the process of water exchange,

$$[\text{M(H}_2\text{O)}_n]^{2+} + \text{H}_2\text{O}^* \rightleftharpoons [\text{M(H}_2\text{O)}_{n-1}(\text{H}_2\text{O}^*)]^{2+} + \text{H}_2\text{O}$$

the rate for $M = Ca^{2+}$ is greater than for Mg^{2+} (perhaps because of the capacity of Ca^{2+} to vary its co-ordination number). The utility of calcium in particular enzymes may reflect such rate-factors.

Conclusions

In this paper, I have tried to illustrate the features of the chemistry of calcium which seem to me likely to prove relevant to its biological function. These are (*a*) its high abundance; (*b*) its high affinity for oxygen ligands; (*c*) its particular size (~ 1 Å) and consequent ionic potential; and (*d*) its capacity to adopt the higher co-ordination numbers. The interactions between calcium and magnesium in three separate biological situations might be interpreted as follows.

(*a*) Where both Ca^{2+} and Mg^{2+} are effective: this might suggest that, in the particular context, stereochemistry is less important than the mere presence of a positively charged ion.

(*b*) Where Mg^{2+} is effective and Ca^{2+} is inhibitory: here, the stereochemistry leading to a 6-fold co-ordination, or the need for a high ionic potential, may be more important than mere positive ionic charge.

(*c*) Where Ca^{2+} is effective and Mg^{2+} is inhibitory: here, it seems likely that a site of high co-ordination number is available, or that a low ionic potential is required. I might add here an item of pure speculation in connection with the highly poisonous nature of soluble barium salts. Barium ions are even more suited to eight co-ordination than are calcium ions, and it could well be that their toxicity stems from their preferential binding by a site of high co-ordination number which would normally contain calcium.

It can truly be said of alkaline earth metal complex chemistry that suitable models have hardly been investigated. The increasing awareness among inorganic chemists of the fascinating problems of specificity and mode of action of metal ions in biological systems, however, gives hope that this situation will change.

REFERENCES

CHATT, J., AHRLAND, S. & DAVIES, N. (1957). *Quart. Rev.*, **11**, 320

GREENWOOD, N. N. (1968). *Ionic Crystals, Lattice Defects and Non-Stoichiometry*, p. 6. Butterworths

LONSDALE, K. (1968). *Sci. Amer.*, December, 104

OPARIN, A. I. (1969). *R.I.C. Rev.*, **2**, 1

PARKER, K. D. & RUDALL, K. M. (1955). *Biochim. biophys. Acta*, **17**, 287

WARNER, B. E. & BRAGG, W. L. (1928). *Z. Krist.*, **69**, 168

WELLS, A. F. (1962). *Structural Inorganic Chemistry*, 3rd ed., p. 788. Oxford University Press

CALCIUM CHELATION AND BUFFERS

P. C. CALDWELL
Department of Zoology, University of Bristol

An approach which has proved very useful in studies of the role of calcium in biological systems, in particular in muscle, is the use of substances which selectively bind this ion. One of the best known compounds of this type is ethylene-diamine-tetra-acetic acid (EDTA) which binds calcium about 100 times more effectively than it binds magnesium. In recent years, however, EDTA has been superseded in many studies involving calcium by ethylene-glycol *bis* (β-amino-ethyl-ether)-N,N'-tetra-acetic acid (EGTA) which can bind calcium over 10^5 times more effectively than it binds magnesium. This article is concerned with some of the uses which have been made of EGTA in studying the role of calcium in the contraction of the large single muscle fibres which can be obtained from certain crustacea, in particular crabs (*Maia squinado*) and barnacles (*Balanus nubilus*).

Calculating the concentration of free calcium in a mixture of calcium, magnesium, and EGTA or EDTA
The logarithms to the base 10 of the association constants for the interactions of EGTA and EDTA with H^+, Ca^{2+}, and Mg^{2+} are given in Table 1. Normally only the binding of Ca^{2+} and Mg^{2+} by the forms of EGTA and EDTA with three negative charges (HL^{3-}) and four negative charges (L^{4-}) are considered. The total amount (MeL_{total}) of a divalent cation (Me^{2+}) which is bound is the sum of the amounts ($MeHL^- + MeL^{2-}$) bound to these two forms, (Me^{2+}) in this case being either calcium or magnesium. $MeHL^-$, MeL^{2-}, Me^{2+}, HL^{3-}, and L^{4-} are related to the association constants given in Table 1 by

$$K_{MeHL} = [MeHL^-]/[Me^{2+}][HL^{3-}]$$

and
$$K_{MeL} = [MeL^{2-}]/[Me^{2+}][L^{4-}].$$

At a given pH apparent association constants, K'_{MeHL} and K'_{MeL} can be defined as follows:

$$K'_{MeHL} = [MeHL^-]/[Me^{2+}][L_{total}]$$

and
$$K'_{MeL} = [MeL^{2-}]/[Me^{2+}][L_{total}]$$

where L_{total} is the total ligand not complexed. The relationships between the various constants, HL^{3-}, L^{4-}, L_{total}, and hydrogen ion concentration $[H^+]$ are given by:

$$K_{MeHL}/K'_{MeHL} = [L_{total}]/[HL^{3-}]$$
$$= 1 + 1/[H^+]K_1 + [H^+]K_2 + [H^+]^2K_2K_3 + [H^+]^3K_2K_3K_4$$

and

$$K_{MeL}/K'_{MeL} = [L_{total}]/[L^{4-}]$$
$$= 1 + [H^+]K_1 + [H^+]^2K_1K_2 + [H^+]^3K_1K_2K_3$$
$$+ [H^+]^4K_1K_2K_3K_4$$

The values for K'_{MeHL} and K'_{MeL} can be calculated from these equations. A combined apparent association constant $K'_{MeL_{total}}$ can now be defined and calculated where

$$K'_{MeL_{total}} = K'_{MeHL} + K'_{MeL} = [MeL_{total}]/[Me^{2+}][L_{total}]$$

This last equation can be used to calculate the concentration of free divalent ion from the calculated value for $K'_{MeL_{total}}$ and values of MeL_{total} and L_{total}. If only one cation is present, the calculation can often be done quite simply by successive approximation, MeL_{total} being taken as approximately equal to the total cation present and the free ligand L_{total} as approximately equal to the total ligand less MeL_{total}. The logarithms of K'_{MeHL} and K'_{MeL} to the base 10 for calcium and magnesium at pH 7·1 and 6·6 are given in Table 1 and the actual constants are obtained by

TABLE 1. *True* (K) *and apparent* (K') *association constants for the ligands EGTA and EDTA for an ionic strength of 0·1 (KCl) and a temperature of 20°C.*
(From Portzehl, Caldwell, and Rüegg, 1964)

Abbreviation for constant in the text	Cation	Ligand	EGTA			EDTA	
				log K' at			log K' at
			log K^*	pH 7·1	pH 6·6	log K^*	pH 7·1
K_1	H^+	L^{4-}	9·46†	—	—	10·26	—
K_2	H^+	HL^{3-}	8·85	—	—	6·16	—
K_3	H^+	H_2L^{2-}	2·68	—	—	2·67	—
K_4	H^+	H_3L^-	~2·00	—	—	1·99	—
K_{MeL}	$\begin{cases} Ca^{2+} \\ Mg^{2+} \end{cases}$	L^{4-} L^{4-}	11·00 5·21	6·882 1·092	5·89† 0·098	10·7 8·69	7·491 5·483
K_{MeHL}	$\begin{cases} Ca^{2+} \\ Mg^{2+} \end{cases}$	HL^{3-} HL^{3-}	5·33 3·37	3·572 1·612	3·077 1·118	3·51 2·28	3·464 2·234

* From Bjerrum, Schwarzenbach, & Sillén (1957).
† Weber & Winicur (1961) give 9·43 and 5·92.

P. C. Caldwell

raising 10 to the power of these logarithms. $K'_{MeL_{total}}$ is calculated at a pH of 7·1 to be $10^{6·88}$ for calcium and $10^{1·73}$ for magnesium. If both calcium and magnesium are present, then the concentration of free calcium and magnesium ions can be calculated by a process of successive approximation with the last equation given above, starting with a calculation for calcium and then continuing with alternating calculations for magnesium and calcium until reasonably steady values are reached.

Use of EGTA in the study of muscular contraction
These calculations show that the solutions which contain mixtures of

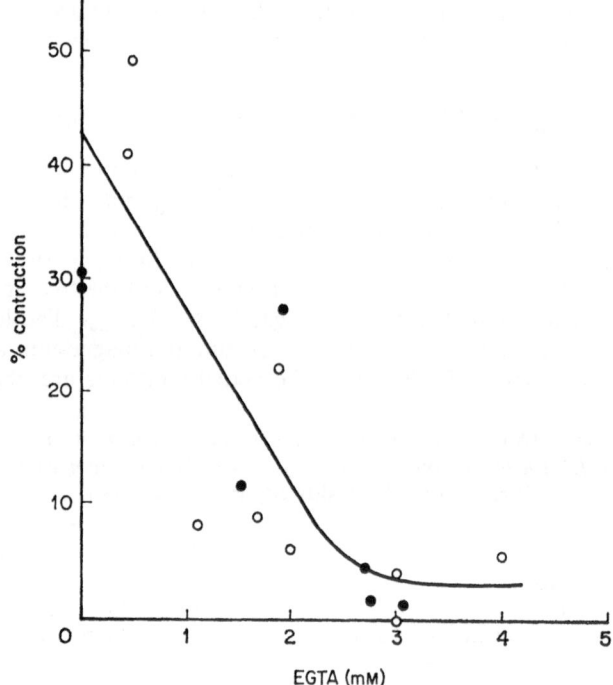

FIG. 1. Contraction (as % of resting length) of single *Maia* muscle fibres induced by 400 mM KCl about 30 min after the injection of EGTA. Estimated concentration of EGTA in each fibre shown as abscissa. Resting potentials before application of KCl, −45 to −60 mV; 22°–23°C. (From Ashley, 1967.)

EGTA and its calcium complex will also contain stabilized concentrations of free calcium ions which are changed only slightly either by dilution at a given pH or by the addition of extra calcium or magnesium. A mixture of EGTA and its calcium complex therefore acts as a calcium buffer. Calcium/ EGTA buffers have been used in experiments to determine the threshold concentration of free calcium ions needed to activate the contractile mechanism in intact muscle fibres. Portzehl, Caldwell, & Rüegg (1964) found that the injection of calcium/EGTA buffers into single muscle fibres from the crab, *Maia squinado*, caused contraction if the ionized calcium concentration in the fibre was raised to more than about 10^{-6} M. This

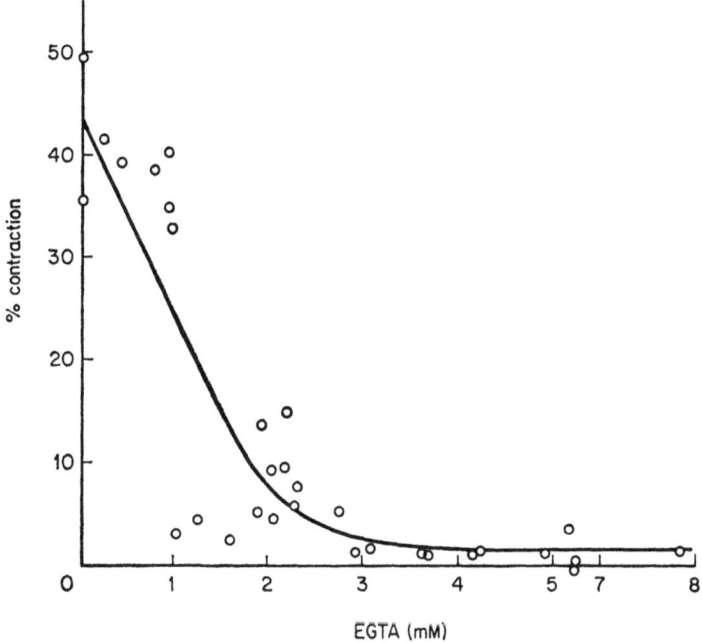

FIG. 2. Contraction of single *Maia* muscle fibres induced by crab saline containing 5 mM caffeine about 30 min after the injection of EGTA. Estimated concentration of EGTA in each fibre shown as abscissa. Resting potentials, -45 to -60 mV; 22°–23°C. (From Ashley, 1967.)

result has been confirmed for barnacle muscle fibres by Hagiwara & Nakajima (1966). This concentration of ionized calcium is similar to that required for the activation of isolated actomyosin systems and myofibrils.

Certain other experiments can be done which show that EGTA can bind calcium in intact muscle fibres in such a way as to keep the ionized calcium concentration of the sarcoplasm in excitation below that needed to activate the contractile mechanism. Ashley, Caldwell, Lowe, Richards, & Schirmer (1965) found that the injection of EGTA into *Maia* muscle fibres could suppress their contractile response. Figures 1, 2, and 3 (taken from Ashley, 1967) are from this work and show that the contractile responses to high potassium, caffeine, and intense electrical stimulation are almost completely suppressed if the concentration of EGTA inside the fibre after the injection is in excess of about 3mM. This concentration of EGTA just exceeds the total concentration of calcium in *Maia* muscle fibres, which is about 2 mM (Ashley *et al.*, 1965). This suggests that if the EGTA present in the fibres is sufficient to bind all the calcium and keep the ionized calcium below about 10^{-6} M then contraction is suppressed. On the other hand, the fact that some contractile response can still be elicited when the EGTA concentration in the fibres is as high as 1·5 mM suggests that as much

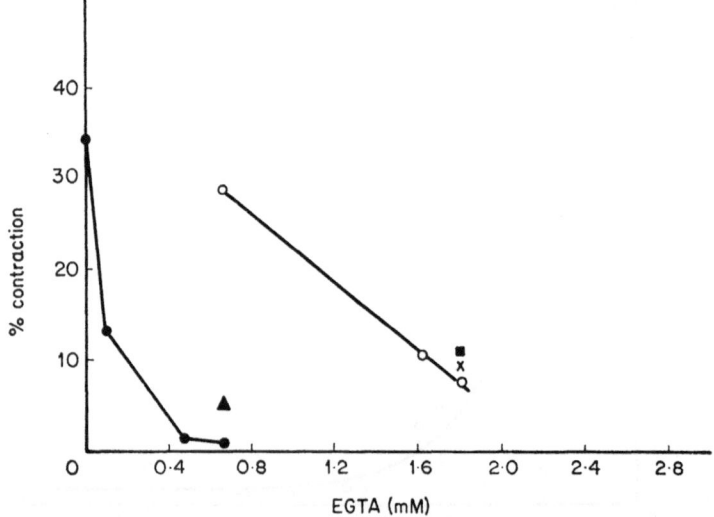

FIG. 3. Effect of a.c. shocks of 2 sec duration at varying current strengths and high potassium saline on the contraction of a *Maia* muscle fibre as increasing amounts of EGTA were injected to bring the estimated internal concentration to the values shown as abscissa. All points are the mean of two values except for the contraction in high potassium saline. Resting potential about -58 mV; 22°–24°C. ●, 46–48 mA; ▲, 55 mA; ○, 91 mA; ×, 175 mA; ■, high potassium saline. All 2 sec shocks at 50 c/s. (From Ashley, 1967.)

as 1·5 mM of the fibre calcium can be released into the sarcoplasm during intense stimulation, the release of this amount being necessary in the presence of 1·5 mM EGTA to bring the concentration of ionized calcium in the sarcoplasm above the threshold value of 10^{-6} M. It is possible that the release of this amount of calcium does not occur in the absence of EGTA, but these experiments indicate that it is possible to bring about the release of the bulk of the fibre calcium during intense stimulation, presumably from the sarcoplasmic reticulum.

Another type of experiment shows that injected EGTA can reduce the concentration of free ionized calcium which can be reached in muscle fibres. Figure 4 (taken from Caldwell, 1964) shows the changes in calcium efflux which are observed when ^{45}Ca is injected into a single *Maia* muscle fibre as a solution of 40 mM $CaCl_2$. The injected calcium causes a contraction of the fibre lasting about 30 sec and this is associated with a high rate of calcium loss from the fibre. After relaxation, when the sarcoplasmic ionized calcium has fallen to a low concentration, this rate of loss is greatly diminished. If the contractile mechanism is then reactivated, in

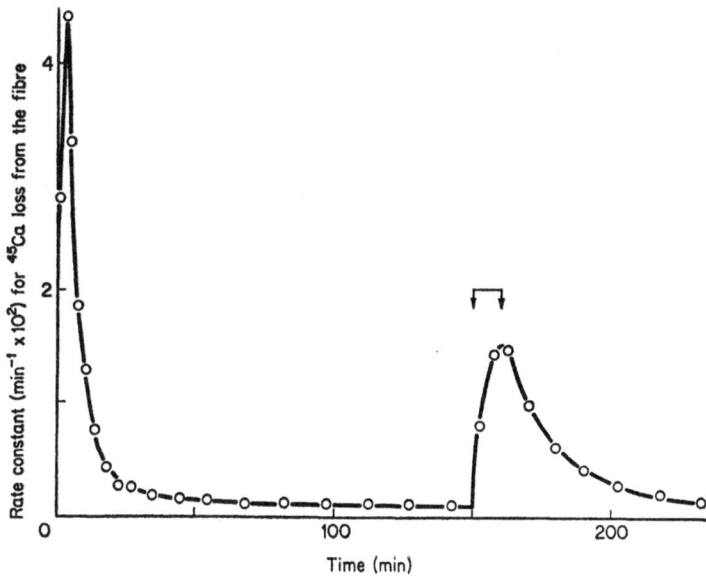

FIG. 4. The efflux of ^{45}Ca from a *Maia* muscle fibre. The ^{45}Ca was injected at the beginning of the experiment as 40 mM $CaCl_2$. The fibre contracted and relaxed after 2 min. 2 mM caffeine was applied externally in the crab saline for the period indicated (⌐⌐). The fibre remained contracted during this period and relaxed slowly when the caffeine was removed. Resting potential throughout −50 to −57 mV; 20°C. (From Caldwell, 1964; from unpublished work by Caldwell & Lowe.)

this case with caffeine, the contraction is associated with a marked rise in the efflux of radioactive calcium, presumably as a result of the increase in the sarcoplasmic ionized calcium to a concentration above 10^{-6} M. A different version of this experiment can be carried out in which the ^{45}Ca is injected as a calcium/EGTA buffer rather than as a solution of calcium chloride. Figure 5 shows the results of such an experiment. The ionized calcium in this buffer was so low (4 × 10^{-9} M) that on injection into the fibre no contraction was observed and there was no initial rapid loss of ^{45}Ca. Treatment with caffeine produced no contraction even though the

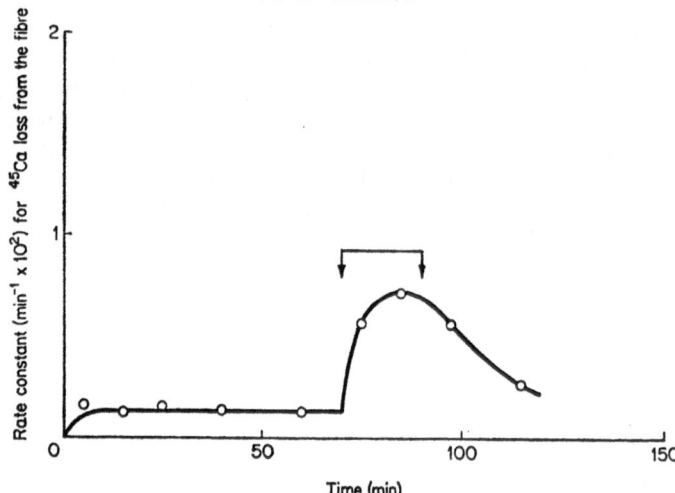

Time (min)

FIG. 5. The efflux of ^{45}Ca from a *Maia* muscle fibre. In this case the ^{45}Ca was injected at the beginning of the experiment as a calcium/EGTA buffer containing about 4×10^{-9} M of ionized calcium. No contraction was observed. 5 mM caffeine was applied externally in the crab saline for the period indicated (⌐¬). No contraction was observed. Resting potential, -50 to -57 mV; 20°C.

caffeine was able to release sufficient calcium to increase the ionized calcium concentration in the sarcoplasm, this increase being reflected in an increase in calcium efflux. The increase in sarcoplasmic ionized calcium induced by caffeine in this experiment was presumably prevented (by the EGTA) from reaching the threshold needed for contraction ($\sim 10^{-6}$ M).

These few experiments by no means exhaust the possibilities for the use of calcium binding agents in the study of the contraction of crab muscle, but they illustrate the type of experiment which can be carried out.

Figs. 1, 2, and 3 are reproduced with permission of the *American Zoologist* and Fig. 4 is reproduced with permission of the Royal Society.

REFERENCES

ASHLEY, C. C. (1967). *Am. Zool.*, **7**, 647
ASHLEY, C. C., CALDWELL, P. C., LOWE, A. G., RICHARDS, C. D. & SCHIRMER, H. (1965). *J. Physiol., Lond.*, **179**, 32P
BJERRUM, J., SCHWARZENBACH, G. & SILLEN, L. G. (1957). *Stability Constants, Part 1: Organic Ligands*, pp. 76 and 90. London: The Chemical Society
CALDWELL, P. C. (1964). *Proc. Roy. Soc., B*, **160**, 512
HAGIWARA, S. & NAKAJIMA, S. (1966). *J. gen. Physiol.*, **49**, 807
PORTZEHL, H., CALDWELL, P. C. & RÜEGG, J. C. (1964). *Biochem. biophys. Acta*, **79**, 581
WEBER, A. & WINICUR, S. (1961). *J. biol. Chem.*, **236**, 3198

BINDING OF CALCIUM TO PHOSPHOLIPIDS

R. M. C. DAWSON

Biochemistry Department, A.R.C. Institute of Animal Physiology, Babraham, Cambridge

H. HAUSER

Unilever Research Laboratories, The Frythe, Welwyn Garden City, Hertfordshire

It has often been assumed that phospholipids play an important part in the binding of calcium to cell membranes. The calcium contained in the cell is largely concentrated in the phospholipid-rich membrane fractions, while the concentration of ionized calcium in the intracellular fluid is minimal (Thiers & Vallee, 1952; Hofer & Kleinzeller, 1963; Hodgkin & Keynes, 1957; Harris, 1957).

When membrane fragments which are capable of binding radioactive calcium are extracted by organic solvents, the protein-containing residue will not bind calcium whereas the extracted lipids do (Koketsu, Kitamura & Tanaka, 1964). Since phospholipids constitute the main bulk of the polar lipids extracted from the membranes, extrapolation of such results suggests that these are mainly responsible for the binding of the metal. Thus it has been suggested that the phospholipids of mitochondrial membranes constitute the initial receptor sites for calcium taken up by these organelles, at least in the absence of inorganic phosphate (Slater & Cleland, 1953; Chappel, Cohn & Greville, 1963; Peachey, 1964).

As a result, nearly every paper published on the binding of calcium ions to pure phospholipids contains a rather plaintive justification for the work which states that organized phospholipid structures such as bimolecular leaflets and monolayers act as simplified models for the biological structure. Although the study of interactions between pure phospholipids and Ca^{2+} is of importance in understanding the binding of the metal ion to biological membranes it has to be borne in mind that this type of binding would be greatly influenced by the perturbation of other macromolecular components in the membrane and that the non-phospholipid parts of the membrane would almost certainly possess some, although maybe limited, affinity for calcium ions.

It was realized by early workers that isolated phospholipids possessed a high affinity for calcium (for example, Drinker & Zinsser, 1943; Folch, 1949a,b), but it is only in the last decade that this has been put on a more quantitative basis. The reason lies in the development of isolation procedures which have allowed the preparation of phospholipids which are homogeneous; at least as far as the hydrophilic regions of their molecules are concerned. Also important has been the development of techniques for estimating calcium binding.

Binding of calcium to phospholipid particles

In contact with water, isolated phospholipids form hydrated polymolecular structures in which the hydrophobic regions (usually fatty acyl residues) are held together by cohesive forces (van der Waals) and the polar head regions are orientated to the aqueous phase because they are hydrophilic. The type of structure formed depends largely on the shape and dimensions of the phospholipid molecule and on the properties of its polar head group. At a physiological pH the bulk of the phospholipids which constitute the cell membrane consists of roughly cylindrically shaped molecules in which the polar end is a zwitterion (for example, phosphatidylcholine or lecithin, phosphatidylethanolamine, choline plasmalogen, sphingomyelin). Due to the absence of a net charge on the head groups and also to the cylindrical shape of the molecules which allows easy side by side stacking, the phospholipids exist in the hydrated state as bimolecular liquid crystals. Such crystals are in the smectic mesophase form in which successive sheets of bimolecular layers of phospholipids are separated from one another by layers of water. Alternatively, certain pure phospholipids can exist in an aqueous environment in the form of classical micelles, either because the molecule is wedge-shaped (lysolecithin) or because it possesses a highly negatively charged polar region (triphosphoinositide). Neither property is conducive to the molecules stacking in a continuous bimolecular lamella.

Aqueous suspensions of such liquid crystals or micelles have been used recently to examine the affinity between phospholipids and calcium. The binding can be assessed by a number of different techniques. That most frequently used depends on the ability of metallic salts to liberate hydrogen ions from phospholipid interfaces; this property was originally described by Christensen & Hastings (1940) and Dervichian (1955). When calcium salts are added to ultrasonically dispersed suspensions of acidic phospholipids (for example, phosphatidylserine) the pH change can be appreciable, and the suspension is then titrated back to the original pH generally using a base, for example, tetramethylammonium hydroxide, whose bulky cation is assumed to have little affinity for the phosphate moiety of the phospholipid (Abramson et al., 1964a,b; 1966, 1968). A variation of this method is to titrate the acidic phospholipid suspension both in the presence and absence of the added divalent metal salt and to calculate the binding of the calcium from the displacement of the titration

TABLE 1. *Comparison of* $^{45}Ca^{2+}$ *adsorbed on monolyers of various phospholipids with differing ionic structures on their head group regions. Ca concentration 0·21 µg atoms/l.*

Phospholipid			Monolayer/cm² at collapse		
Ionic portion	Name	Nett ⊖ve charges	Ca atoms adsorbed ($\times 10^{-14}$)	Molecules Lipid ($\times 10^{-14}$)	⊖ve charges /Ca atom
$-\overset{\overset{O}{\|\|}}{\underset{\underset{O_{\ominus}}{\|}}{P}}-OCH_2CH_2\overset{\oplus}{N}(CH_3)_3$	Phosphatidyl choline (lecithin)	0	0	–	–
	Sphingomyelin	0	0	–	–
$-\overset{\overset{O}{\|\|}}{\underset{\underset{O_{\ominus}}{\|}}{P}}-OCH_2CH_2\overset{\oplus}{N}H_3$	Phosphatidyl ethanolamine	0	0	–	–
$-\overset{\overset{O}{\|\|}}{\underset{\underset{O_{\ominus}}{\|}}{P}}-OCH_2-\underset{\underset{\oplus NH_3}{}}{CH}-\underset{\underset{O_{\ominus}}{}}{C}=O$	Phosphatidyl serine	1	0·39	2·3	5·9
(inositol ring structure)	Phosphatidyl inositol (Mono-phosphoinositide)	1	0·38	2·0	5·3
$-\overset{}{\underset{\underset{O_{\ominus}}{\|}}{P}}=O$	Dicetyl phosphoric acid	1	0·50	2·8	5·6
$-\overset{\overset{O}{\|\|}}{\underset{\underset{O_{\ominus}}{\|}}{P}}-O_{\ominus}$	Phosphatidic acid	2	0·71	1·9	5·3
(triphosphoinositide ring structure)	Triphosphoinositide	5	0·78	1·7	10·8
			0·41	0·55 (2·5 dyne/cm)	6·6
			0·045	0·061 (95 % mole of lecithin)	6·8

curve produced (Hendrickson & Fullington, 1965). The affinity of the calcium ion can also be assessed by the change in turbidity of the phospholipid suspension produced by its addition and also by the concentration required to produce flocculation of the phospholipid particles (Abramson *et al.*, 1965, 1967, 1968), or reversal of their zeta potential (Barton, 1968).

When these methods are applied to ultrasonicated suspensions of phospholipids (lecithin, phosphatidylserine, phosphatidic acid, phosphatidylinositol, triphosphoinositide, Table 1 gives structure of ionic portion) it becomes clear that acidic phospholipids bind the calcium ion much more strongly than do zwitterionic lecithin particles. Also, the binding of calcium ions to phospholipid surfaces is shown to be somewhat stronger than Mg^{2+} binding and very much more so than that of Na^+ and K^+. Precise quantification of the results in terms of the calculation of binding constants is rendered difficult because of certain assumptions that have to be made. Thus it is assumed that all anionic sites on the phospholipid particles are free to react with added calcium ions since all the theoretical sites can be titrated with univalent cations. There is evidence, however, that at least with phosphatidic acid (Abramson *et al.*, 1964b) certain of the anionic sites are masked in the presence of calcium and become unavailable for titration. Furthermore, it is often assumed in these calculations that the binding of cation to the phosphate moiety does not alter the ionization characteristics of the phospholipid, although it might be expected that the reduction in the surface charge would effect the dissociation of the surface ionic groups. In spite of these uncertainties, the determinations of the association constants for calcium binding to phosphatidic acid and phosphatidylserine particles by various methods show remarkable agreement (Barton, 1968).

Binding of calcium to phospholipid monolayers

The reaction between calcium and phospholipids has also been studied using unimolecular layers of the lipids oriented at the air–water interface rather than polymolecular particulate suspensions. Although the presence of air above the film can cause some perturbation of the phospholipid molecules (Haydon & Taylor, 1963), the orientation of the lipid molecules is more precise and better understood in a monolayer than in a polymolecular aggregate. Furthermore, the spacing of the molecules at the interface and the areas occupied by the polar head groups on which the calcium exchange occurs can be varied by careful compression or expansion of the film. In measurements on dispersed phospholipid particles the parameters which are related to intermolecular spacing and state of compression are not known.

Although early workers assessed the binding of calcium to fatty acid monolayers by direct analyses after skimming them off the water surface (Langmuir & Schaefer, 1936), two other methods for following the interaction are now in general use; these concern interfacial potential measurements and surface radioactivity determinations with ^{45}Ca. Because the latter method has been criticized recently (Shah & Schulman, 1967a), and this criticism reviewed favourably by Bangham (1968), it is perhaps as well to consider the merits of the two techniques in some detail.

The surface potential of a lipid monolayer is measured as the change in interfacial potential which occurs when a lipid film is spread at the air–water interface. This potential (measured for example with an air ionizing electrode) results from the sum of the vertical (to the surface plane) components of permanent dipoles in the lipid molecule, those of the charged ionic groups in the same molecule, and those due to water molecules orientated by the presence of the film. Thus in lecithin, for example, there are probably three permanent dipoles, those of the -o- ester links of the fatty acyl bonds, of the ketonic groups, and of :o- ester links to the glycerol carbons (Shah & Schulman, 1965). The two charged groups in the phosphorylcholine moiety form another large ionic dipole component.

When calcium ions bind with the film, it is presumed that they are located near to these charged ionic groupings (presumably the phosphate) and consequently change the moment and the interfacial potential. Although at first sight it would appear that the surface potential change reflects a simple specific electrostatic interaction between the calcium ion and the phosphate group, this is not necessarily so. The magnitude and sign of the change will depend on the precise position of the interacting calcium ions along an axis vertical to the plane of the monolayer and thus its relationship to the charged ionic groups. In addition the presence of calcium ions could cause perturbation of the orientation of the permanent dipoles in the monolayer and thus change their vertical components. They could also bring about dipole changes by displacing other counter ions which may be reacting with the film or altering the orientation of water dipoles adjacent to the film molecules. It is not surprising therefore that the attraction of Ca^{2+} to negatively charged phosphate groups on monolayers of dicetylphosphoric acid increases the surface potential whereas, in contrast, the equivalent binding to the negatively charged carboxyl groups of fatty acid monolayers decreases the potential (Shah & Schulman, 1965).

Although it can be generally assumed that any change in the surface potential brought about by an addition of calcium ions to the subphase means that the metal ions are binding, it is theoretically possible for small surface potential changes to be brought about by the presence of calcium in the surface phase at a concentration no greater than that in the bulk subphase (no effective binding). What is important is that although measurements of surface potential do probably reflect a specific reaction of the Ca^{2+} with the film, they cannot be used to quantify such an interaction. In addition interpretation of such potential changes in terms of the precise position of the calcium in relation to the orientated lipid molecule must be regarded as highly speculative.

The measurement of calcium binding by the change in surface radioactivity brought about by adding ^{45}Ca to the subphase below a lipid monolayer is an extension of the technique introduced by Salley, Weith, Argyle & Dixon (1950) for studying the adsorption of ^{35}S-labelled-dioctyl sodium sulphosuccinate at an air–water interface. ^{45}Ca emits β rays of very low energy which in the absence of a monolayer are mainly absorbed by the subphase and do not reach a counter (mica-windowed Geiger–Müller or flow type) mounted just above the surface. When

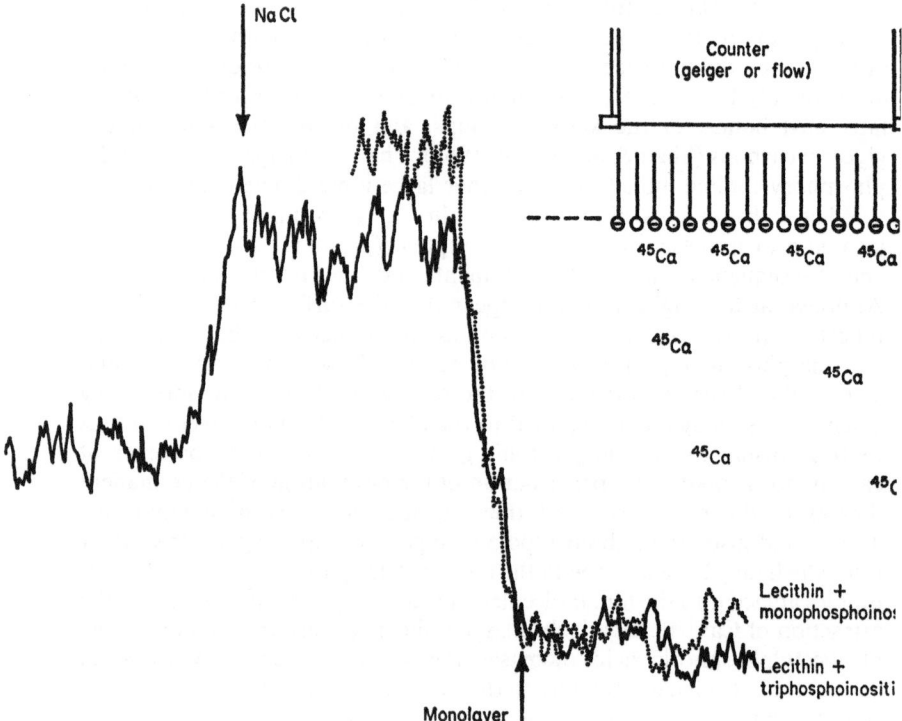

Fig. 1. Diagram showing how the spreading of a monolayer containing an acidic phospholipid attracts ^{45}Ca ions to the surface and produces an increment in the surface radioactivity measured with a counter. On adding sodium chloride to the subphase a proportion of the ^{45}Ca is displaced.

calcium ions are attracted to a lipid monolayer at the interface the radio-activity coming from the surface is increased above that obtained when the ^{45}Ca is distributed evenly throughout the subphase (Fig. 1). It is comparatively easy, therefore, to estimate precisely the amount of calcium which is associated with the lipid film because this is directly related to the increment in surface radioactivity. The calcium binding measured repre-sents not only the calcium ions immediately adjacent to the charges in the lipid head groups; that is, in the fixed portion of the electrical double layer, but also the excess concentration present in the outer diffuse double layer. This diffuse layer of Ca ions is due to the electrostatic field which is set up close to the surface by the interfacial charges (Fig. 2). The contribution of the diffuse double layer to the radioactivity can, of course, be negative; in other words, a repulsion of calcium ions from the interface can occur so that their concentration becomes less than in the bulk subphase. This can only happen with an acidic lipid, however, when the concentration of bulk

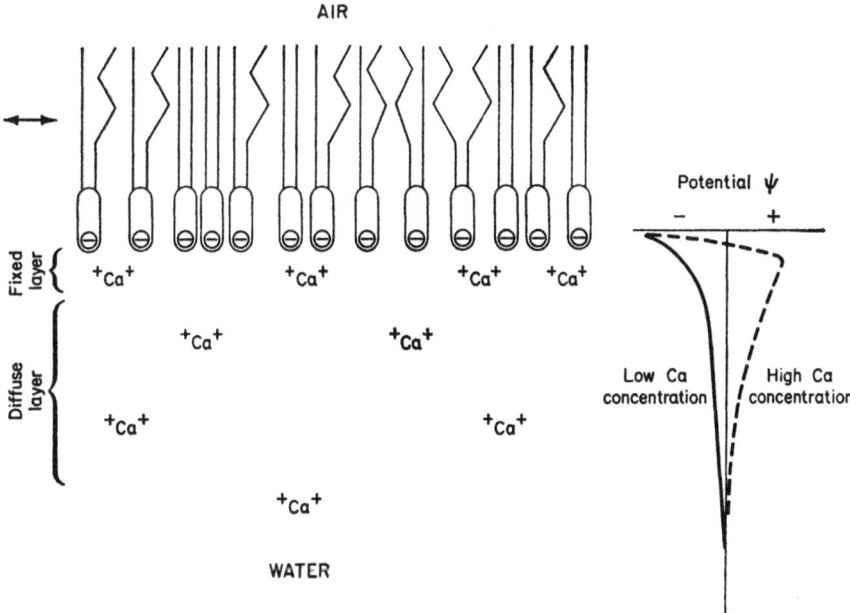

FIG. 2. Diagram showing a monolayer of an acidic phospholipid, whose molecules have differing degrees of unsaturation in their fatty acyl chains. Ca ions are attracted from the bulk phase forming the fixed and diffuse electrical double layers. The potential (ψ) near the surface plane varies according to whether the Ca ions are present in low concentration or in sufficient concentration to cause charge reversal of the surface.

calcium is high (> 10 mM), so that the layer of 'fixed' calcium ions immediately adjacent to the interface is sufficient not only to neutralize the negative sites on the lipid but also because of the divalent nature of the cation to bring about a charge reversal of the interfacial potential (Fig. 2). Measurements are therefore usually made with calcium concentrations below 1 mM.

Surface radioactivity measurements provide an estimate of the increased concentration of Ca ions in the total surface phase; that is, those ions bound close to the surface and held by ionic and perhaps non-ionic forces, and those in the diffuse double layer held more loosely by long range electrostatic forces. This enables the binding of Ca to the film in various circumstances to be assessed and allows investigation of how this binding is influenced by the presences of other substances which react with either the film or calcium ions or both. Such ternary systems are very difficult to evaluate from surface potential changes. Like the surface potential measurements the technique can only give limited evidence concerning the distribution of calcium ions along an axis vertical to the surface plane. The two methods of assessing Ca binding are therefore complementary—ideally both parameters should be measured in the same experiment but up to now such a course has not been systematically adopted.

Binding of calcium to phospholipid monolayers as a function of calcium concentration

When ^{45}Ca is introduced into the subphase below monolayers of acidic phospholipids (for example, phosphatidylinositol or phosphatidylserine) the surface radioactivity (calcium bound) initially increases with the increasing concentration of calcium added (Hauser & Dawson, 1967a; Hauser, Chapman & Dawson, 1969) (Fig. 3). At low concentrations the relationship is not linear but initially followed a Langmuir-type adsorption isotherm: surface radioactivity $= K_1 C/(K_2 + C)$, where C is the concentration and K_1 and K_2 are constants.

The adsorption data can be analysed in terms of the apparent association constant K_a for the reaction $Ca^{2+} + lipid \rightleftharpoons Ca$ surface with K_a being calculated from the mass equation

$$K_a = \frac{[Ca]_s}{[Ca]_f [L]_f}$$

where $[Ca]_s$ is the Ca adsorbed on surface, $[Ca]_f$ the Ca in subphase, and $[L]_f$ the concentration of free binding sites on phospholipid surface.

Treatment by three theoretical methods shows that the binding of calcium to monolayers of phosphatidylserine and phosphatidylinositol follows the mass equation rigorously only at low film pressures (or low charge densities), and with increasing film pressures deviations are apparent (Hauser et al., 1969). This deviation may be due to electrostatic repulsion between adjacent calcium ions bound to the surface.

As the calcium concentration is raised to high levels the binding sites become completely saturated and adsorption ceases. This occurs with phosphatidylserine monolayers at physiological pH values when approximately one calcium atom is bound to one phospholipid molecule; this ratio may be reduced in the presence of NaCl at physiological concentrations (Rojas & Tobias, 1965). Using surface potential measurements Bangham & Papahadjopoulos (1966), Shah & Schulman (1967a), and Papahadjopoulos (1968) found that with increasing concentrations of calcium, plots of the change in surface potential of phospholipid monolayers against log Ca concentration were linear but often showed a biphasic pattern. Above 1·0 mM Ca (where there is 1 equivalence of bound Ca per phospholipid molecule) they rose at a faster rate as the calcium concentration increased, suggesting a greater affinity of the film for calcium. Similar biphasic changes were also observed in plots of zeta potentials of phospholipid particles when increasing amounts of calcium were added. While the interpretation of such curves is at present uncertain, the authors

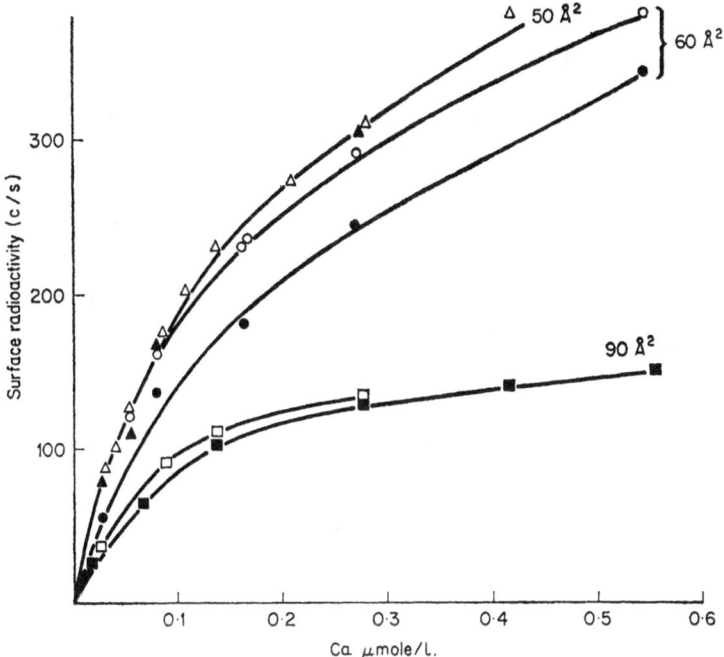

FIG. 3. Adsorption of $^{45}Ca^{2+}$ to monolayers of phosphatidylserine (black symbols) and phosphatidylinositol (open symbols) at various surface pressures as a function of calcium concentration. The area occupied by each phospholipid molecule in $Å^2$ is indicated. (From Hauser, Chapman & Dawson, 1969.)

suggest that the effect may be due to a different type of binding occurring at higher Ca concentrations. What is certain is that at this calcium concentration (1·0 M) there is a dramatic increase in the permeability of phosphatidylserine bilayers to univalent cations, and this implies that the calcium is causing a profound physical change in the membrane (Papahadjopoulos & Bangham, 1966).

Binding of calcium to individual phospholipids
A single low concentration of ^{45}calcium chloride can therefore be used to compare the adsorption of ^{45}Ca to phospholipid films of differing chemical

composition. The calcium adsorbed at this one concentration will presumably represent a point on the Langmuir adsorption isotherm for the particular phospholipid involved. The results of such a study are indicated in Table 1 (Hauser & Dawson, 1967a) for which all the observations were obtained at a single pressure (collapse). A number of conclusions are immediately apparent. First of all, at this Ca concentration (0·22 μM) there appears to be little adsorption of calcium on phosphatidylcholine sphingomyelin, and phosphatidylethanolamine monolayers. At the subphase pH (5·5) which was used these phospholipids would presumably exist as zwitterions with no net charge on their head groups. Rojas & Tobias (1965), using a similar surface radioactivity technique, also found little binding of ^{45}Ca (1·0 mM) to lecithin monolayers at this pH although there was limited adsorption onto phosphatidylethanolamine. With the same concentration of calcium (1·0 mM), Papahadjopoulos (1968) could find no change in the surface potential of egg lecithin and phosphatidylethanolamine monolayers. These results are in complete contrast to those of Kimizuka, Nakahara, Uejo & Yamauchi (1967), who found ready adsorption of ^{45}Ca to lecithin and phosphatidylethanolamine at an ambient concentration of 10^{-4} M CaCl$_2$. Although these workers were adding phospholipids at a far greater concentration than was necessary to form a monolayer on the available surface, such results are still difficult to understand unless the material was impure.

It is to be expected that at higher concentrations of calcium some affinity for the zwitterionic phospholipids might be apparent, as calcium would then be able to compete successfully with the quaternary ammonium or protonated amino groups of the phospholipid for the anionic site on its phosphate moiety. Bangham & Dawson (1962) showed that calcium produced a small positive zeta potential on lecithin particles examined by microelectrophoresis; this is indicative of a layer of 'fixed' calcium ions near to the interface (Fig. 2). Shah & Schulman (1965) obtained an increase of surface potential when calcium (10 mM) was added to monolayers of dipalmitoyl, egg, and yeast lecithins. The change was, however, related to the unsaturation of the fatty acid chains being highest in dipalmitoyl lecithin, less so in egg lecithin (medium unsaturation); the very unsaturated yeast lecithin only showed a change in surface potential at fairly high pressures. Shah & Schulman (1967a) have suggested, from studying surface potential-pH plots of the three lecithins, that this is due to the zwitterion structure being more readily broken as the unsaturation of the fatty acid chains is decreased. The separation of the phosphate and quaternary N in the lecithin molecule by two methylene groups presumably sterically helps the calcium competition because a mixed monolayer of equal molecular ratios of dicetylphosphate and eicosanyltrimethylammonium showed no surface potential change on calcium addition.

As might be expected, negatively charged acidic phospholipids are much more effective at concentrating ^{45}Ca in the surface phase (Table 1). It is perhaps surprising to find that with most of the acidic phospholipids the amount of calcium bound at collapse pressure is related to the number of net negative sites on the film if complete ionization of the ionic groups in the polar region is assumed. With triphosphoinositide, however, where there are five possible phosphate ionizations, the calcium adsorption per negative site is less than anticipated. Here, it seems that the very high density of negative sites on the monolayer is sufficient to reduce the surface pH by attracting protons from the bulk and so partially suppressing ionization. Thus the calcium adsorption is less than would be predicted from the other acidic phospholipids. When the packing density of tri-phosphoinositide in a monolayer is decreased by dilution with lecithin or by reducing the surface pressure to 2·5 dyne/cm, the surface charge density is reduced, the suppression of ionization is removed and as a result the adsorption of calcium per triphosphoinositide molecule increases and approaches that predicted from other phospholipids (Table 1). Micro-electrophoresis studies confirm that at the pH used (5·5) all the anionic sites are ionized when triphosphoinositide is diluted out with lecithin.

These results indicate therefore that under the conditions used the affinity of calcium for the phospholipid interface is largely controlled by Coulombic forces. This affinity is directly related to the net excess charge on the phospholipid molecule and is independent of its chemical nature.

Dependence of calcium binding to phospholipid monolayers on surface pressure

Using monolayers of phosphatidylinositol, Hauser & Dawson (1968) showed that the binding of ^{45}Ca was dependent on the density of packing of the molecules in the film. As the film was compressed the affinity of calcium per molecule of phospholipid increased appreciably and reached a maximum between 25 and 30 dyne/cm; above these pressures the binding of calcium decreased (Fig. 4). Similar observations have recently been obtained for phosphatidylserine films (Hauser *et al.*, 1969). Rojas & Tobias (1965) found a somewhat different binding of ^{45}Ca to films of phosphatidylserine, with the affinity increasing steadily as the surface pressure was increased to collapse; however, these experiments were performed in the presence of higher concentrations of calcium and appreciable competing concentrations of univalent metal cations in the subphase. In order to observe the maximal binding at intermediate pressures it is perhaps necessary to choose a low enough metal ion con-centration so that the anionic sites are not quite saturated. Although the maximal binding of calcium to phospholipid films at intermediate pressures is imperfectly understood, it is perhaps significant that the calculated average spacing between the negative charges on the film at pressures of 25–30 dyne/cm reaches a value which is equivalent to the diameter of the hydrated calcium ion. It seems possible therefore that the similarity in the

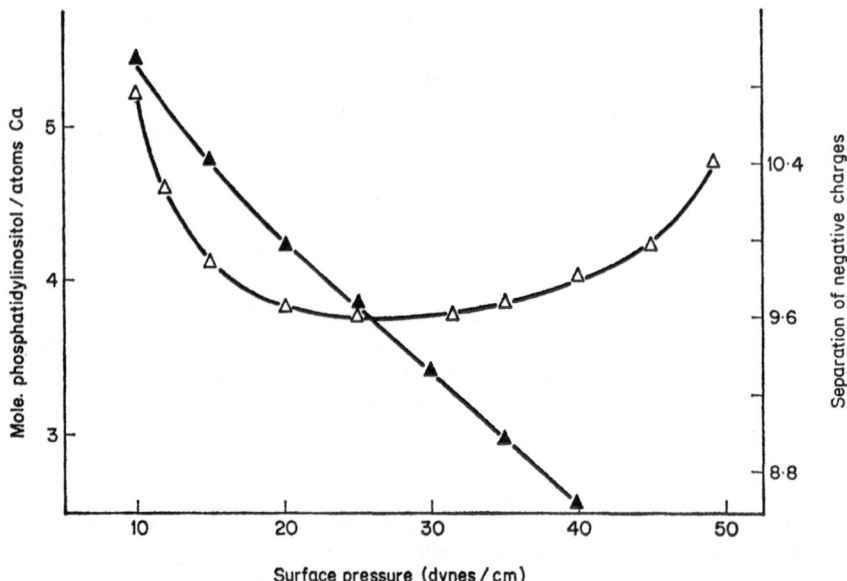

Surface pressure (dynes / cm)

FIG. 4. Variation of the adsorption of $^{45}Ca^{2+}$ on a phosphatidylinositol mono-layer (Δ) with surface pressure. $\blacktriangle$ shows the calculated average spacing (Å) between the negative charges on the film. (From Hauser, Chapman & Dawson, 1969.)

dimensions of the components encourages this two-point electrostatic attachment of calcium to the film.

As well as showing a variation in affinity for an individual phospholipid depending on its molecular spacing in the film, differences between the various phospholipids become apparent at lower film pressures. Thus Fig. 3 shows that although the adsorption isotherms for phosphatidyl-serine and phosphatidylinositol are identical at their collapse pressures, at lower pressures the former phospholipid exhibits a stronger binding of calcium. However, calculation of the association constant (K_a) from Hughes–Klotz plots (Hauser *et al.*, 1969) indicates that the difference is less than an order of magnitude. This small difference is probably accounted for by the different chemical structures of the polar groups of the phospholipids which allow a different type of binding at pressures lower than collapse. This is not necessarily non-ionic, and it could be that with the wider spacing of the molecules the Ca binds more readily to the carboxylic acid groups of the serine residues. At interfaces these can show a somewhat lower affinity for calcium than do the anionic sites on the phosphate groups of phospholipids (Hauser & Dawson, 1967a).

Dependence of calcium binding to phospholipid monolayers on pH

Because the binding of calcium depends mainly on whether a net negative charge is available on the phospholipid polar region, it is perhaps not surprising that the adsorption is dependent on the pH of the subphase. The situation is complicated because the pH of the surface phase adjacent to a negatively charged interface is somewhat lower than that of the bulk subphase, and this is due to proton attraction.

$$\text{pH surface} = \text{pH bulk} + \frac{\psi \varepsilon}{2 \cdot 3kT}$$

(ψ is surface potential, ε is electronic charge, and k is Boltzmann's constant). Moreover, at high pH values the results can be disturbed by chemical instability of the film (Hauser & Dawson, 1967b; Papahadjopoulos, 1968) and by precipitation of calcium carbonate unless CO_2 is rigorously excluded from the subphase. In addition, the presence of calcium can itself change the ionization characteristics of a phospholipid.

The ionization of the film is usually assessed by plotting its surface potential against the bulk pH ($\Delta V/\text{pH}$ plots) (Schulman & Hughes, 1931). A graph which is a true indication of surface ionization can often be of quite a different character to that which would be expected from a soluble molecule. For example, the $\Delta V/\text{pH}$ plot of dicetylphosphoric acid shows a gradual ionization from pH 2–7 (Shah & Schulman, 1967a).

The $\Delta V/\text{pH}$ plots of lecithin are straight lines with zero slopes between pH 3 (when the phosphate group becomes ionized) and pH 11 (Shah & Schulman, 1967a; Papahadjopoulos, 1968). The high pK of the quaternary nitrogen of the choline residue keeps the head group as a zwitterion so that there is little Ca^{2+} binding in this pH range. The same is true with phosphatidylethanolamine between pH 3 and 7 as both the amino group and phosphate are fully ionized and calcium adsorption is minimal. Above pH 7 the amino group is deprotonated and calcium adsorption occurs (Hauser & Dawson, 1967a; Papahadjopoulos, 1968). Phosphatidylserine shows a $\Delta V/\text{pH}$ plot which indicates that the carboxylic acid group is fully ionized at pH 5 and the amino group depolarized at pH 9. The binding of ^{45}Ca to such films (Rojas & Tobias, 1965) almost exactly corresponds to the ionization indicated by this plot (Fig. 5). Thus at saturation approximately one molecule of calcium is bound as the carboxylic acid group is ionized and a second as the amino group is depolarized. However, changes in the surface potential (ΔV) of phosphatidylserine films in the presence of $CaCl_2$ show an increased Ca binding with initial acid ionization but give no evidence of extra binding on the depolarization of the amino group. This phenomenon might be due to multiple binding of Ca^{2+} with both the acidic groups and the deprotonated amino group (Papahadjopoulos, 1968).

With the non-nitrogen containing phospholipids, $\Delta V/\text{pH}$ curves for phosphatidylinositol in the presence and absence of $CaCl_2$ indicate constant Ca binding as the bulk pH is increased above 3·5 and the single anionic site becomes fully dissociated. Also, $\Delta V/\text{pH}$ curves of phosphatidic

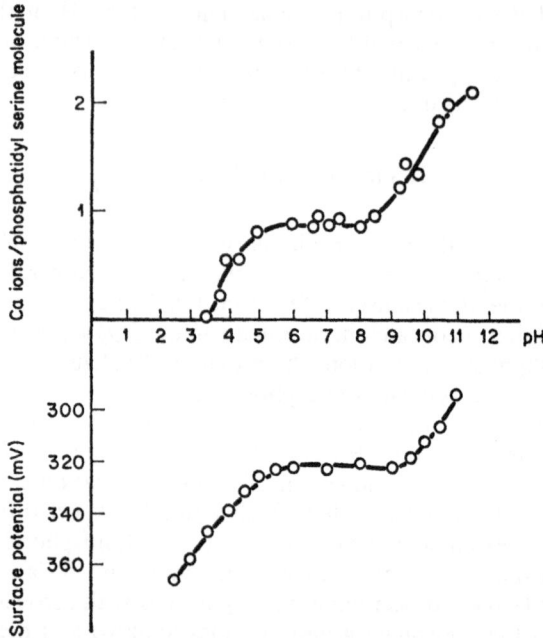

FIG. 5. Variation with pH of $^{45}Ca^{2+}$ adsorption on a phosphatidylserine monolayer (upper curve, Rojas & Tobias, 1965) and interfacial potential of a phosphatidylserine monolayer (lower curve, Papahadjopoulos, 1968).

acid show two apparent acid ionizations one of which is complete at pH 3·5 and one with an apparent pK_2 of 8·1 (Papahadjopoulos, 1968). In the presence of Ca^{2+}, pK_2 is shifted downwards to pH 7.1. It seems likely that the presence of calcium encourages the ionization of the phosphatidic acid at a lower pH. The ionizations

$$PA \rightleftharpoons HPA^- + H^+$$
$$HPA^- \rightleftharpoons PA^{2-} + H^+$$

would be displaced by the reaction

$$PA^{2-} + Ca^{2+} \rightleftharpoons CaPA$$

(Abramson *et al.*, 1964b, 1966). This displacement has been used to explain why a monolayer of phosphatidic acid on distilled water (pH 5·5) could adsorb ^{45}Ca as if it were completely dissociated (Hauser & Dawson, 1967a).

Position of calcium atoms in a phospholipid–water interface

Shah & Schulman (1965) interpret the surface potential changes observed when calcium is added under lecithin monolayers as indicating that the adsorbed Ca ions are lying just behind the interfacial plane of the phosphate charges. However, in view of what has previously been said concerning the interpretation of surface potential changes, and the fact that there is still disagreement concerning whether the phosphorylcholine moiety is oriented coplanar or vertically to the surface (Pethica, 1965; Hanai, Haydon & Taylor, 1965; Shah & Schulman, 1967a), such conclusions must be assessed with extreme caution.

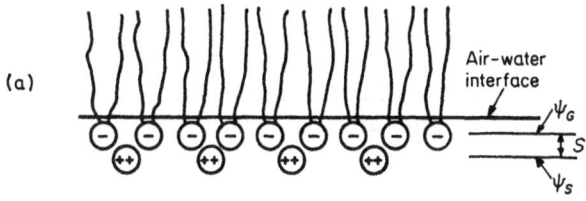

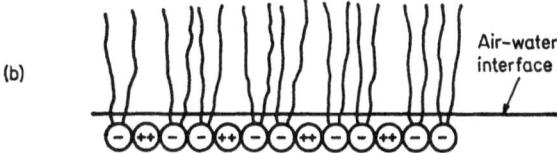

FIG. 6. Possible positions of Ca^{2+} beneath a monolayer of acidic phospholipid. (a) Ca^{2+} forming a Stern layer; (b) Ca^{2+} ions penetrating into the plane of the negative charges of the monolayer (Gouy). The plane of the Gouy potential and the plane of the Stern potential are marked by ψG and ψS respectively. (From Hauser, Chapman & Dawson, 1969.)

Hauser *et al.* (1969) have calculated how the surface potential ψ of a phosphatidylinositol monolayer changes as the ratio of the calcium bound/total lipid is varied. Two methods were used to calculate ψ: first, the Gouy equation which assumes that the calcium ions move into the plane of the phosphate groups (Fig. 6), and second, the Stern equation which assumes that the calcium ions lie outside the plane and towards the aqueous phase. The relationships obtained with the Stern equation fitted the experimental data best, and this would suggest that with this phospholipid the calcium atoms are located as shown in Fig. 6a.

Effect of calcium on the physical state of phospholipid monolayers
Monolayers of phospholipids at low pressures are generally in a liquid expanded state so that the fatty acid chains are liquid in a two-dimensional sense and can oscillate, rotate, and vibrate within the surface plane. As the pressure on the film is increased the molecules pack more closely and there is a tendency for them to gel and finally solidify to the condensed state.

The presence of saturated fatty acyl groups in the phospholipids encourages this solidification: for example, films of completely saturated lecithins and phosphatidylethanolamines will condense well below their collapse pressures (Watkins, 1968) whereas unsaturated phospholipids isolated from natural sources often remain in the liquid expanded state up to collapse pressures. Presumably the absence of an unsaturation 'kink' in the hydrocarbon chain allows better and more symmetrical packing of the molecules so that cohesive forces can produce a two-dimensional 'crystallization'. Such changes can be detected by surface viscosity measurements, or more simply by sprinkling talc particles on the film and observing their behaviour when they are blown by a light current of air.

The presence of calcium can cause a change in this surface rheology of phospholipid monolayers. Thus monolayers of dipalmitoyl lecithin on a subphase of 0·02 M NaCl usually solidify at pressures of 35–40 dyne/cm, whereas if 0·01 M $CaCl_2$ is added the films condense at 30–35 dyne/cm (Shah & Schulman, 1967b). This is presumably due to Ca^{2+} forming a bridge between two phosphate groups of adjacent lecithin molecules and thus restricting their degrees of freedom. If cholesterol is added to dipalmitoyl lecithin monolayers it becomes liquefied at high pressures and the effect of calcium is reversed. Presumably the cholesterol interferes with interactions between the fatty acid chains and prevents easy bridging of calcium between adjacent phosphate groups (Shah & Schulman, 1967c). On the other hand, films of the more unsaturated egg lecithin are in a liquid expanded state at all pressures up to collapse on subphases containing calcium (Shah & Schulman, 1967c). Presumably solidification does not occur because the cohesive forces are decreased by unsaturation and the reduced affinity of calcium. The unsaturated 'kink' in the fatty acid carbon chain causes a steric hindrance to the formation of a calcium bridge between adjacent molecules and a rigid structure.

Although the surface rheology of monolayers of acidic phospholipid has not been systematically examined it is to be expected that the strong interaction of calcium might cause a solidification. This has been described with monolayers of dicetylphosphoric acid (10^{-2} M Ca in subphase) but here again the presence of a 10–15 molecular percentage of cholesterol in the film is sufficient to bring about liquefaction. Solidification with monolayers of acidic phospholipid would require spacing between the head groups such that the calcium ions would bridge and form a transient lattice structure on the surface. Thus, calcium does not change the surface viscosity of phosphatidic acid or phosphatidylserine monolayers at a surface pressure of 5 dyne/cm where the spacing between the head groups is rather large (Deamer & Cornwell, 1966). Whether Ca solidification occurs with phospholipid films which are more compressed will depend on their degree of unsaturation and whether the temperature is low enough for a transition to occur below the collapse pressure.

If the fluidity of phospholipid structures is essential for the correct functioning of a biological membrane (Chapman & Wallach, 1968), then the presence of unsaturation in the fatty acid chains, the substantial dilution of the acidic phospholipids with cholesterol and zwitterionic phospholipids and homotherms in the elevated temperatures are all factors which would tend to prevent solidification. If calcium does produce its effect on the permeability of membranes by such a solidification process it would have to be in some region where these factors are minimized, for example, where the phospholipid in the membrane is already almost gelled by interaction with protein.

The solidification of a phospholipid film often leads to a change in the way in which the area of a film varies with the applied pressure (force–area curve). Condensed monolayers show steep surface pressure-area curves whereas liquid-expanded films are much more compressible and the pressure does not rise very steeply as the area is reduced. Force–area curves, however, are not a good indication of surface rheology. For example, whereas mixed monolayers of dicetylphosphoric acid and cholesterol are very incompressible on a subphase containing calcium ions, the film is nevertheless in the liquid state. The force–area curve of dipalmitoyl lecithin is not affected by calcium (0·01 M) in the subphase in spite of this producing solidification at much lower film pressures (Shah & Schulman, 1967c).

Calcium ions are known to produce a contraction of monolayers of certain acidic phospholipids; for example, in the presence of 10^{-2} M $CaCl_2$ and at constant pressure which may be high or low, the area occupied by each cardiolipin molecule is reduced by about 10–13% (Shah & Schulman, 1965). Similar contractions have been observed with phosphatidylserine and phosphatidic acid monolayers on adding 10^{-3} M $CaCl_2$ (Rojas & Tobias, 1965; Papahadjopoulos, 1968). Generally the contraction produced by calcium chloride is reversed by high concentrations of sodium chloride although the effect of the latter can vary with the investigator (Papahadjopoulos, 1968). It is usually assumed that the calcium counter ions reduce the electrostatic repulsion between the negatively charged head groups and cause contraction of the film, but a bridging effect may also be necessary. The importance of steric factors is suggested by the lack of contraction when $CaCl_2$ is introduced below a dicetylphosphoric acid monolayer (Shah & Schulman, 1965) although here the condensed state of monolayer produced by the saturated hydrocarbon chain may prevent the molecules moving closer together.

Displacement of calcium from monolayers
When ^{45}Ca is adsorbed on a monolayer of phospholipid, it can be completely displaced by adding excess unlabelled calcium chloride (Hauser & Dawson, 1967a; Kimizuka *et al.*, 1967). This shows that the calcium ions bound to a phospholipid surface even by direct two-point electrostatic attachment are not 'fixed' in the electrical double layer in an absolute

sense but can rapidly exchange with bulk phase Ca. This displacement finds a parallel in the way that ^{45}Ca bound to muscle tissue is lost more rapidly in a Ringer solution containing unlabelled calcium than in a solution which is calcium-free (Shanes & Bianchi, 1959).

^{45}Ca adsorbed onto phospholipid can also be displaced by adding other metallic cations to the subphase. Mg^{2+} is very much more efficient than the monovalent Na or K ions; even so the surface ^{45}Ca is only reduced by 50% when magnesium is added at a concentration 5 to 21 times greater than the calcium present, depending on the nature of the phospholipid (Table 2). Sodium and potassium are required at a concentration many thousand times greater than calcium to produce 50% ^{45}Ca displacement, and again

TABLE 2. *Displacement of $^{45}Ca^{2+}$ adsorbed on lipid/water interfaces by Na^+, K^+, and Mg^{2+}*

Lipid monolayer	Ratio of ion concentrations[a] required to reduce ^{45}Ca adsorption by 50%		
	$K^+/Ca^{2+} \times 10^{-2}$	$Na^+/Ca^{2+} \times 10^{-2}$	Mg^{2+}/Ca^{2+}
Phosphatidylserine	82	82	8·0
Phosphatidylinositol	50	50	5·2
Phosphatidic acid	78	80	16
Triphosphoinositide	85	86	21
Dicetylphosphoric acid	98	98	

[a] These ratios refer to the total ion concentrations present in the system. All unimolecular films at collapse pressure.

the precise amount is dependent on the chemical nature of the phospholipid film. There is, however, no difference in the displacing ability of Na^+ and K^+ at final equilibrium (Rojas & Tobias, 1965; Hauser & Dawson, 1967a; Kimizuka et al., 1967). As a biological parallel it has been shown by Koketsu et al. (1964) that Na^+ and K^+ were equally effective at impeding the binding of ^{45}Ca to the membrane fragments from bullfrog skeletal muscle fibres. Kimizuka et al. (1967) and Yamauchi, Matsubara, Kimizuka & Abood (1968) claim that Na^+ displaced ^{45}Ca from lecithin and stearic acid films faster than K^+ although the final equilibrium displacement was the same. In view of the fact that in the experiments with lecithin these authors were not using monolayers, and in contrast with previous work, the phospholipid was found to adsorb appreciable ^{45}Ca at pH values above 3, these results must await confirmation. With monolayers of acidic phospholipids we have not observed any difference between the rate at which Na^+ or K^+ displaces ^{45}Ca from monolayers. The displacement of Ca by other metallic cations may be regarded as a direct competition of the two metal ions for the available anionic sites on the surface. A theoretical form of the displacement curve can be calculated from:

$$Ca_b + Li_e \rightleftharpoons Ca_s$$
$$Na_b + Li_e \rightleftharpoons Na_s$$

where subscript b and s represent the surface and bulk concentrations of Na^+ and Ca^{2+}, respectively, and Li_e is the number of anionic sites per cm^2 lipid surface. Then

$$Li_e = \frac{Ca_s}{Ca_b}K_1 = \frac{Na_s}{Na_b}K_2$$

$$\frac{Ca_s Na_b}{Na_s Ca_b} = \frac{K_2}{K_1} = K$$

Assuming that at high Na^+ concentrations virtually all the anionic sites are occupied by sodium ions, and that the calcium ions are bound by two-point electrostatic attachment, then

$$Li_e - 2Ca_s = Na_s$$

Figure 7 shows the displacement of adsorbed ^{45}Ca from films of phosphatidylserine by Na^+ and K^+ and also the theoretical curve drawn from the mean constant K for all points ($K = 1 \cdot 51 \pm 0 \cdot 03$). The latter agrees well with the experimentally determined displacement.

It was observed by Hauser & Dawson (1968) that ^{45}Ca adsorbed onto acidic phospholipid monolayers could be easily removed by adding small amounts of pharmacologically active bases to the subphase. Drugs such as chlorpromazine or tetracaine were effective at displacing calcium in concentrations many hundreds of times less than Na^+ or K^+. Such drugs also produced an increase in the surface pressure ($\Delta\Pi$) of the monolayer which was kept at constant area, and this indicated that the drugs were penetrating into the film and occupying space at the interface. The variation of $\Delta\Pi$ and ^{45}Ca displacement from films of phosphatidyl inositol on adding the local anaesthetic tetracaine has been systematically examined (Hauser & Dawson, 1968). At a constant initial film pressure (10 dyne/cm) and on varying the amount of tetracaine added, $\Delta\Pi$ was directly proportional to the calcium displacement (Fig. 8). As the initial pressure of the film was varied the pressure increment on adding a constant amount of tetracaine was again related to the calcium displaced at all starting pressures above 4 dyne/cm (Fig. 9). As the pressure was increased from 5 to 40 dyne/cm the percentage calcium displaced was reduced as the increased packing of the phospholipid molecules prevented the penetration of the tetracaine into the film. When the starting pressure was less than 5 dyne/cm, the value of $\Delta\Pi$ fell and the relationship was lost. At such low pressures it is probable that the phospholipid fatty acyl chains are much more flopped over than those in the more densely packed films at higher pressures. Consequently, penetration of tetracaine molecules would presumably not register as such a marked surface pressure increment.

Although the surface concentration of tetracaine was not measured in these experiments it seems reasonable to assume that the $\Delta\Pi$ value at film pressures above 5 dyne/cm gives a reasonable assessment of this parameter.

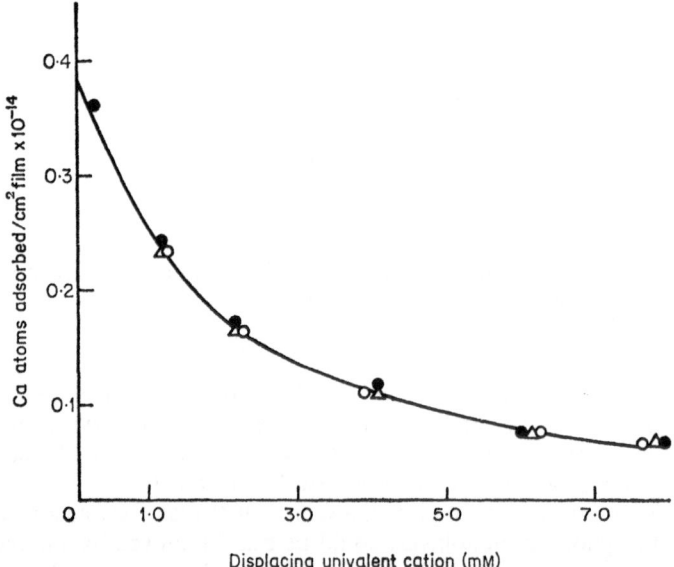

FIG. 7. Final displacement of ^{45}Ca from a monolayer of phosphatidylserine by NaCl (●) and KCl (△). The theoretical curve (○) has been drawn from the mean constant K calculated for all experimental points. (From Hauser and Dawson, 1967a.)

Displacement of calcium would therefore be due to the drug in the cation form competing with Ca^{2+} for the anionic sites on the phospholipid. In this competition it is presumably assisted by the amphipathic nature of the drug molecule so that its affinity for the film is greatly increased by non-ionic bonding (for example, van der Waals forces). This conclusion received confirmation from measurements made of the ability of cetyltri-methyl-ammonium ions (CETA+) added to the subphase to displace ^{45}Ca ions from phosphatidylinositol monolayers. In equivalent experiments the surface concentration of ^{14}C-labelled CETA+ in the phospholipid film was measured by a surface radioactivity technique, and this clearly showed that the displacing ability was related to the surface concentration of CETA+ (Hauser & Dawson, 1968).

In further experiments the ability of a series of simple straight chain

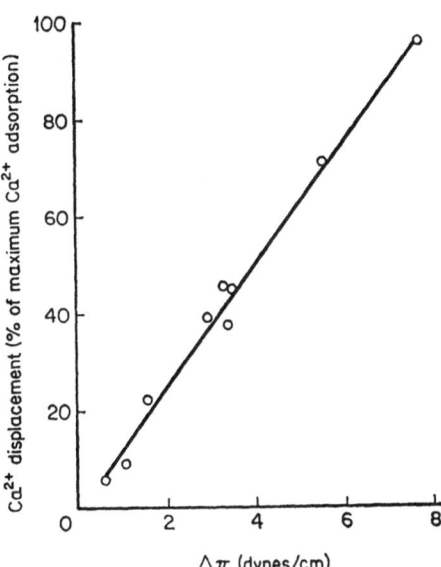

FIG. 8. ⁴⁵Ca²⁺ displacement plotted against the increase in surface pressure ($\Delta\Pi$) obtained on adding tetracaine in increasing amounts to the subphase below a monolayer of phosphatidylinositol at an initial starting pressure of 10 dyne/cm. (From Hauser and Dawson, 1968.)

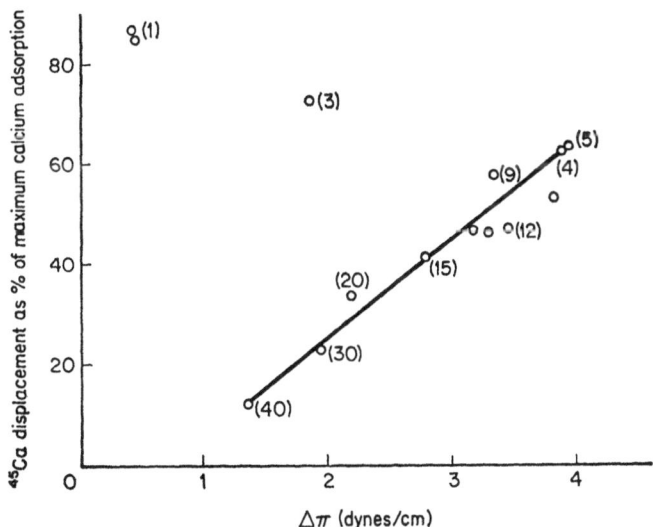

FIG. 9. Increase in surface pressure ($\Delta\Pi$) plotted against the displacement of ⁴⁵Ca²⁺ produced by adding tetracaine (0·1 μmole) to the subphase below phosphatidylinositol monolayers at various starting pressures (indicated in dynes/cm by the number in parentheses). (From Hauser and Dawson, 1968.)

aliphatic amines to displace ^{45}Ca from phospholipid monolayers was examined. Plotting the log of the concentration of amine required to produce 50% ^{45}Ca displacement against the number of methylene groups in the amine gave a linear relationship with a clear lateral shift between C_6 and C_7 amines (Fig. 10). This indicates that heptylamine can produce displacement more effectively than could be predicted from the behaviour

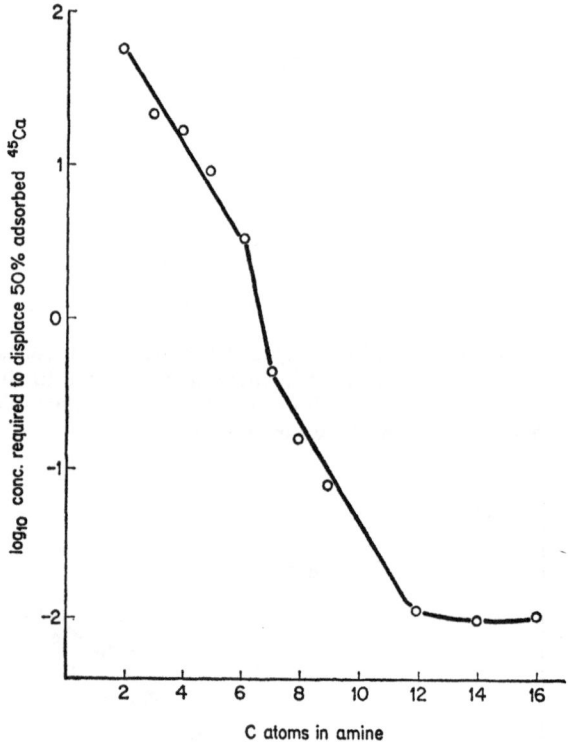

FIG. 10. Displacement of $^{45}Ca^{2+}$ from phosphatidylinositol monolayers (10 dyne/cm) by adding straight chain aliphatic amines to the subphase.

of the C_2–C_6 amines. Calculations and the building of molecular models show that if it is assumed that the protonated amino group of the displacing amine is in the same plane as the interfacial phosphate groups of the phospholipid, the C_7 chain would just reach into the hydrophobic fatty acyl chain region of the phospholipid assuming that the head group of the latter has a vertical orientation.

Although it has been suggested that the displacement of calcium from phospholipid interfaces may be of importance in the initiation of anaesthesia (Feinstein & Paimre, 1966; Blaustein, 1967) it is perhaps premature to extrapolate these results in this direction. It would be necessary first to explain why certain organic bases which are not anaesthetics can displace ^{45}Ca from phospholipid monolayers and why this property is also shown by excitory drugs, for example, brucine, strychnine, or amphetamine.

Summary

The monolayer technique for measuring Ca^{2+} binding to phospholipids has certain advantages over other methods in that the orientation of the molecules of phospholipid is known and the spacing between them can be varied. The Ca^{2+} bound can be assessed either by the change in interfacial potential or by an increment in surface radioactivity when ^{45}Ca is added. The former method is largely dependent on the change in ionic dipole produced by the Ca ions immediately adjacent to the monolayer polar head groups (fixed layer); but it is difficult to interpret and quantitate. The latter method accurately measures the net movement of calcium ions into the entire surface phase—fixed and diffuse electric double layers. The methods are therefore complementary.

At low calcium concentrations little binding occurs on zwitterionic phospholipids (for example, lecithin, phosphatidylethanolamine, sphingomyelin) in the physiological pH range, but some association takes place at higher concentrations. Monolayers of acidic phospholipids (for example, phosphatidylserine, phosphatidylinositol) avidly attract Ca^{2+}, provided the dissociations of the ionic groups on the polar head region produce a net negative charge. The adsorption isotherms are of the Langmuir type although deviations become apparent at high film pressures and Ca concentrations. At high pressures, the binding of calcium appears to be entirely Coulombic in nature and is directly related to the net negative charge on the acidic phospholipid rather than its chemical nature. The binding is strongest at intermediate film pressures when the average distance between the negative charges on the films approximates to the dimensions of a hydrated Ca ion and allows favourable two point electrostatic attachment. Under these conditions small variations in the binding to individual phospholipids become apparent. The binding of calcium to a phospholipid monolayer can cause a contraction of the film at constant pressure, and may change the surface rheology, encouraging gelling or solidification from the liquid state.

Bound ^{45}Ca can readily be displaced from phospholipid films by inactive Ca^{2+}, Mg^{2+}, Na^+, K^+ and pharmacologically active bases. To displace 50% of the adsorbed ^{45}Ca, Mg^{2+} is required at 5–20 times the calcium concentration and Na^+ and K^+ 5–10 thousand times more, depending on the chemical nature of the phospholipid. Na^+ and K^+ are always equally effective at displacing Ca^{2+}. The displacing ability of pharmacological bases is directly proportional to their penetration into the monolayer which can be measured as an increase in surface pressure. Non-ionic interactions (probably hydrophobic) between the drug and the phospholipid help the cationic group of the former to compete with calcium for the negative site on the polar head group of the phospholipid.

REFERENCES

ABRAMSON, M. B., KATZMAN, R. & GREGOR, H. P. (1964a). *J. biol. Chem.*, **239**, 70

ABRAMSON, M. B., KATZMAN, R., WILSON, C. E. & GREGOR, H. P. (1964b). *J. biol. Chem.*, **239**, 4066.

ABRAMSON, M. B., KATZMAN, R., CURCI, R. & GREGOR, H. (1967). *Biochemistry*, **5**, 2207

ABRAMSON, M. B., COLACICCO, G., CURCI, R. & RAPPORT, M. M. (1968) *Biochemistry*, **7**, 1692

ABRAMSON, M. B., KATZMAN, R. & CURCI, R. (1965). *J. colloid Sci.*, **20**, 777.

BANGHAM, A. D. (1968). *Progr. Biophys. mol. Biol.*, **18**, 29

BANGHAM, A. D. & DAWSON, R. M. C. (1962). *Biochim. biophys. Acta*, **59**, 103

BANGHAM, A. D. & PAPAHADJOPOULOS, D. (1966). *Biochim. biophys. Acta*, **126**, 181

BARTON, P. G. (1968). *J. biol. Chem.*, **243**, 3884

BLAUSTEIN, M. P. (1967). *Biochim. biophys. Acta*, **135**, 653

CHAPPELL, J. B., COHN, M. & GREVILLE, G. D. (1963). In *Energy Linked Functions of Mitochondria*, ed. Chance B., p. 219. New York: Academic Press

CHAPMAN, D. & WALLACH, D. F. H. (1968). In *Biological Membranes*, ed. Chapman, D. p. 125. New York: Academic Press

CHRISTENSEN, H. N. & HASTINGS, A. B. (1940). *J. biol. Chem.*, **136**, 387

DEAMER, D. W. & CORNWELL, D. G. (1966). *Biochim. biophys. Acta*, **116**, 555

DERVICHIAN, D. G. (1955). In *Biochemical Problems of Lipid*, ed. POPJAK, G. and LEBRETON, E. New York, N.Y.: Interscience

DRINKER, N. & ZINSSER, H. H. (1943). *J. biol. Chem.*, **148**, 187

FEINSTEIN, M. B. & PAIMRE, M. (1966). *Biochim. biophys. Acta*, **115**, 33

FOLCH, J. (1949a). *J. biol. Chem.*, **177**, 497

FOLCH, J. (1949b). *J. biol. Chem.*, **177**, 505

HANAI, T., HAYDON, D. A. & TAYLOR, J. (1965). *J. gen. Physiol.*, **48**, 59

HARRIS, E. J. (1957). *Biochim. biophys. Acta*, **23**, 80.

HAUSER, H., CHAPMAN, D. & DAWSON, R. M. C. (1969). *Biochim. biophys. Acta*, **183**, 320.

HAUSER, H. & DAWSON, R. M. C. (1967a). *Eur. J. Biochem.*, **1**, 61

HAUSER, H. & DAWSON, R. M. C. (1967b). *Biochem. J.*, **105**, 401

HAUSER, H. & DAWSON, R. M. C. (1968). *Biochem. J.*, **109**, 909

HAYDON, D. A. & TAYLOR, J. (1963). *J. theoret. Biol.*, **4**, 281

HENDRICKSON, H. S, & FULLINGTON, J. G. (1965). *Biochemistry*, **4**, 1599

HODGKIN, A. L. & KEYNES, R. D. (1957). *J. Physiol., London.*, **138**, 253

HOFER, M. & KLEINZELLER, A. (1963). *Physiol. Bohemslov.*, **12**, 405

KIMIZUKA, H., NAKAHARA, T., UEJO, H. & YAMAUCHI, A. (1967). *Biochim. biophys. Acta*, **137**, 549

KOKETSU, K., KITAMURA, R. & TANAKA, R. (1964). *Am. J. Physiol.*, **207**, 509

LANGMUIR, I. & SCHAEFER, U. J. (1936). *J. Am. chem. Soc.*, **58**, 284
PAPAHADJOPOULOS, D. (1968). *Biochim. biophys. Acta*, **163**, 240
PAPAHADJOPOULOS & BANGHAM (1966). *Biochim. biophys. Acta*, **126**, 185
PEACHEY, L. D. (1964). *J. cell. Biol.*, **20**, 95
PETHICA, B. A. (1965). *Soc. Chem. Ind. (Lond.) monog.*, **19**, 85
ROJAS, E. & TOBIAS, J. M. (1965). *Biochim. biophys. Acta*, **94**, 394
SALLEY, D. J., WEITH, A. J., ARGYLE, A. A. & DIXON, J. K. (1950). *Proc. R. Soc., A*, **203**, 42
SCHULMAN, J. H. & HUGHES, A. H. (1931). *Proc. R. Soc., A*, **138**, 430
SHAH, D. O. & SCHULMAN, J. H. (1965). *J. lipid Res.*, **6**, 341
SHAH, D. O. & SCHULMAN, J. H. (1967a). *J. lipid Res.*, **8**, 227
SHAH, D. O. & SCHULMAN, J. H. (1967b). *Lipids*, **2**, 21
SHAH, D. O. & SCHULMAN, J. H. (1967c). *J. lipid Res.*, **8**, 215
SHANES, A. M. & BIANCHI, C. P. (1959). *J. gen. physiol.*, **42**, 1123
SLATER, E. C. & CLELAND, K. W. (1953). *Biochem. J.*, **55**, 566
THIERS, R. E. & VALLEE, B. L. (1952). *J. biol. Chem.*, **226**, 911
WATKINS, J. C. (1968). *Biochim. biophys. Acta*, **152**, 293
YAMACHI, A., MATSUBARA, A., KIMIZUKA, H. & ABOOD, L. G. (1968). *Biochim. biophys. Acta*, **150**, 181

AEQUORIN—CALCIUM LUMINESCENCE AND ITS APPLICATION TO MUSCLE PHYSIOLOGY

C. C. ASHLEY
Department of Zoology, University of Bristol

E. B. RIDGWAY
Department of Biology, University of Oregon

There is much evidence to suggest that small transient changes in intracellular calcium play an important intermediate role in the complex process of excitation–contraction (E–C) coupling in skeletal muscle (Weber, Herz & Reiss, 1964; Portzehl, Caldwell & Rüegg, 1964; Jöbsis & O'Connor, 1966; Hellam & Podolsky, 1969). These changes in intracellular calcium can now be monitored during the contraction of a single muscle fibre by a photometric method (Ridgway & Ashley, 1967; Ashley & Ridgway, 1968, 1969, in preparation; Ashley, 1969). The method uses the remarkable properties of the photoprotein, aequorin, which is extracted from the photogenic organs of the hydromedusa, *Aequorea forskalea*.

Properties of aequorin
Aequorin under physiological conditions emits light solely in the presence of calcium ions and appears to be the simplest bio-luminescent system so far described (Shimomura, Johnson & Saiga, 1962, 1963). The reaction expressed in its most simple form is as follows:

$$\text{Aequorin} + \text{Ca}^{2+} \longrightarrow \text{Products} + \text{Ca}^{2+} + \text{Light}$$

No other factors except for calcium appear to be required for the reaction and the protein in its isolated state has been aptly described as a pre-charged molecule (Hastings, 1968). The protein has a high solubility (ca. 10 mg/ml.) and a relatively low molecular weight of 32,000 (Shimomura, & Johnson, 1969). Thus only small volumes of a concentrated solution of aequorin are needed and the photoprotein should be readily distributed throughout the fibre sarcoplasm.

Rapid mix experiments have indicated that the reaction of aequorin with excess calcium follows pseudo-first order kinetics with a half rise-time of about 6 msec at 20°C (van Leeuwen & Blinks, 1969; Hastings, Mitchell, Mattingly, Blinks & van Leeuwen, 1969). Additional experiments suggest that the response time of the aequorin may be limited by the presence of an intermediate in the luminescence reaction which has a $t_{1/2}$ of about 7 msec at 20°C (Hastings *et al.*, 1969). In these crustacean fibres, however, peak tension is reached in 400–600 msec (Fig. 2); because the response time of the aequorin is in the 5–10 msec range the sarcoplasmic calcium events associated with this slow contraction can be followed with some degree of precision.

Experiments with barnacle muscle fibres
After extraction and purification according to the scheme of Shimomura, Johnson & Saiga (1962), the aequorin was dissolved in potassium phosphate buffer (pH 6·1) and 0·4–0·5 μl. of this solution was injected into the large single muscle fibres from the barnacle *Balanus nubilus* (Hoyle & Smyth, 1963). The cannulation and micro-injection of these fibres has already been described. Tensions as high as 15 g can be recorded without the fibre pulling away from the cannula (Ashley, 1967). The fibre was stimulated by a dual, axial stimulating-recording electrode similar to that used for experiments on squid giant axons (Hodgkin, Huxley & Katz, 1952), except that the uninsulated region of the stimulating electrode (50 μ diameter silver) extended the length of the fibre (1·9 cm), while the recording electrode (25 μ diameter platinum) was not insulated over a region of 2–3 mm some 0·9 cm from the tendon end of the fibre. The output from the recording electrode was connected to the input stage of a cathode follower.

The light emission caused by the interaction between aequorin and the intracellular calcium released by the stimulus was monitored by a sensitive end-window photomultiplier tube (RCA 6342 A); the resulting signal has been termed the *calcium transient* (Ridgway & Ashley, 1967). The 2-inch diameter face-plate of the photomultiplier was optically coupled to a thin glass plate which formed part of the front surface of the muscle-fibre chamber (Fig. 1). When in position in the chamber the whole length of the single fibre was monitored during the course of the experiment, and the face-plate of the photomultiplier was only 3–4 mm from the front surface of the muscle fibre. Tension from these single fibres on stimulation was monitored using a RCA 5734 mechano-electric transducer via a small lever system (Fig. 1). One end of the lever was attached to the tendon of the fibre by a stainless steel hook, and the other end was attached to the anode peg of the transducer by a length of fine gold chain. The experiments were performed at 11°–12°C which is close to the ambient temperature experienced by the barnacles in their natural environment.

With this experimental arrangement the relationships between perhaps the three most important processes in E–C coupling can be examined during a single contraction. The time course of these three processes, namely the membrane depolarization (electrical event), calcium release

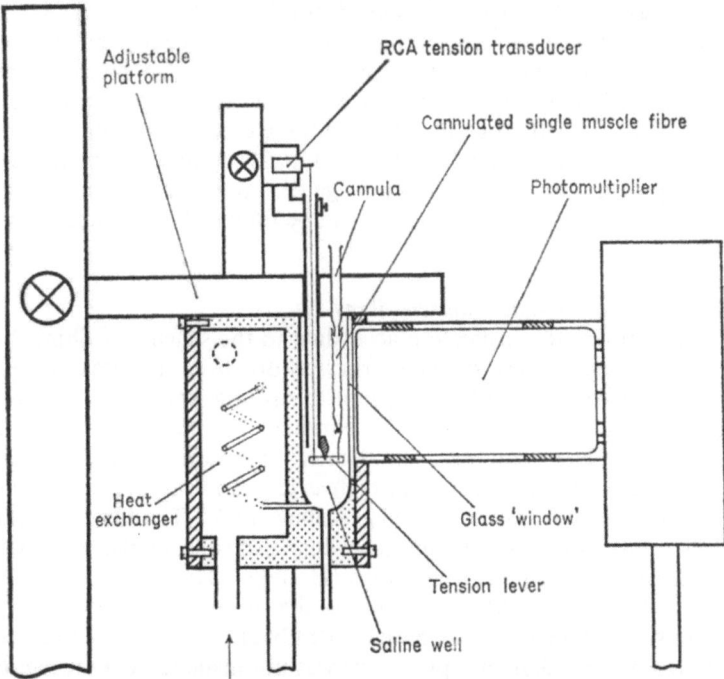

FIG. 1. Diagrammatic representation of the single muscle-fibre chamber. The fibre was firmly attached to the glass cannula by silk ligatures (not shown) and immersed in the well containing constant temperature, circulating, crab saline (Fatt & Katz, 1953). The cannula mount and tension transducer were arranged on a separate adjustable platform which was readily raised or lowered. A light-tight box containing thermal insulating material surrounded the chamber. The face-plate of the RCA 6342A photomultiplier was optically coupled (Dow-Corning 20-057) to the thin glass 'window'. (From Ashley & Ridgway, 1969b.)

and reaccumulation (the calcium transient; chemical event) and isometric tension (mechanical event) are illustrated in the traces shown in Fig. 2. The intermediate event in the sequence is obviously the calcium transient (trace 2) whose rapid rising phase, representing an increase in the intra-cellular calcium concentration, is dependent on the membrane depolariza-tion. The rapid falling phase of the transient, initiated during membrane repolarization, represents a decrease in the intracellular calcium concen-tration. This rapid falling phase can be approximated to an exponential having a time constant of 60–80 msec. The calcium transient begins after the onset of the membrane response but before there is detectable tension. The peak of the calcium transient coincides with the maximum rate of development of tension while the calcium transient is virtually complete at peak tension. Thus visual inspection of the calcium transient trace suggests

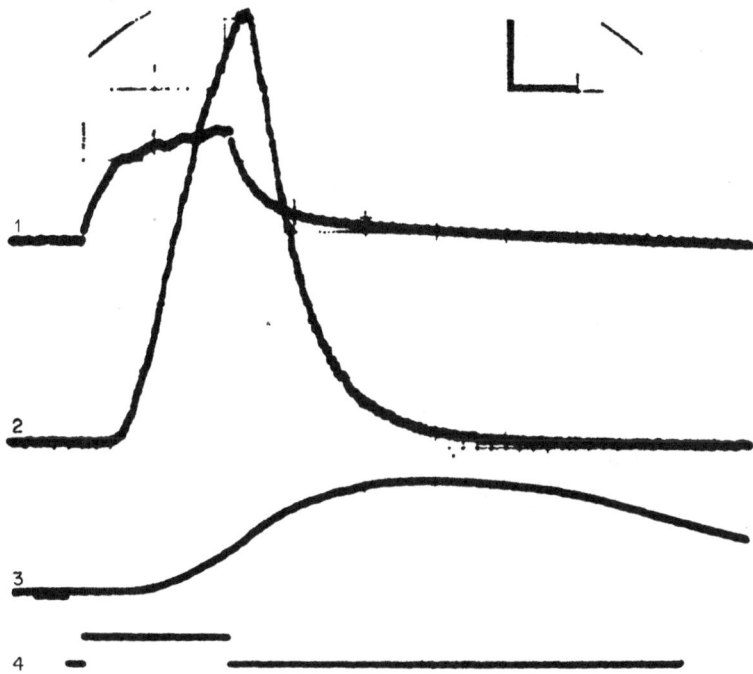

FIG. 2. Result of applying a single depolarizing pulse to the fibre membrane. Trace 1, membrane response; trace 2, calcium-mediated light emission; trace 3, isometric tension; trace 4, calibration pulse and stimulus mark. Nominal strength: 3·5; nominal duration: 200 msec. Calibration: vertical, trace 1, 20 mV cm^{-1}; trace 2, 1·9 × 10^{-9} lumen; trace 3, 5 g; trace 4, 1 V cm^{-1} calibration pulse; horizontal, 100 msec. Temperature 11°–12°C. Resting light emission 0·9 × 10^{-9} lumen.

a first derivative relationship between the calcium transient and the rising phase of tension. Finally, it is of interest to note that relaxation proceeds at what are virtually resting calcium levels.

If the intensity of the stimulus is increased at constant stimulus duration (Fig. 3) there is an increase in the rate of rise and in the maximum value of the membrane response. Associated with the increased membrane response there is an increase in the rate of rise, maximum value and area of the corresponding calcium transient. This in turn produces an increase in the maximum rate of rise and maximum value of the resulting isometric tension response. In each case the peak of the calcium transient coincides with the maximum rate of rise of the corresponding tension response, while the transient is virtually complete at the peak of the tension response. Associated with the increase in the stimulus intensity there is a decrease in the latent period between the onset of the membrane response and the

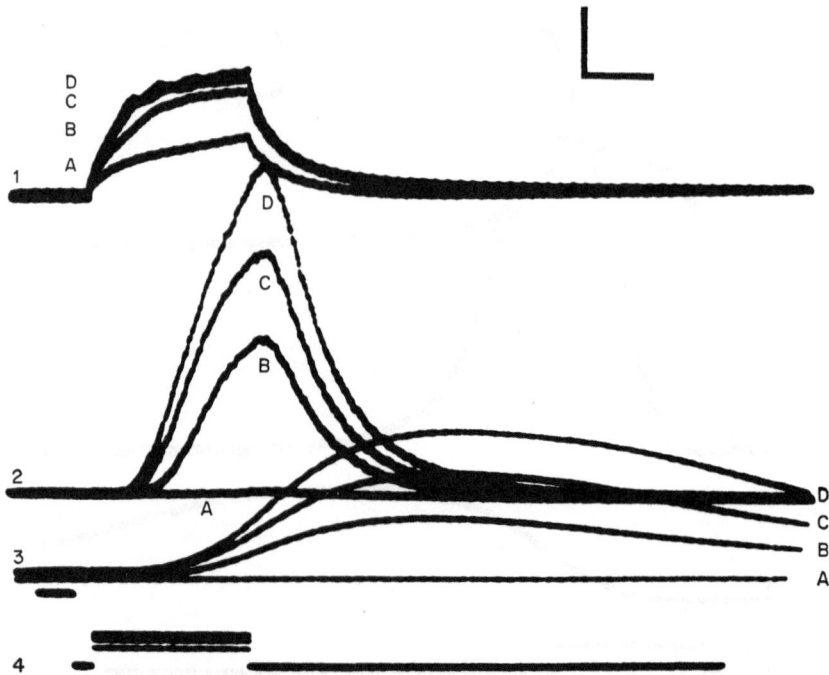

FIG. 3. Effects of increasing the stimulus strength at constant stimulus duration. Trace 1, membrane response; trace 2, calcium-mediated light emission; trace 3, isometric tension; trace 4, calibration pulse and stimulus marks. Nominal strength: A, 2·0; B, 3·0; C, 3·5; D, 4·0. Nominal duration: 200 msec. Calibration: vertical, trace 1, 20 mV cm^{-1}; trace 2, 3·8 × 10^{-9} lumen; trace 3, 5 g; trace 4, 1 V calibration pulse; horizontal, 100 msec. Temperature 11°–12°C. Resting light emission 0·9 × 10^{-9} lumen. (From Ashley & Ridgway, 1968.)

onset of detectable tension. This can be accounted for satisfactorily by a decrease in the delay between the onset of the membrane response and the onset of the calcium transient (Fig. 3).

In an associated experiment, where the stimulus intensity remains constant, increasing the stimulus duration (Fig. 4) prolongs the membrane depolarization and thus prolongs the S-shaped rising phase of the calcium transient. This produces an increase in the maximum value and area of the respective calcium transient and in turn increases the maximum rate of rise and maximum value of the corresponding isometric tension response. As in Figs. 2 and 3, the peaks of the calcium transients coincide with the maximum rate of rise of the corresponding tension responses, and peak tension occurs when the light and hence calcium level is virtually at resting values. At constant stimulus intensity, the delays remain constant between

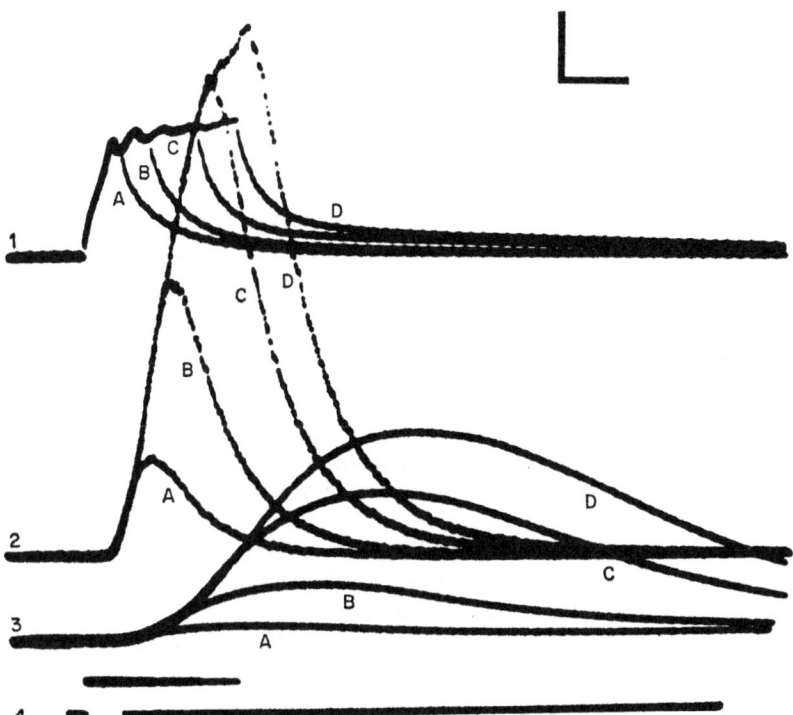

FIG. 4. Effects of increasing the stimulus duration at constant stimulus strength. Trace 1, membrane response; trace 2, calcium-mediated light emission; trace 3, isometric tension; trace 4, calibration pulse and stimulus marks. Nominal durations (msec): A, 50; B, 100; C, 150; D, 200. Nominal strength: 4·0. Calibration: vertical, trace 1, 20 mV cm^{-1}; trace 2, 1·9 × 10^{-9} lumen; trace 3, 5 g; trace 4, 1 V calibration pulse (indistinct); horizontal: 100 msec. Temperature 11°–12°C. Resting light emission ca. 0·9 × 10^{-9} lumen. (From Ashley & Ridgway, 1968.)

the onset of the membrane response, the onset of the calcium transient and the onset of the isometric tension response.

Visual inspection of the calcium transient traces presented in Figs. 2, 3, and 4 suggests a first derivative relationship between the calcium transient, and the rising phase of the corresponding isometric tension response. Moreover, if the *rate* of tension development is proportional to the calcium transient, then peak tension must be proportional to the area of the calcium transient; in particular we can write that

$$P_{max} = k \cdot S_{I_g} + C \qquad (1)$$

where P_{max} is the maximum isometric tension, S_{I_g} is the total area of the calcium transient and C is the constant of integration.

Equation (1) specifically states that P_{max} depends on the total area of the calcium transient and not on the actual shape of the transient. Moreover, graphical integration of each calcium transient presented in Figs. 3 and 4 yields a linear relationship, at least over a range extending from 50–500 g cm^{-2}, when plotted against the respective isometric tension values. Further observations on a series of longer duration transients and from brief tetani confirm this relationship as stated in equation (1). In addition preliminary studies using the large single muscle fibres from the crab *Maia* (Ashley, 1969, and unpublished experiments) indicate that, over a range of tension values, peak tension is linearly related to the area of the respective calcium transient.

At the present time, however, there is no accurate and unequivocal *in vivo* calibration of the calcium transient traces in terms of the absolute calcium concentration. In addition the linear relation between the area of the calcium transient and P_{max} certainly does not prove that both the reaction of the myofibrils and aequorin with calcium ions is linear, although if both reactions are non-linear they must be non-linear in exactly the same way.

When the duration of the stimulus pulse is increased beyond 300 msec there is a distinct change in the shape of the associated calcium transient response (Fig. 5a). The familiar S-shape rising phase is followed by a slow falling phase, and both phases are mediated by the membrane depolarization. Finally during membrane repolarization the rapid falling phase of the transient is initiated. At longer stimulus durations the slow falling phase forms a plateau, whose height above the resting light value is dependent on the intensity of the stimulus pulse. The tension response continues to rise at a slow rate during this slow falling phase of the calcium transient. In the case of a brief tetanus (Fig. 5b), each membrane response produces a definite calcium transient which in turn leads to a change in the rate of development of tension. At this stimulus frequency all the responses are partially fused. It is clear, however, that the calcium transient responses are not all of equal intensity; the third and fourth transients are smaller than the second, even though the membrane responses which initiate these transients are very similar in size and are of identical duration (see legend to Fig. 5b). Moreover, at the stimulus intensity illustrated in Fig. 5b, an envelope bounding the peaks of the calcium transients during the tetanus is very similar to the shape of the transient in the long duration pulse (Fig. 5a).

This decline in the light emission seems to be dependent on the membrane depolarization. At low stimulus intensities, both for the long duration pulse and the tetanus, the slow fall is not observed, although the calcium transient at these intensities is still sufficient to produce a detectable tension response. Whatever its underlying cause the phenomenon is certainly of great interest in the process of E–C coupling, for it results in a decreased concentration of calcium ions in the fibre sarcoplasm, and this produces a

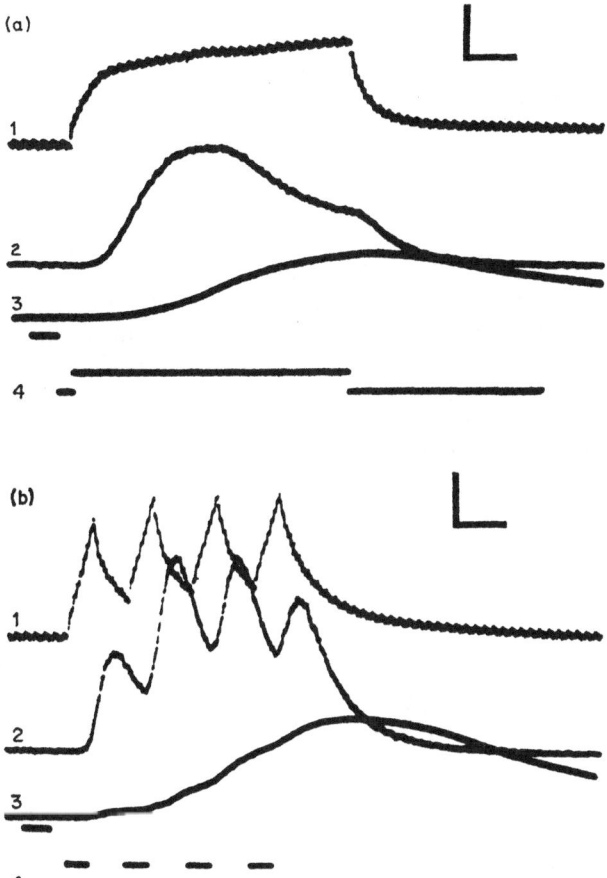

FIG. 5. Effect of applying a long duration (a) and brief tetanic pulse (b) to a fibre. Trace 1, membrane response; trace 2, calcium-mediated light emission; trace 3, isometric tension; trace 4, calibration pulse and stimulus mark. Nominal intensity: 3·0; nominal duration (msec): (a) 450, (b) 40. Calibration: vertical, trace 1, 10 mV cm^{-1}; trace 2, (a) 3·8 × 10^{-9} lumen and (b) 1·9 × 10^{-9} lumen; trace 3, 2·5 g; trace 4, 1 V calibration pulse; horizontal, 100 msec. Temperature 11°–12°C. Resting light emission (a) 1·1 × 10^{-9} lumen and (b) 0·6 × 10^{-9} lumen.

considerable reduction in the rate of rise of tension (see Ashley & Ridgway, 1969, and in preparation).

Intracellular injection of certain compounds, including potassium citrate, converts the normally passive external membrane of single barnacle muscle fibres to one capable of supporting an active all-or-nothing action potential or spike (Hagiwara & Naka, 1964; Hagiwara, Chichibu

& Naka, 1964). When a single fibre is initially injected with potassium citrate (pH 7·3) and then with aequorin, there is a considerable suppression of the tension responses with little change in the calcium transient associated with the active membrane response. At a stimulus duration of about 300 msec and an intensity less than that normally required to produce a transient, the fibre was capable of supporting two active membrane responses within the time course of single depolarizing pulse (Fig. 6a). The first calcium transient rises very steeply after the onset of the first active membrane response, reaches a peak at the end of the response and begins an exponential fall. The second calcium transient occurs soon after the onset of the second active membrane response, which is similar in amplitude to the first, but which initiates a much larger calcium transient. The second calcium transient reaches a peak and starts to fall after the end of the second active membrane response.

Tension responses are much suppressed in both instances although the second transient results in more tension than the first. It is interesting to note that the second active membrane response is initiated while the light, representing the calcium level, is still about 3·8 nlumen above the resting light level of 0·68 nlumen. It appears therefore tl.at potassium citrate is

(a)

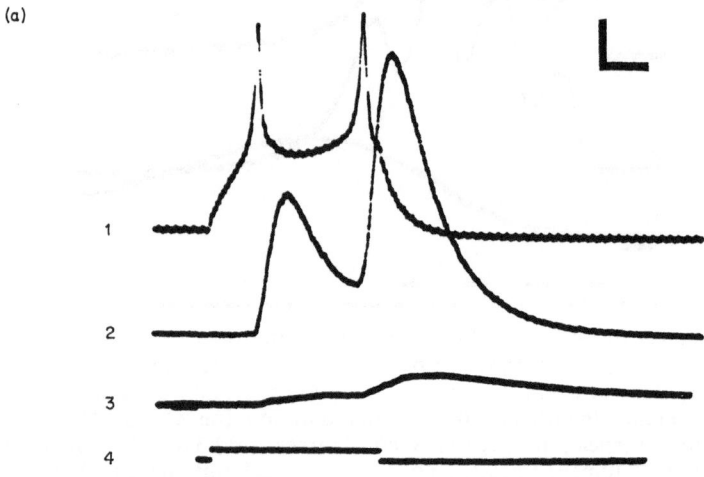

Fig. 6 (a). Effect of potassium citrate applied internally on the fibre responses. (a) Trace 1, membrane response; trace 2, calcium-mediated light emission; trace 3, isometric tension; trace 4, calibration pulse and stimulus mark. Nominal intensity, 1.6; nominal duration, 300 msec. Calibrations: vertical— trace 1, 10 mV cm^{-1}; trace 2, $3\cdot8 \times 10^{-9}$ lumen; trace 3, 0·35 g; trace 4, 1 V calibration pulse; horizontal—100 msec. Citrate concentration estimated assuming uniform dilution inside the fibre as 5–6 mM, external saline contained 20 mM calcium and 100 mM magnesium. Resting light emission: $0\cdot68 \times 10^{-9}$ lumen.

(b)

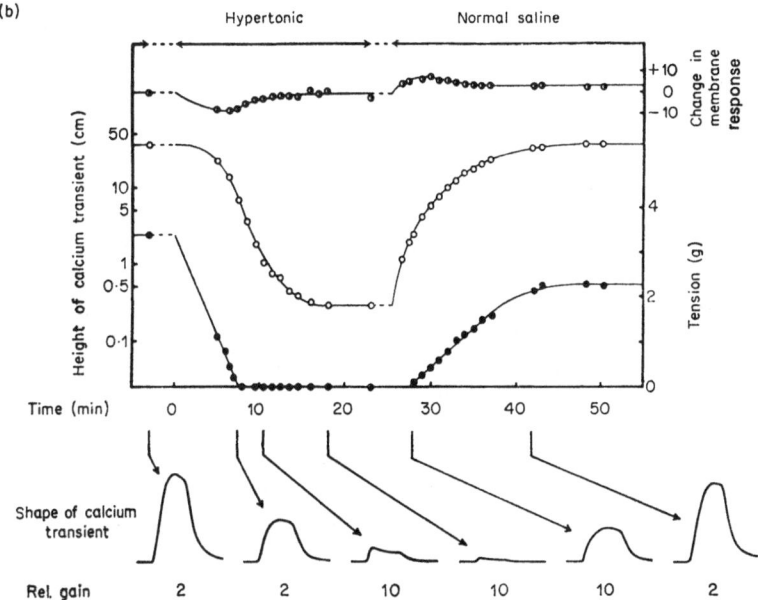

Fig. 6(b). Effect of potassium citrate applied externally on the fibre responses. Top line represents the time in the different salines; second line (◗), the relative change in membrane response (mV); third line (○), the change in the height of the calcium transients on a logarithmic scale, and fourth line (●), the change in the values of peak tension (P_{max}). The last lines illustrate the shape of the calcium transients at various stages during the experiment, together with the relative gains. Saline: 1 M glycerol in addition to the normal concentration of ions. Nominal duration of the pulses: 250 msec, nominal strength: 5·0. Temperature 11°–12°C.

able to interrupt, at least partially, the normal processes of E–C coupling at the level between calcium release and the production of tension. Hypertonic salines, however, appear to suppress both the calcium transient and tension responses but leave the membrane responses virtually unaffected. This is illustrated by the data presented in Fig. 6b, where the tonicity of the bathing saline has been increased by the addition of 1 M glycerol (or 1 M sucrose). Application of the hypertonic saline produces a rapid decline in the calcium transient and tension responses and after about 8 min there is no longer any detectable tension. After about 15 min in the hypertonic saline the calcium transient responses have declined to a small but apparently constant size. Reapplication of normal bathing saline produces a rapid restoration of the calcium transient and tension responses and after about 20 min both sets of responses are restored to 80–90% of their original values. Associated with the decline in the height of the

3—C.C.F.

calcium transient there is also a marked change in shape, the fast rising phase becoming attenuated and the slow falling phase considerably prolonged. The transient is terminated by the familiar faster falling phase initiated during membrane repolarization.

Examination under the electron microscope of frog muscle fibres soaked for short periods in hypertonic salines indicates that, despite an overall decrease in fibre volume, there is a dilation of the central as well as outer elements of the triad (Huxley, Page & Wilkie, 1963). The central element of the triad is part of the transverse (T) tubular system which is continuous with the outer fibre membrane, whilst the outer elements of the triad form part of the lateral cysternae of the longitudinal SR (from two different sarcomeres in frog) which form the main calcium 'sink' inside the fibre. The arrangement of these tubules is slightly simpler in the barnacle (Hoyle, 1965), where the T tubules form a series of diads with the longitudinal SR. Evidence from the local activation experiments on single frog and crab fibres (Huxley & Taylor, 1958) suggests that certain of the T tubules and presumably the associated diads and triads are vital in the linking of membrane depolarization to the release of intracellular calcium. A temporary dilation of the lumina of the tubules forming the diads or triads by a fairly brief application of hypertonic salines could well produce the temporary interruption in E–C coupling observed in Fig. 6b.

The photoprotein aequorin has been shown to be a useful indicator of calcium *in vivo*. Although in the present experiments its usefulness has been restricted to the monitoring of changes in intracellular calcium, there is work in progress on a simple calibration of the light emission in terms of the absolute calcium concentration. With regard to the role played by calcium during the contraction of a single barnacle muscle fibre, the results can best be summarized by saying that the level of calcium controls the *rate* of tension development.

This work was supported in part by Training Grant 2T1 GM 336, P.H.S Grant NBO 381904 and by an M.R.C. Grant G/968/399/B.

REFERENCES

ASHLEY, C. C. (1967). In *Comparative Aspects of Muscle*, ed. HOYLE, G., *Amer. Zool.*, **7**, 647
ASHLEY, C. C. (1969). *J. Physiol. Lond.*, **203**, 32P
ASHLEY, C. C. & RIDGWAY, E. B. (1968). *Nature, Lond.*, **219**, 1168
ASHLEY, C. C. & RIDGWAY, E. B. (1969). *J. Physiol. Lond.*, **200**, 74P
ASHLEY, C. C. & RIDGWAY, E. B. (1969b). (In preparation)
FATT, P. & KATZ, B. (1953). *J. Physiol., Lond.*, **120**, 171
HAGIWARA, S. & NAKA, K. (1964). *J. gen. Physiol.*, **48**, 141
HAGIWARA, S., CHICHIBU, S. & NAKA, K. (1964). *J. gen. Physiol.*, **48**, 163
HASTINGS, J. W. (1968). *A. Rev. Biochem.*, **37**, 597

HASTINGS, J. W., MITCHELL, G., MATTINGLY, P. H., BLINKS, J. R. & VAN LEEUWEN, M. (1969). *Nature, Lond.*, **222**, 1047

HELLAM, D. C. & PODOLSKY, R. J. (1969). *J. Physiol. Lond.*, **200**, 807

HODGKIN, A. L., HUXLEY, A. F. & KATZ, B. (1952). *J. Physiol., Lond.*, **116**, 424

HOYLE, G. (1965). *Science, N.Y.*, **149**, 70

HOYLE, G. & SMYTH, T. (1963). *Comp. Biochem. Physiol.*, **10**, 291

HUXLEY, A. F. & TAYLOR, R. E. (1958). *J. Physiol., Lond.*, **144**, 426

HUXLEY, H. E., PAGE, S. & WILKIE, D. R. (1963). In Appendix of DYDYŃSKA, M. & WILKIE, D. R., *J. Physiol., Lond.*, **169**, 312

JÖBSIS, F. F. & O'CONNOR, M. J. (1966). *Biochem. biophys. Res. Commun.*, **25**, 246

PORTZEHL, H., CALDWELL, P. C. & RÜEGG, C. (1964). *Biochim. biophys. Acta*, **79**, 581

RIDGWAY, E. B. & ASHLEY, C. C. (1967). *Biochem. biophys. Res. Commun.*, **29**, 229

SHIMOMURA, O. & JOHNSON, F. H. (1969). *Biochemistry, N. Y.* **8**, 3991

SHIMOMURA, O., JOHNSON, F. H. & SAIGA, Y. (1962). *J. cell. comp. Physiol.*, **59**, 223

SHIMOMURA, O., JOHNSON, F. H. & SAIGA, Y. (1963). *J. cell. comp. Physiol.*, **62**, 1

VAN LEEUWEN, M. & BLINKS, J. R. (1969). *Fedn Proc. Fedn Am. Socs exp. Biol.*, **28**, No. 571

WEBER, A., HERTZ, R. & REISS, I. (1964). *Proc. Roy. Soc.*, B, **160**, 489

DISCUSSION TO SESSION I

Baker (*Cambridge*)

I should like to ask Dr. Gillard how easily do Ca^{2+} ions become hydrated in aqueous solutions? I ask this because it seems likely that in biological systems Ca ions might normally be hydrated. Second, to what extent would hydration alter any of your arguments—especially with reference to the differences between Ca^{2+} and Mg^{2+} ions?

Gillard (*Canterbury*)

Ca^{2+} ions are, of course, like all small highly charged species, hydrated in solution. I should have made it clear that the equilibrium (stability) constants quoted refer to processes involving hydrated ions of the type

$$[M(H_2O)_n]^{2+} + L^{n-} \underset{k_b}{\overset{k_f}{\rightleftharpoons}} [ML]^{(2-n)+} + nH_2O$$

where M is either Ca or Mg, L is a ligand molecule (itself usually solvated), and n is the hydration number of the metal ion in water. The basic arguments relating to complex formation which I have given are essentially independent of hydration.

I should perhaps add two points. First, there are clearly circumstances in which the size difference between the hydrated Ca^{2+} and Mg^{2+} ions may be important. Secondly, the rate of displacement of water from calcium by L (k_f in the equilibrium above) is commonly greater than k_f for magnesium. In special circumstances, for example, those relating to membrane potentials, both effects may well come into play.

Goodford (*London*)

Can the reactions of calcium be treated as infinitely rapid on a biological time scale of milliseconds?

Hague (*Canterbury*)

In general, complexes of Ca^{2+} are formed approximately three orders of magnitude more rapidly than the corresponding complexes of Mg^{2+}. The second-order rate constants for Ca^{2+} are typically of the order of $10^9 \, l. \, m^{-1} \, sec^{-1}$, indicating that the reactions are close to being diffusion-controlled. The half-time for a reaction involving calcium complex formation will evidently depend on the concentrations of the reactants, but if they are as low as 10^{-6} M the time will be in the millisecond range.

Cowan (Portsmouth)
Can I ask Dr. Ashley the nature of the first order reaction between calcium and aequorin?

Ashley (Bristol)
At the present time we have to rely solely on the original experiments of Shimomura, Johnson & Saiga (1962), who reported that at several different calcium concentrations, the light emission from the aequorin system decayed in a way that suggested first order kinetics. More detailed kinetic analysis of the system has not yet been undertaken. (*Added in proof:* These kinetic experiments have recently been performed by Hastings *et al.* (1969) using rapid mix techniques.)

Birks (Montreal)
Could the diphasic response of calcium release seen during a prolonged depolarization be interpreted in terms of calcium removal by a calcium pump that is only turned on when the membrane of the sarcoplasmic reticulum is repolarized.

Ashley (Bristol)
Additional experiments would suggest that the slower falling phase of the calcium transient, observed during continued membrane depolarization, can be more easily explained in terms of a decreased rate of release rather than by an increased rate of uptake. It is likely, however, that the fast, exponential falling phase of the transient, initiated during membrane repolarization, can at least partially be accounted for by the removal of calcium from the sarcoplasm by the sarcoplasmic reticulum.

REFERENCES

HASTINGS, J. W., MITCHELL, G., MATTINGLY, P. H., BLINKS, J. R. & VAN LEEUWEN, M. (1969). *Nature, Lond.,* **222**, 1047
SHIMOMURA, O., JOHNSON, F. H. & SAIGA, Y. (1962). *J. cell. comp. Physiol.,* **59**, 223

Session II

Calcium and Cell Surfaces

Chairman:

L. WOLPERT

*Department of Biology as Applied to Medicine,
Middlesex Hospital Medical School*

DIVALENT CATIONS AND CELL ADHESION

D. GINGELL & D. R. GARROD
Department of Biology as Applied to Medicine,
Middlesex Hospital Medical School, London, W.1

J. F. PALMER
Department of Physiology,
Middlesex Hospital Medical School, London, W.1

Divalent cations have long been known to play a crucial part in the adhesive functions of cells. As long ago as 1894 Roux showed that media lacking Ca^{2+} ions aided the disaggregation of frog embryos. Herbst (1900) discovered that the blastomeres of echinoderm embryos fall apart in Ca^{2+} free seawater, though it is now known that the principal effect is dissolution of the hyaline capsule. Later workers frequently used chelating agents, most commonly sodium salts of ethylenediaminetetracetic acid (EDTA), to help the removal of divalent cation and bring about disaggregation. Its effectiveness has been demonstrated by Anderson (1953) on mammalian cells, by Zwilling (1954) on chick embryos, by Coman (1954) in adult rat liver, and by Curtis (1967) on amphibian embryos. Because EDTA chelates divalent cations in the preferential sequence

$$Ca^{2+} > Mg^{2+} > Sr^{2+} > Ba^{2+}$$

however, disaggregation in EDTA might be connected with chelation of less strongly bound cations, as well as Ca^{2+}. EGTA which is potentially capable of distinguishing between Ca^{2+} and Mg^{2+}, since the respective stability constants of Ca^{2+} and Mg^{2+} complexes are 10^{11} and 10^{5}, does not seem to have been utilized for this purpose.

Where a number of divalent cations are capable of promoting adhesion, it is difficult to decide which are of importance in natural adhesion. Curtis (1957) and Steinberg (1958, 1962) have shown that Ca^{2+}, Mg^{2+}, and Sr^{2+} promote reaggregation of amphibian embryonic cells, while Armstrong (1966) found Mg^{2+} to be more effective than Ca^{2+} in the case of embryonic chick cells. Weiss (1960), however, found that Ca^{2+} is specifically required for the adhesion of sarcoma cells to glass. Armstrong & Jones (1968) have similarly found that only Ca^{2+} can prevent disaggregation of *Xenopus* embryos by EDTA, whereas Lipman, Dodelson & Hays (1966) report that although Ca^{2+}, Mg^{2+}, or Ba^{2+} bind to the surface of toad bladder cells, only Ca^{2+} is capable of promoting adhesion. This work is also of great interest with regard to the mechanism of cell adhesion and will be referred to again.

Although there is no doubt that divalent cations, and Ca^{2+} in particular, are involved in adhesive processes, their mode of action is much disputed and there are basically two schools of thought regarding their role. Steinberg (1958, 1962), Pethica (1961), and Moscona (1968) have all supported the concept of direct or indirect ionic bridging between adjacent cell surfaces by divalent cations. It has been calculated by Curtis (1967) that only six bonds of 20 kcal./mole per 0.1 μ^2 would give an adhesive energy of 8×10^{-5} erg/cm and account for most adhesive energies reported (Brooks, Millar, Seaman & Vassar, 1967). Curtis (1962, 1967), in contrast, has maintained that cell adhesion is a colloidal phenomenon. The colloid theory states that a stable separation is achieved when the van der Waals–London attractive forces between the surfaces of particles are equal to the electrostatic repulsive forces due to surface charges Because all cells investigated have net negative surface charges, the removal of divalent cations from the double layer should increase the surface potential and thus increase the force of repulsion between cells. This would tend to decrease the energy of adhesion and lead to disaggregation. The theory thus embodies a clear prediction that the surface potentials of cells should be more negative when sufficient EDTA to inhibit stable adhesion is present.

We decided to examine the action of EDTA on the electrophoretic properties of cells of the cellular slime mould, *Dictyostelium discoideum*, at the pre-aggregation stage of the life cycle in relation to intercellular adhesiveness (Gingell & Garrod, 1969). Previous studies by De Haan (1959) and Gerisch (1961) have shown that EDTA reduces the adhesiveness of these cells and also of cells at later stages in the cycle.

A group of cells was incubated in a buffered salt containing 0·5 mg/ml. $MgCl_2$ while a second group were incubated without divalent cations but with 10^{-2} M EDTA. After 2 hr the cells were washed in distilled water and resuspended for electrophoresis in 10^{-2} M phosphate buffer, pH 7.05. Cells treated with EDTA were suspended in this medium containing additional 10^{-3} M EDTA. Electrophoresis was performed at 23°C in a Zeiss Cytopherometer with a current of 1 mA. The mean zeta potential of cells in EDTA buffer was lower than the controls, as shown in Table 1. Because cells in buffered EDTA are in a medium of higher ionic strength than those in buffer alone, the surface charge densities of cells in each group are not simply proportional to their respective zeta potentials. The mean surface charge density of EDTA-treated cells is slightly but

TABLE 1. *Effect of EDTA on surface properties of slime mould cells*

	EDTA	Control
Zeta potential ($\pm$ S.D.)	$-18\cdot32 \pm 1\cdot54$ mV (131)	$-18.95 \pm 1\cdot50$ mV (64)
Surface charge density ($\pm$ S.D.)	$-1\cdot96 \pm 0\cdot16 \times 10^3$ e.s.u./cm^2	$-1\cdot84 \pm 0\cdot15 + 10^3$ e.s.u./cm^2

The number of measurements is shown in parentheses.

significantly higher than that of controls. In a parallel series of experiments cells were treated as for electrophoresis, then left 10 min, when comparable samples were examined microscopically. Those in EDTA remained essentially separate whereas control cells formed large clumps indicating that aggregation is prevented by EDTA in these conditions. These experiments show that the ability to form stable mutual contacts in the presence of 1 mM EDTA is greatly reduced while the overall electrostatic surface potential is not increased. This is in apparent conflict with the colloid theory of cell adhesion, though if there are significant local differences in potential undetectable by electrophoresis, the strength of this conclusion might be lessened.

Our electrophoretic experiments do not, of course, prove that the anti-adhesion effect produced by EDTA is due to chelation of divalent cations. Experiments on the kinetics of aggregation do, however, suggest that this is the probable mode of action. Garrod & Born (unpublished data) have shown that after treatment with 2×10^{-3} M EDTA, cells could be aggregated by 10^{-3} M solutions of Ba^{2+}, Ca^{2+}, and Mg^{2+} chlorides, but not by higher concentrations of monovalent ions alone. There is thus no evidence of a high degree of divalent cation specificity in *Dictyostelium* adhesion.

In view of our results it seems necessary to seek an explanation for the effect of EDTA on cell adhesion in terms other than those of the colloid theory. It is possible that a clue to the mechanism whereby EDTA affects slime mould adhesion may lie in the fact that it increases the cell surface negative charge density. It is also possible that EDTA is having an effect on the cytoplasm as well as on the surface membrane. These possibilities will be discussed in turn.

Surface effects of EDTA

It is possible that the increase in surface charge density is due simply to the removal of divalent cations bound to the cell surface within 20 Å or so of the electrophoretic plane of shear. Lipman *et al.* (1966) have found that 10^{-3} M Ca^{2+} and 10^{-3} M Mg^{2+} bind to the surface of toad bladder cells, reducing the surface negative charge. They point out that in 10^{-4} M Ca^{2+}, which is five times greater than the minimal concentration necessary to maintain cell adhesion in the intact bladder, the change in surface charge density is too small to measure. Again, divalent cation action is clearly not a simple colloidal effect. Weiss (1967) has shown that calcium can bind to the surface of R.P.M.I. No. 41 cells, and can be removed by EDTA, but without change in adhesiveness, assayed by agglutination test. Calcium binding to the surface of various chick embryo cell types has also been demonstrated by Collins (1966). Addition of 1.8×10^{-3} M Ca^{2+} decreased the surface charge density by 5-10%, but no tests on adhesion were reported. An effect of EDTA on adhesion may therefore be the removal of surface-bound divalent cations, conceivably involved in intercellular ionic bridging. Similarly, it is possible that the removal of divalent cations inhibits adhesion via a calcium-sensitive intercellular ligand. Evidence for such a mechanism has been provided by Moscona (1968) who has isolated an adhesion-promoting macromolecular substance from sponge cells. Alternatively EDTA might react with groups within the membrane, resulting in conformational changes manifested by an increase in surface charge density. This does not immediately suggest why adhesion is affected. It is clear from these considerations that if the action of EDTA is limited to removal of divalent cations from the surface, then some sort of intercellular molecular bonding is implicated in slime mould cell adhesion.

Cytoplasmic effects of EDTA
It is by no means certain that the significant effects of EDTA on adhesion are limited to the cell surface membrane. It is likely that the ability of cells to put out and withdraw pseudopodia is affected by EDTA treatment. Because adhesiveness may be affected by motility, as emphasized by Gustafson & Wolpert (1967), we may anticipate an indirect action of EDTA on adhesion due to cytoplasmic changes. It is also possible that having to come into contact, active cell movement is necessary in order to spread the area of contact and increase the energy of adhesion. EDTA causes violent pseudopodal activity in cells of the slime mould grex (De Haan, 1959; Gerisch, 1961; Whitefield, 1964) and Whitefield has suggested that such activity may bring about disaggregation by pushing the cells apart.

It is perhaps not obvious how pseudopodal activity of cells in suspension or out of contact with a solid substratum might cause disaggregation or prevent the formation of stable adhesions. It may, however, occur in the following way. Having made contact, protrusion of pseudopodia at the free surface would tend to reduce the area of contact by causing the body of the cell to round up, if the cell surface area remains constant (Wolpert & Gingell, 1968). We have also found that treatment of slime mould amoebae in aggregation streams with the anionic and cationic detergents sodium dodecyl sulphate and hexadecyltrimethylammonium bromide at concentrations less than 1 mM causes a very striking increase in pseudopodal activity and disaggregation of cells in aggregation streams. The mechanism by which EDTA, as well as ionic detergents, might affect cytoplasmic motile processes may involve membrane permeability changes. It is well known that Ca^{2+} is essential in maintaining normal cell membrane permeability, and that EDTA increases membrane permeability. It would, therefore, be surprising if motility were not affected by prolonged EDTA treatment, for a cytoplasmic contractile mechanism must become exposed to different ionic concentrations. EDTA treatment of amoeba cytoplasm *in vitro* causes the assembly of fibrous filaments which may be involved in locomotion (Morgan, Fyfe & Wolpert, 1967). We have also shown that cytoplasmic contraction *in vivo* is dependent on Ca^{2+} in amphibian embryos (Gingell, 1967; Gingell & Palmer, 1968).

There is a Ca^{2+} activated contractile mechanism located immediately beneath the cell membrane of amphibian blastula cells which responds to the influx of Ca^{2+} when membrane permeability is increased by adsorption of polycations or when a glass microelectrode is used to iontophorese ions just beneath the membrane. A contractile response is initiated by Ca^{2+} but not by Mg^{2+}, K^+, Na^+, or Cl^-. Response is localized and capable of spontaneous relaxation, suggesting that a Ca^{2+} sequestering apparatus is operating as in muscle cells. This system causes wound healing and it seems likely that it is responsible for changes in cell shape which occur in gastrulation and possibly in cytoplasmic cleavage.

We therefore have evidence, from what may be regarded as rather special cells, that local calcium concentration beneath the membrane can regulate motile processes. It is possible that EDTA is somehow affecting a system of this kind in adhesion experiments, either by the entry of EDTA as permeability increases, or by other ionic exchanges across the membrane.

In addition to Whitefield's suggestion (1964), we have considered the possibility that cellular motile behaviour might determine the ability of the cells to protract or withdraw micro-spikes, which may be important in making initial adhesive contacts (Bangham & Pethica, 1961; Pethica, 1961; Weiss, 1964) or in drawing the cells together once adhesion at the tip is established. We have examined *Dictyostelium discoideum* amoebae treated as in the electrophoresis experiments by Nomarski light microscopy and electron microscopy. It appears from the study of cells fixed in suspension that initial contacts between control cells are sometimes made by micro-spikes. However, these are also present in EDTA-treated cells and the EM axial fibrillar elements are evident in the spikes of both groups, suggesting a potential contractile function.

In conclusion it is not possible to be sure at present how EDTA inhibits cell adhesion although the chelation of divalent cations is probably basic to its action. Our experiments are consistent with the possibility that divalent cations may be involved in intercellular bridging. It seems possible,

64   *D. Gingell, D. R. Garrod & J. F. Palmer*

however, that the inhibition of adhesion may be due not only to changes at the cell surface but also to effects on a cytoplasmic motile system. We feel that both these possibilities should be taken into account in considering the results of adhesion experiments.

REFERENCES

ANDERSON, N. G. (1953). *Science, N.Y.*, **117**, 627
ARMSTRONG, P. B. (1966). *J. exp. Zool.*, **163**, 99
ARMSTRONG, P. B. & JONES, D. P. (1968). *J. exp. Zool.*, **167**, 275
BANGHAM, A. D. & PETHICA, B. A. (1961). *Proc. R. Soc. Edin.*, **28**, 43
BROOKS, D. E., MILLAR, J. S., SEAMAN, G. V. F. & VASSAR, P. S. (1967). *J. cell. Physiol.*, **69**, 155
COLLINS, M. (1966). *J. exp. Zool.*, **163**, 39
COMAN, D. R. (1954). *Cancer Res.*, **14**, 519
CURTIS, A. S. G. (1957). *Proc. R. Phys. Soc. Edin.*, **26**, 25
CURTIS, A. S. G. (1962). *Biol. Rev.*, **37**, 82
CURTIS, A. S. G. (1967). *The Cell Surface: Its Molecular Role in Morphogenesis.* New York: Academic Press
DE HAAN, R. L. (1959). *J. Embryol. exp. Morph.*, **7**, 335
GERISCH, G. (1961). *Exp. cell. Res.*, **25**, 535
GINGELL, D. (1967). Ph.D. thesis, University of London
GINGELL, D. & GARROD, D. R. (1969). *Nature, Lond.*, **221**, 192
GINGELL, D. & PALMER, J. F. (1968). *Nature, Lond.*, **217**, 98
GUSTAFSON, T. & WOLPERT, L. (1967). *Biol. Rev.*, **42**, 442
HERBST, C. (1900). *Arch. Entwickmech.*, **9**, 424
LIPMAN, K. M., DODELSON, R. & HAYS, R. M. (1966). *J. gen. Physiol.*, **49**, 501
MORGAN, J., FYFE, D. & WOLPERT, L. (1967). *Exp. cell Res.*, **48**, 194
MOSCONA, A. A. (1968). *Dev. Biol.*, **18**, 250
PETHICA, B. A. (1961). *Exp. cell Res.*, suppl., **8**, 123
ROUX, W. (1894). *Arch. Entwickmech.*, **1**, 43
STEINBERG, M. S. (1958). *Am. Nat.*, **92**, 65
STEINBERG, M. S. (1962). In *Biological Interactions in Normal and Neoplastic Growth*, ed. BRENNAN, M. J. and SIMPSON, W. I. Boston: Little, Brown & Co.
WEISS, L. (1960). *Exp. cell Res.*, **21**, 71
WEISS, L. (1964). *J. theor. Biol.*, **6**, 275
WEISS, L. (1967). *J. cell Biol.*, **35**, 347
WHITEFIELD, F. E. (1964). *Exp. cell Res.*, **36**, 62
WOLPERT, L. & GINGELL, D. (1968). *Symp. Soc. exp. Biol.*, **22**, 169
ZWILLING, E. (1954). *Science, N.Y.*, **120**, 219

CALCIUM AND CELLULAR FUNCTION AFTER FREEZING AND THAWING

J. FARRANT

Clinical Research Centre Laboratories, National Institute for Medical Research, Mill Hill, London, N.W.7

Since the discovery that glycerol protects living cells against the damage caused by freezing and thawing (Polge, Smith & Parkes, 1949), many advances have been made towards the storage of different cells and tissues at very low temperatures ($-80°C$ or below). Damage caused by freezing and thawing is thought to be linked with the excessively high concentrations of electrolytes, particularly sodium chloride, that occur around the cells as ice forms (Lovelock, 1953). The part of the cell most susceptible to this stress appears to be the cellular membranes (Lovelock, 1957). Smooth muscle preparations have been used to investigate the recovery of an organized tissue after freezing and thawing in the presence of another protective substance (dimethyl sulphoxide) (Farrant, 1964a,b, 1965; Farrant, Walter & Armstrong, 1967), and the present experiments show how calcium ions are involved in the responses of the muscle to drugs after thawing.

Dose-response curves, either to histamine or acetylcholine, have been obtained from isometric tension responses of isolated guinea-pig uteri suspended in Krebs solution at 37°C. The normal calcium concentration of this solution was 1·4 mM. Low Ca^{2+} conditions were achieved by simply omitting the calcium salt from the solution. Some solutions were made containing 1·4 M dimethyl sulphoxide (DMSO) without altering the electrolyte concentration.

Each tissue was assigned to one of two experimental groups after the initial sensitivity of the tissue had been tested in Krebs solution at 37°C containing 1·4 mM $CaCl_2$. These sensitivities were taken as the 100% level of response. The two experimental groups were:

(*A*) Low Ca^{2+} conditions at 37°C together with a period of incubation at 37°C in the low calcium Krebs solution containing DMSO (1·4 M). Drug sensitivities were tested in both these conditions.

(*B*) Low Ca^{2+} conditions and incubation in the DMSO at 37°C followed by freezing and thawing.

After each of these procedures, uterine horns were re-examined at 37°C for responses to histamine (or acetylcholine) in Krebs solution without DMSO but with varying concentrations of Ca^{2+}.

The procedure for freezing and thawing was as follows: After the incubation in the DMSO solution at 37°C, uteri in the second group were carefully removed from the organ bath and placed into 2 ml. of the same DMSO solution in an ampoule at 0°C. The ampoule was then sealed, cooled to -79°C and rewarmed. Cooling was carried out by placing the ampoule in alcohol in a small thermally lagged container. This container was suspended in a bath of alcohol maintained at -79°C by solid carbon dioxide. The thermal lagging was arranged so as to allow a reproducible cooling rate of 1°C per min from 0°C to -40°C. Temperatures were recorded from a similar unsealed ampoule using copper–constantan thermocouples and a potentiometric recorder. Thawing was carried out by agitating the ampoule in a water bath at 45°C until the last particle of ice had vanished. The ampoule was then broken open and the uterine horn tipped out into a low Ca^{2+} solution without DMSO at 37°C before being resuspended in the same organ bath. The recovery of the responses to drugs after treatments *A* or *B* was tested.

Graded amounts of calcium ions (as chloride) were added to the bath 1 min before the next dose of drug. The recorder was switched on at this time so that any effects of calcium itself might be observed. Both the drug and the added calcium ions were washed out of the bath and replaced with the low Ca^{2+} solution after each dose. Frequent checks to see that all the calcium ions had been washed out of the bath were made by interspersing doses of drugs without previously adding calcium.

Spontaneous contractions after low calcium conditions

Uteri remained relaxed when transferred back into Krebs solution containing Ca^{2+} (1·4 mM) after exposure to low-Ca^{2+} conditions for up to 60 min at 37°C. In contrast the inclusion during the low-Ca^{2+} conditions of a 20 min period of incubation in the protective non-electrolyte dimethyl sulphoxide (DMSO, 1·4 M) produced a transient spontaneous contraction of the uterus on return to a DMSO-free Krebs solution with calcium ions (1·4 mM). The effect of delaying the addition of calcium ions (1·4 mM) after removal of the DMSO was also examined. A contraction occurred only when the calcium ions were added within 12 min of the removal of the DMSO.

After exposure to the low Ca^{2+} incubation with DMSO followed by freezing and thawing, uteri always contracted when Ca^{2+} (1·4 mM) was added to the organ bath, no matter how long (up to 4 hr) this addition of Ca^{2+} was delayed.

Uteri frozen and thawed without prior incubation in the protective substance (DMSO) never showed any responses after thawing whether calcium ions were present during freezing or not.

Responses to histamine after low Ca²⁺ conditions

As soon as a regular submaximal response to histamine (0·5 to 1·0 μM) was established the transfer to low Ca^{2+} conditions was carried out and responses were quickly reduced to a negligible level. Responses were also completely absent during the period of incubation in DMSO (1·4 M). The reintroduction of varying concentrations of calcium ions after the removal

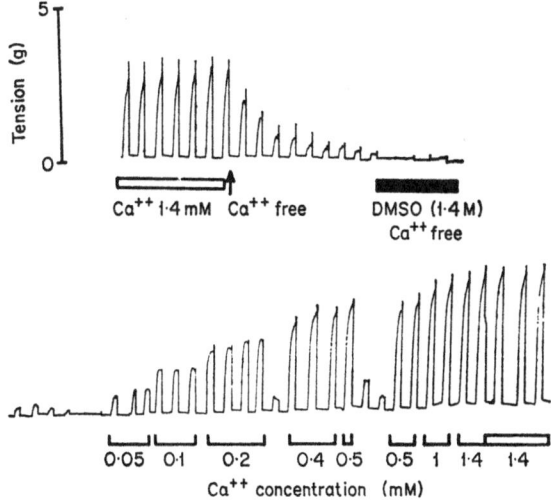

FIG. 1. Isometric responses of an isolated guinea-pig uterus to a repeated submaximal dose of histamine (0·5 μM) at +37°C. Initially (open rectangle) the solution contained Ca^{2+} (1·4 mM). At the arrow, a solution without added Ca^{2+} was substituted and the responses to histamine declined. The solid rectangle represents exposure for 20 min to the same solution without Ca^{2+} but with added dimethyl sulphoxide (DMSO, 1·4 M). After the removal of the DMSO, the regular doses of histamine were continued, often with added Ca^{2+} (concentration shown) that is present both during the 30 sec exposure to the histamine and for the 1 min preceding it. Finally (open rectangle) the muscle was returned to a solution containing Ca^{2+} (1·4 mM).

of the DMSO allowed histamine responses to return. Figure 1 shows these effects in a single experiment where, for clarity, the reintroduction of calcium ions was carried out in ascending order of concentration. Finally the uterus was returned to a Krebs solution which contained calcium ions (1·4 mM). Providing the reintroduction of calcium was delayed for at least 12 min after the removal of the DMSO, there was no response from the muscle to calcium itself.

Figure 2 shows an identical experiment, except that the uterus was frozen to −79°C and thawed at the end of the DMSO incubation. In these circumstances the muscles contracted during the 1 min period of exposure

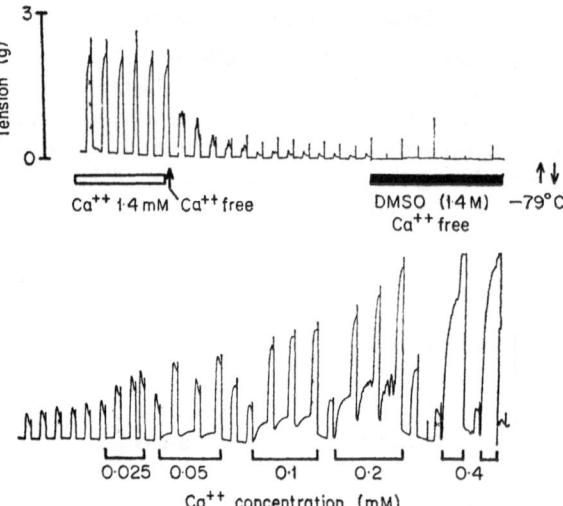

FIG. 2. As Fig. 1, except that at the end of the incubation in the solution containing DMSO (1·4 M) the muscle was frozen to −79°C and thawed before being resuspended in the organ bath. It can be seen that the thawed preparation contracts to Ca^{2+} alone. In addition, this figure shows the dependence of the histamine response to the concentration of Ca^{2+}. In this experiment the standard dose of histamine was 1 μM.

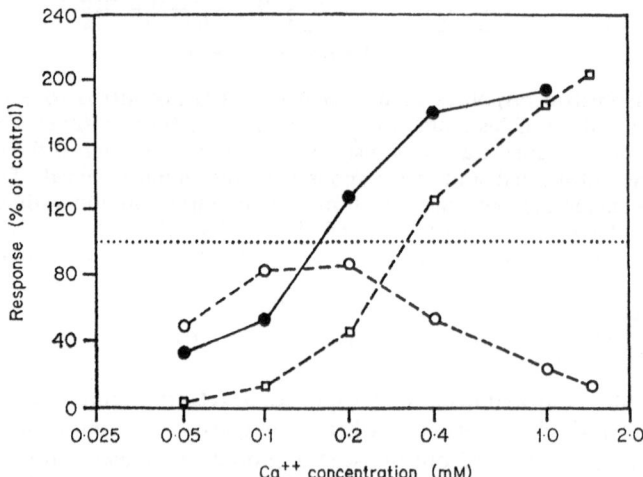

FIG. 3. Mean percentage isometric responses of guinea-pig uteri as a function of Ca^{2+} concentration (mM). The 100% response level is taken to be the mean response to the standard dose of histamine (0·5 to 1·0 μM) in Ca^{2+} (1.4 mM) before the treatment. The solid circles show the mean responses to the dose of histamine after low Ca^{2+} conditions and incubation in DMSO. The open circles and open squares show the mean responses to histamine and Ca^{2+} respectively in thawed preparation.

to calcium ions before the addition of the histamine. The histamine then caused a second contraction.

Dose response relationships are shown in Fig. 3. The response (percentage of control) is linked at the 100 per cent level to the mean isometric response which is stimulated by the repeated submaximal dose of histamine in Ca^{2+} (1·4 mM) at the start of the experiment. It can be seen that the

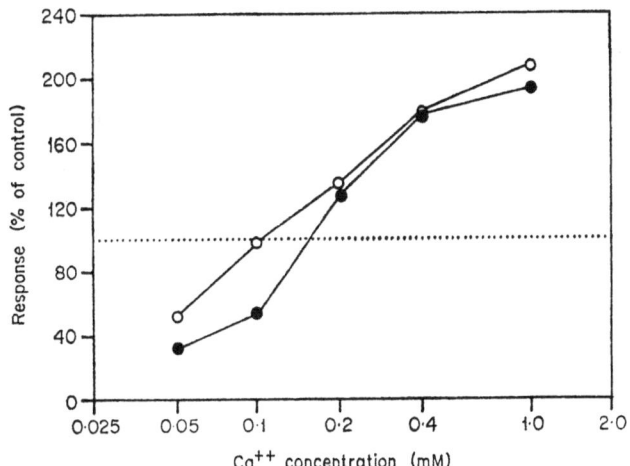

FIG. 4. Mean percentage isometric responses of guinea-pig uteri as a function of Ca^{2+} concentration. The solid circles show the mean responses to histamine after low Ca^{2+} conditions without freezing, and the open circles show the sum of the histamine and Ca^{2+} responses in thawed uteri.

recovery of the histamine response, after the low Ca^{2+} conditions and incubation in DMSO, is a function of the Ca^{2+} concentration present. In addition there seems to have been a more powerful response to histamine in Ca^{2+} (1·4 mM) after the exposure to low Ca^{2+} conditions.

Figure 3 also shows the mean response of uteri after freezing and thawing, both to Ca^{2+} itself and also to the dose of histamine in the presence of Ca^{2+}. The higher the concentration of Ca^{2+} the greater the response to Ca^{2+} itself, but at higher concentrations of Ca^{2+} (0·4 mM and above) the subsequent response to the histamine declines.

The summed effect of the contractions to Ca^{2+} itself and to histamine by thawed uteri is shown in Fig. 4, and these are compared with the calcium-dependent histamine responses of unfrozen muscles.

Responses to acetylcholine after low Ca^{2+} conditions
Similar results were obtained when acetylcholine (1 μM) was used instead of histamine. Figure 5 shows responses to the repeated dose of acetyl-

choline as a function of Ca^{2+} concentration for both unfrozen and thawed muscles. It also includes responses by thawed uteri to Ca^{2+} itself.

How may these results with uterine muscle be interpreted? The transient contraction observed in unfrozen uteri after return to Ca^{2+} (1·4 mM) within 10 min of removal from the solution containing DMSO could be associated with the osmotic swelling of the smooth muscle cells that occurs

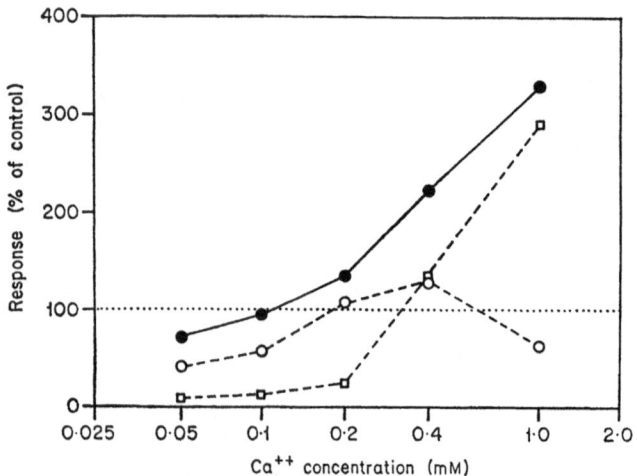

FIG. 5. Mean percentage isometric responses of guinea-pig uteri as a function of Ca^{2+} concentration. The solid circles show the responses to acetylcholine after low Ca^{2+} conditions but without freezing. The open circles and open squares record the responses to acetycholine and Ca^{2+} respectively in thawed preparations.

after removal from DMSO (Farrant, 1964a,b). It seems clear that the use of DMSO and the freezing technique described allows partial recovery of the contractile apparatus of guinea-pig uterine muscle after it has been frozen to $-79°C$. Protection is not complete because the readmission of Ca^{2+} to the solution after thawing induces a contraction of the muscle. Smooth muscle that has been depolarized in low Ca^{2+} conditions will contract when Ca^{2+} is added (Edman & Schild, 1962; Isojima & Bozler, 1963); moreover several authors have reported that depolarization alters the calcium flux in smooth muscle (Briggs, 1962; Edman & Schild, 1962; Van Breeman & Daniel, 1966). An increase in the rate of Ca^{2+} influx could explain the calcium contraction seen in these thawed preparations but the phenomenon differs from that of a depolarized preparation in the following respects: (a) The Ca^{2+} induced contraction of thawed muscle occurs in a solution with a high Na^+ concentration; (b) The electron microscope also provides evidence that smooth muscle cells frozen and thawed by a technique similar to that described here have a very much higher frequency of cellular membrane breaks than either control material or material which is frozen and thawed by methods that do not produce Ca^{2+} contractions on thawing (Farrant, 1965; Farrant, Armstrong & Walter, 1967).

This evidence suggests that the technique used protects the contractile apparatus during freezing but does not protect the cellular membranes. The contractions induced by Ca^{2+} after thawing could thus be due to direct access to the contractile proteins. It is interesting that drugs such as histamine and acetylcholine still cause a further contraction in the presence of added Ca^{2+}. More experiments are needed to distinguish whether the responses to histamine and acetylcholine in these partially damaged thawed muscles are mediated by some direct action at the contractile apparatus or whether they are mediated indirectly perhaps by competition with Ca^{2+} at sites on the disrupted membrane fragments. Some indications favouring the second possibility are that competition of Ca^{2+} with car-bamyl choline has been suggested for a different preparation (Nastuk & Liu, 1966) and also that Van Breeman & Daniel (1966) have postulated that contraction could be induced by the release of Ca^{2+} from inside the cell membranes.

REFERENCES

BREEMAN, van C. & DANIEL, E. E. (1966). *J. gen. Physiol.*, **49**, 1299

BRIGGS, A. H. (1962). *Am. J. Physiol.*, **203**, 849

EDMAN, K. A. P. & SCHILD, H. O. (1962). *J. Physiol., Lond.*, **161**, 424

FARRANT, J. (1964a). *J. Physiol., Lond.*, **170**, 33P

FARRANT, J. (1964b). *J. pharm. Pharmac.*, **16**, 472

FARRANT, J. (1965). *Nature, Lond.*, **205**, 1284

FARRANT, J., WALTER, C. W. & ARMSTRONG, J. A. (1967). *Proc. R. Soc., B*, **168**, 293

ISOJIMA, C. & BOZLER, E. (1963). *Am. J. Physiol.*, **205**, 681

LOVELOCK, J. E. (1953). *Biochim. biophys. Acta*, **10**, 414

LOVELOCK, J. E. (1957). *Proc. R. Soc., B*, **147**, 427

NASTUK, W. L. & LIU, J. H. (1966). *Science, N.Y.*, **154**, 266

POLGE, C., SMITH, A. U. & PARKES, A. S. (1949). *Nature, Lond.*, **164**, 666

DISCUSSION TO SESSION II

Jones (Aberystwyth)
Dr. Gingell, as chelation by EDTA would undoubtedly effect ATPase activity, have you considered the possibility that the effect of EDTA may relate indirectly to a consequent elevation of ATP inducing conformational changes at or below the cell membrane? I ask this because whereas non-adhesive *Dictyostelium myxamoebae* have a high ATP level, in adhesive amoebae ATP is low.

Have you measured ATP levels?

Gingell (London)
We have not investigated this possibility, but it is perhaps of interest that Rasmussen & Tenenhouse (1968) have pointed out the importance of reversible Ca^{2+} binding to ATP at the membrane. This may provide a link between EDTA, Ca^{2+}, and ATP.

Gillard (Canterbury)
Dr. Farrant, the protective agents used in your freezing experiments are all notorious modifiers of the structure of water. Is this relevant to a possible mechanism in the freezing experiments, rather than one based purely on calcium?

Farrant (London)
Although substances like glycerol and dimethyl sulphoxide do modify water structure, it has been possible to explain their actions in protecting living cells from damage during freezing and thawing simply by invoking their effects on the bulk properties of the solution. In their presence the proportion of the system that is converted to ice at any temperature during freezing is reduced, thus maintaining at a higher level the volume of liquid phase available for the dissolved solutes. This reduction in the concentration level of solutes (particularly electrolytes) during freezing has been correlated with protection against damage. The phenomena involving calcium that I described are a consequence of cellular damage rather than a cause.

REFERENCE

RASMUSSEN, H. & TENENHOUSE, A. (1968). *Proc. natn Acad. Sci. U.S.*, **59**, 1364

Calcium Permeability and Transport

Chairman:
D. H. SMYTH
Department of Physiology, University of Sheffield

ACTIVATION AND INHIBITION OF THE SARCOPLASMIC CALCIUM TRANSPORT

W. HASSELBACH, M. MAKINOSE & W. FIEHN

Max-Planck-Institut f.med.Forschung, Abt. Physiologie, Heidelberg/Germany

Many observations suggest that all cells possess transport systems for the elimination of ionized calcium from the cytoplasm. Presumably no cell can tolerate high concentrations of free calcium ions intracellularly, but so far the concentration of free calcium ions which can be tolerated is known for only one cell with any degree of accuracy. In the resting giant muscle fibres of *Maia* (Portzehl, Caldwell & Rüegg, 1964) and *Balanus* (Hagiwara, 1966) the free calcium concentration has been found not to exceed 5×10^{-7} M. On the other hand, these muscles like any other muscle need relatively large amounts of calcium ions for the activation of their contractile machinery (Weber & Herz, 1963; Jöbsis, 1967; Ashley & Ridgway, 1969). These quantities are presumably liberated in a few milliseconds during excitation. The removal of these ions from the cytoplasm brings activation to a halt, and at room temperature the time required for this process is in the range of 10–100 msec. In skeletal muscles these rapid calcium movements take place mainly across the membranes of the sarcoplasmic reticulum. The sarcoplasmic membranes are especially suitable objects for studying the mechanism of and the structural basis for the active transport of calcium ions.

During muscle relaxation 10 μmoles of calcium per sec must be transported across the surface of 2–5 mg membranal protein which is the amount present in 1 ml. muscle fibre. This is an extremely high transport rate. Even more astonishing is the fact that this high transport rate is brought about at extremely low concentrations of ionized calcium. A great number of functional units must therefore be present in the sarcoplasmic membranes. Their number has been estimated from the data mentioned before, and the rate constant for the formation of calcium chelates was determined by Eigen & Meyer (1963) to be $3 \times 10^{11}/\text{cm}^2$ (Hasselbach & Seraydarian, 1966).

Studies on isolated sarcoplasmic vesicles

Studies of the calcium movements across the membranes and the energy yielding reaction were made by a further unique property of the sarcoplasmic tubules and cysternae. If the tubules are isolated according to the procedure shown in Fig. 1, they form closed vesicles which are perfectly sealed. These vesicles take up small amounts of calcium from solutions

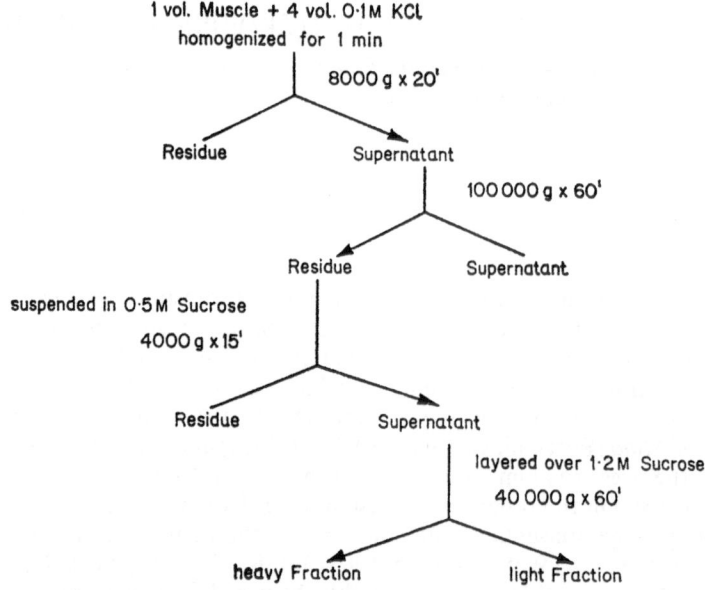

FIG. 1. Scheme for the isolation and purification of sarcoplasmic membranes from skeletal muscle. (From Makinose, 1969.)

which contain ATP and magnesium in addition to calcium (Table 1) and the vesicles are filled to capacity in a very short time (Weber, Herz & Reiss, 1966). If, however, calcium precipitating agents like oxalate or phosphate are also added, huge amounts of calcium oxalate or calcium phosphate—10 μmoles calcium oxalate/mg vesicular protein—are stored inside the vesicles (Table 1). The formation of these calcium precipitates does not occur spontaneously in the solutions because sufficient concentrations of calcium chelating agents are also added. The vesicles need about 2–5 min at 20°C to accumulate these large quantities of calcium (Fig. 2) and during this time the rate of uptake can be measured with great accuracy. Because an active accumulation of oxalate or phosphate has been excluded, an active calcium transport must have given rise to the formation of calcium precipitates (Hasselbach & Makinose, 1961, 1963). The presence of anions which precipitate calcium causes an increase in calcium storing

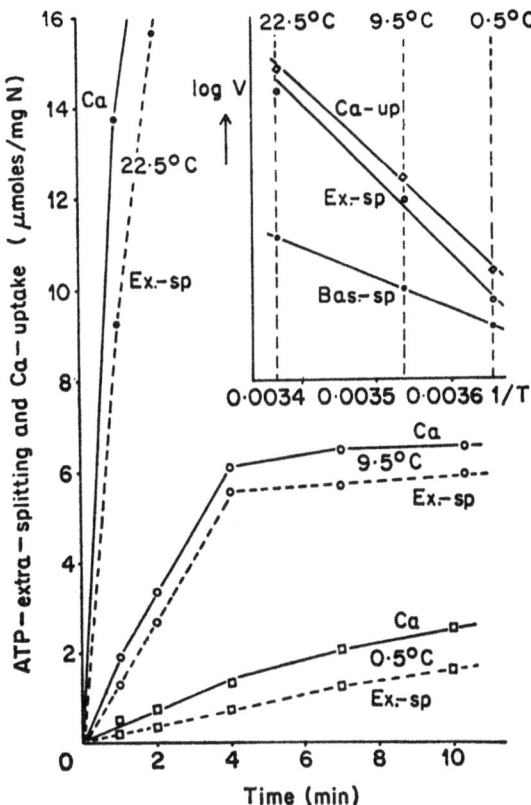

FIG. 2. Calcium uptake and the accompanying calcium dependent ATP splitting at different temperatures. A preparation whose transport ratio Ca uptake/ATP split is approximately 1, has been used.

capacity and also provides information on the energetics of the pump. In the presence of oxalate the product of $[Ca^{2+}]$ and $[oxalate^2]$ inside the vesicles is known and this product is identical with the solubility product of calcium oxalate as soon as the very first calcium oxalate crystal is formed. The value of $[Ca^{2+}][oxalate^{2-}]$ for the solution outside can be determined. From the ratio of the two ion products, the energy requirement of the pump can be obtained. Its maximal value reaches 5000 cal/mole when at the cessation of net calcium uptake calcium influx equals calcium efflux (Table 2) (Hasselbach & Makinose, 1963).

The membrane can use all nucleoside triphosphates as energy sources. They are hydrolysed in a calcium and magnesium dependent reaction which accompanies calcium transport (extra NTPase). Even though the different NTPs are hydrolysed at quite different rates, the transport ratio

TABLE 1. *Composition of solutions from which the isolated sarcoplasmic vesicles take up calcium*

	Concentration total (M)	Concentration free (M)	Concentration total (M)	Concentration free (M)
ATP	5×10^{-3}	7×10^{-4}	5×10^{-3}	10^{-3}
Mg	5×10^{-3}	7×10^{-4}	5×10^{-3}	4×10^{-4}
EGTA	—	—	2×10^{-4}	10^{-4}
Oxalate	—	—	5×10^{-3}	4×10^{-3}
Ca^{2+}	10^{-4}	$2 \cdot 7 \times 10^{-5}$	10^{-4}	2×10^{-7}
Calcium storage (binding?)	$0 \cdot 1 - 0 \cdot 2$ μmoles/mg		10 μmoles/mg	
Initial concentration ratio $\dfrac{Ca_i}{Ca_o}$	~ 0?		25	

TABLE 2. *Performance of the sarcoplasmic calcium pump driven by ATP or UTP. Two terms contribute to the energy requirement during the initial phase of net calcium uptake:*

$$E = RT \ln \frac{\text{influx}}{\text{efflux}} + RT \ln \frac{(\text{Ca.Ox})_i}{(\text{Ca.Ox})_o}$$

The first term represents the energy which is required to maintain net calcium uptake (Heinz, 1959)

Performance of the calcium pump			ATP	UTP
Initial	Rates (20°C) [pmoles/cm² × min]	Ca influx	250	70
		Ca efflux	15	4
	Concentration ratio	$\left[\dfrac{Ca_iOx_i}{Ca_oOx_o}\right]$	25	25
	Energy requirement	cal/mol	~ 4000	~ 4000
Steady state	Rates (20°C)	pmoles/cm² × min	$0 \cdot 12$	$0 \cdot 04$
	Concentration ratio	$\left[\dfrac{Ca_iOx_i}{Ca_oOx_o}\right]$	2500	2100
	Energy requirement	cal/mol	~ 5000	~ 5000

of calcium taken up/NTP split remains unchanged, as does the maximal concentration ratio (Makinose & Hasselbach, 1965).

Dependence on calcium concentration

The calcium concentration outside and inside the vesicles is important for the mechanism of calcium translocation. When the calcium concentration outside the vesicle is changed at a constant oxalate concentration we can assume that only the calcium concentration outside changes. When, on the other hand, the oxalate concentration is changed at a constant free calcium concentration in the medium, only the inside concentration of calcium changes. The results of such experiments are shown in Fig. 3.

Half maximal activation occurs at a calcium concentration of 2×10^{-7} M at the outer surface and half maximal inhibition occurs at a calcium concentration of 5×10^{-5} M at the inner surface of the vesicles. The assumption that oxalate determines the concentration of free calcium inside the vesicles is supported by the observation that the apparent dependence on oxalate disappears when the vesicles are disrupted or punctured by ultrasonication or other treatments (Makinose, 1969).

Role of membrane —SH groups

The first interaction between the soluble reaction participants and the membranes has been found to be the binding of ATP to the vesicular membranes. This binding of ATP is inferred mainly from the effect of ATP on the reactivity of the membranal SH groups. In the presence of ATP one SH equivalent/100,000 g of vesicular protein is protected against substitution by thiol reagents (NEM, Mersalyl). As long as these SH groups remain free all functions remain intact, because these essential SH groups are localized at the outer surface of the vesicles as demonstrated by the attachment of Hg-azophenyl ferritin (Hasselbach & Elfvin, 1967). We must assume that the interaction of ATP with the vesicular membranes takes place at the outer surface. Subsequently if magnesium and calcium ions are present and the SH groups mentioned before are intact, the terminal phosphate of ATP is transferred to the membranal protein (Makinose, 1966a,b, 1969) (Fig. 4).

The phosphate groups transferred from any NTP to the protein can be transferred back to ADP or to any other NDP (Makinose, 1966a,b), and the enzyme therefore behaves like a nucleoside diphosphatekinase. The P-exchange rates between the respective NTP and NDP differ considerably. The exchange reaction is catalysed by calcium ions in the solution outside, but the P-transfer and exchange reactions, in contrast to P-liberation and calcium translocation, are not influenced by the ionized calcium at the inner surface of the vesicles. This can be inferred from the observation that oxalate does not affect the P-transfer reaction nor the P-exchange reaction.

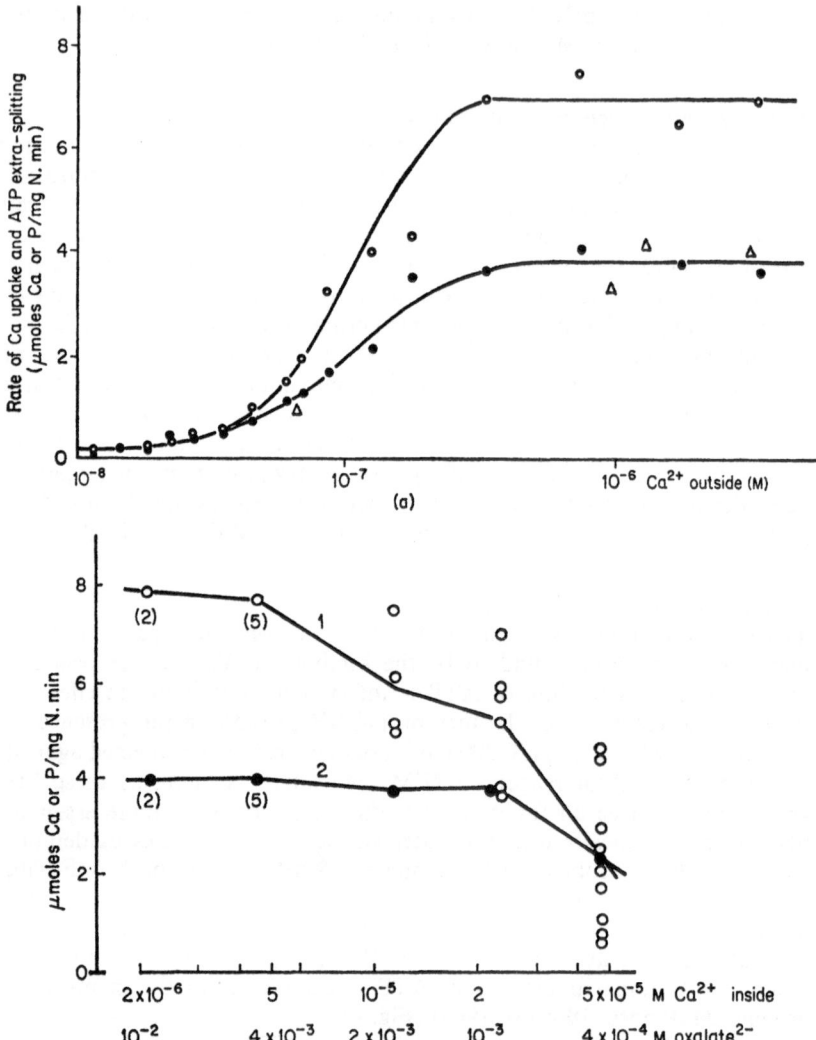

FIG. 3. (a) Activation of calcium transport (○) and ATP extra-splitting (●, △) by increasing the calcium concentration outside the vesicles. Half-maximal activation occurs at a Ca^{2+} ion concentration of 2×10^{-7} M. (b) Inhibition of calcium transport (1) and ATP extra-splitting (2) by increasing the calcium concentration inside the vesicles. Half-maximal inhibition occurs at a Ca^{2+} ion concentration of 5×10^{-5} M.

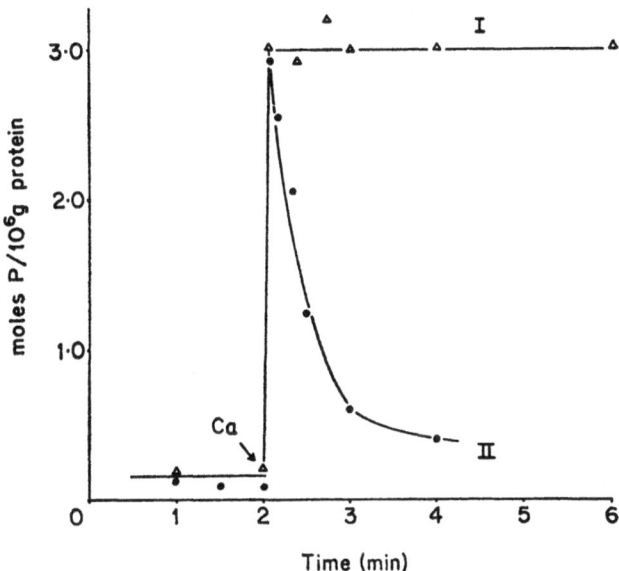

FIG. 4. Experiments illustrating the formation and decomposition of a phosphorylated intermediate (E ~ P) (Makinose, 1966a). The formation of E ~ P was studied under conditions given in Table 2 (right column). In (I) the calcium was added in excess so that the concentration remained high and the level of E ~ P remained constant. In (II) calcium was added in small amount and after the calcium had been removed from the solution the level of E ~ P fell.

Effect of drugs on calcium transport

The reaction sequence which has been discussed is further supported by the action of drugs, such as reserpine, prenylamine, chlorpromazine, and imipramine, which inhibit calcium transport in the sarcoplasm (Balzer, Makinose & Hasselbach, 1968; Balzer, Makinose, Fiehn & Hasselbach, 1968). The way in which these drugs interfere with the reaction sequence is summarized in Table 3. It shows that the reactions thought to be the initial steps in the chain are unimpaired, while calcium accumulation and P-liberation are considerably reduced. The drugs obviously interfere with the change that takes place in the membrane when the calcium ions are translocated and do not influence the initial steps at the outer surface. This type of selective inhibition is in contrast to the effect of other inhibitors, like NEM or ADP, which suppress all reactions equally because they interfere with the initial reaction at the outer surface. We have shown that the latter inhibitor reacts with the protein moiety of the membrane, while selective inhibitors like reserpine, prenylamine, chlorpromazine, and imipramine exert their effect through the unsaturated fatty acids in the membrane lipids. The main evidence for this statement is given in Table 4 and Fig. 5.

TABLE 3. *Effect of prenylamine (10^{-4} M) on the activity of the sarcoplasmic calcium pump* (Balzer, Makinose & Hasselbach, 1968; Balzer, Makinose, Fiehn & Hasselbach, 1968.)

20°C	Control	Prenylamine 10^{-4} M
Ca uptake μmoles/mg × min	1·40	0·44
$\dfrac{\text{Ca uptake}}{\text{ATP splitting}}$	1·7	1·7
$\dfrac{Ca_iOx_i}{Ca_oOx_o}$	2000	2000
Ca exchange (steady state)	0·12	0·04
$E \sim P$ μmoles/g	3	3
$E + ATP \rightleftharpoons$ $EP + ADP$ μmoles/mg × min	3	3

TABLE 4. *Activation and inhibition of the vesicular transport ATPase and the P-exchange reaction produced by modifications of the vesicular lipids*

Preparation 20°C	ATPase μmoles P/mg × min		P ~ Exchange μmoles P/mg prot. × min
	Normal assay	$+10^{-4}$ M Prenylamine	
(1) Normal vesicles	1·2	0·4	3·0
(2) Phospholipase A digestion	1·2	0·1	3·0
(3) Phospholipase A + Albumin	0·1 (basic ATPase)	0·1	≲0·5
(4) 3 + Stearate	0·1	—	≲0·5
(5) 3 + Oleate	0·6	0·06	1·5
(6) 3 + Lysolecithin	0·5	0·4	—
(7) 3 + Oleate + Lysolecithin	1·0	0·3	2·5

Phospholipase A digestion, which like phospholipase C digestion abolishes calcium uptake, affects neither the transport ATPase nor the P-transfer reactions. When the hydrolysis products are removed from the protein by albumin, both ATPase and P-exchange disappear. Both activities can be restored by *cis*-unsaturated fatty acids and lysolecithin. The activation brought about by lysolecithin and *cis*-unsaturated fatty acids differ in at least one respect. While prenylamine strongly inhibits the action of oleic acid, the action of lysolecithin is much less sensitive to prenylamine.

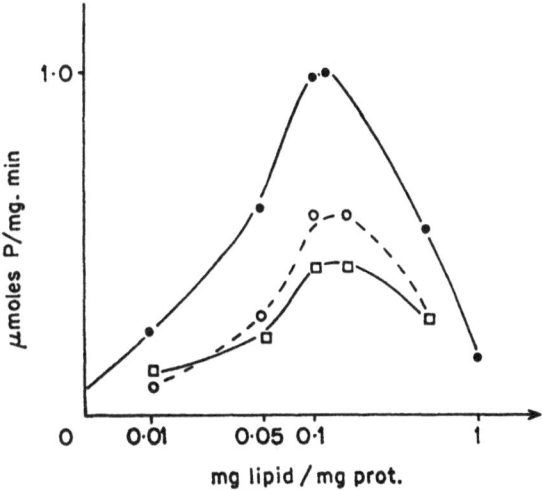

FIG. 5. Restoration of ATPase activity in phospholipase A treated vesicles by oleate (○), lysolecithin (□) and a combination of the two (●).

The results discussed in this paper can be summarized as follows. The sarcoplasmic membranes are characterized by an extremely high transport rate and an extremely high concentrating power. The net rate of calcium uptake is practically identical with the rate of calcium influx during the initial phase of calcium uptake, and as a consequence a close coupling between ATP splitting and calcium uptake has been found. In the presence of oxalate, the energy requirement of the calcium transport can be determined.

The formation of large quantities of a phosphorylating intermediate indicates a large number of active sites in the membranes. An independent estimate has been obtained by labelling the SH groups which are functionally essential. The area corresponding to one site amounts to 5000 Å^2 or to a molecular weight of 100,000. The functional relationship between the membrane lipids and the protein moiety is revealed by the inhibition produced by a number of drugs which combine with the unsaturated fatty

4—C.C.F.

acids in the membrane. When these fatty acids are removed, drug binding disappears and the calcium transport system is completely inactivated. On readdition of unsaturated fatty acids, the P-transfer reaction and the calcium dependent ATPase are restored. The drugs which are bound by the reconstituted preparation produce ATPase inhibition.

REFERENCES

ASHLEY, C. C. & RIDGWAY, E. B. (1969). *J. Physiol., Lond.*, **200**, 74P
BALZER, H., MAKINOSE, M. & HASSELBACH, W. (1968). *Naunyn-Schmiede-berg's Arch. exp. Path. Pharmak.*, **260**, 456
BALZER, H., MAKINOSE, M., FIEHN, H., & HASSELBACH, W. (1968). *Naunyn-Schmeiderberg's Arch. exp. Path. Pharmak.*, **260**, 454
EIGEN, M. & DE MEYER, L. (1963). *Relaxation Methods. Techn. Organic Chem.*, VIII, II, p. 1041. New York: Interscience
HAGIWARA, S. (1966). *Ann. N.Y. Acad. Sci.*, **137**, 1015
HASSELBACH, W. & ELFVIN, L.-G. (1967). *J. ultrastruct. Res.*, **17**, 598
HASSELBACH, W. & MAKINOSE, M. (1961). *Biochem. Z.*, **333**, 518
HASSELBACH, W. & MAKINOSE, M. (1963). *Biochem. Z.*, **339**, 94
HASSELBACH, W. & SERAYDARIAN, K. (1966). *Biochem. Z.*, **345**, 159
HEINZ, E., cited by NETTER, H. (1959). *Theoretische Biochemie*, p. 725. Berlin: Springer-Verlag
JÖBSIS, F. F. (1967). *Symp. Biol. Hung.*, **8**, 151
MAKINOSE, M. (1966a). *Biochem. Z.*, **345**, 80
MAKINOSE, M. (1966b). *2nd Int. Biophys. Congr. Wien*, p. 276
MAKINOSE, M. & HASSELBACH, W. (1965). *Biochem. Z.*, **343**, 360
MAKINOSE, M. (1969). *Europ. J. Biochem.*, **10**, 74
PORTZEHL, H., CALDWELL, C. P. & RÜEGG, J. C. (1964). *Biochem. biophys. Acta*, **79**, 581
WEBER, A. & HERZ, R. (1963). *J. biol. Chem.*, **238**, 599
WEBER, A., HERZ, R. & REISS, I. (1966). *Biochem. Z.*, **345**, 329

TRANSMEMBRANE CALCIUM MOVEMENTS IN RESEALED HUMAN RED CELLS

H. J. SCHATZMANN

Veterinär-pharmakologisches Institut, Universität Bern, Switzerland

There is little doubt that in fresh human red cells there is a low concentration of intracellular calcium. In a careful study, Harrison & Long (1968) and Harrison, Long & Sidle (1968), using a dry ashing technique and absorption flame photometry, found only 0·0158 μmole Ca/ml. of whole cells and were able to show that calcium located in the membrane accounts completely for this figure. This means that the cell interior has a calcium concentration below 0·003 μmole/ml., assuming that, between cells and membranes, a difference of two standard errors could have been detected. In our experiments cells from defibrinated blood were washed in a solution of 145 mM NaCl and 20 mM Tris-Cl at 4°C and were haemolysed in a five-fold volume of water. The haemolysate was extracted with an equal volume of 10% trichloroacetic acid after removal of the stromata, and we found a concentration of 0·03 μmole Ca/ml. cells after corrections were made for possible contamination of the water and reagents with calcium (Schatzmann & Vincenzi, 1969). The two figures differ by a factor of 10. The possibility exists that the temperature of the washing solution might be responsible for the discrepancy. In any event, the value is far below the free ionic calcium concentration in the plasma which surrounds the cells. Because 1 ml. plasma contains 1·5 μmole Ca^{2+}, the ratio $[Ca^{2+}]$ plasma/$[Ca^{2+}]$ cell might be larger than 50. If the potential across the membrane (inside negative) is assumed to be 9 mV, then this ratio should be 0·51 provided that calcium could passively distribute across the membrane. From this it follows that the cells are not at equilibrium with respect to calcium. The fact that the cells keep the intracellular calcium concentration low is in keeping with the finding that exposure of the internal membrane surface to moderate calcium concentrations (0·1–0·3 mM) leads to an inhibition of the Na–K-activated membrane ATPase which is responsible for Na–K transport (Dunham & Glynn, 1961).

Calcium uptake in red cells

It has been known for some time that the passive permeability of the red
cell membrane for calcium is low (Passow, 1963; Rummel, Seifen &
Baldauf, 1962). We suspended washed cells from defibrinated blood in a
medium composed of: 130 mM Na, 5 mM K, 2 mM Mg, 1 mM Ca, 1 mM
phosphate, 20 mM Tris and 161 mM Cl containing ^{45}Ca during 6 days at
4°C. At the end of this time the concentration of ^{45}Ca per ml. cells was
only about 2% of the concentration per ml. medium. After 6 hr at 37°C
it was 0·2%, regardless of the presence or absence of glucose. These
experiments show that, even if the energy supply of the cells is suppressed,
equilibration of calcium across the membrane proceeds slowly. Experimental conditions can be created, however, that increase the permeability
to calcium. One such condition seems to be the presence of very high
calcium concentration in the medium (Rummel *et al.*, 1962; Ponder, 1953);
another is storage of the cells at 37°C in the absence of glucose until they
are thoroughly depleted of their energy stores (Weed, 1968). Figure 1
describes an experiment in which whole blood was incubated under sterile
conditions at 37°C. After 10–15 hr such cells swell and gain sodium. The
figure shows that after 12 hr the calcium concentration in the serum began
to fall, indicating that calcium started moving into the cells (Fuhrmann
& Schatzmann, unpublished). Correction of the calcium concentration

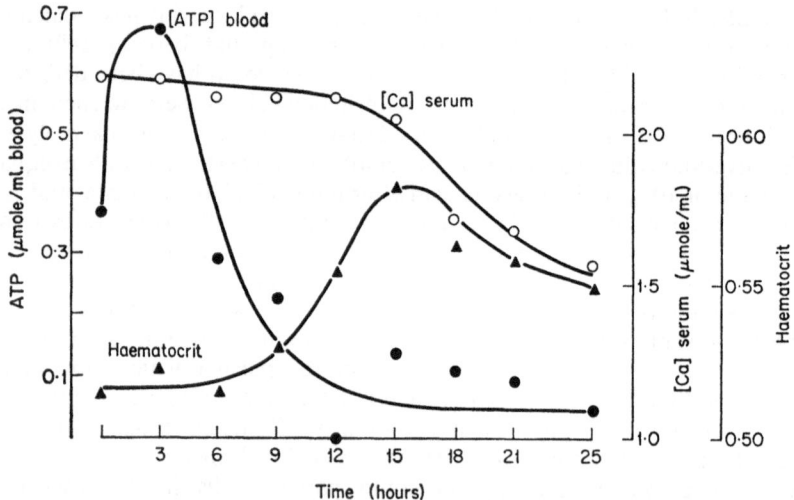

FIG. 1. Whole defibrinated human blood incubated at 37°C in sterile conditions. Single experiment. Ca determined after trichloroacetic acid precipitation in the serum by absorption flame photometry. ATP measured in 1 ml.
blood after perchloric acid precipitation by enzymatic method. Haematocrit:
18,000 *g* for 5 min, in 3·5 × 60 mm Perspex tubes. Haemolysis after 25 hr,
2·3%. Notice entry of Ca into cells after 12 hr.

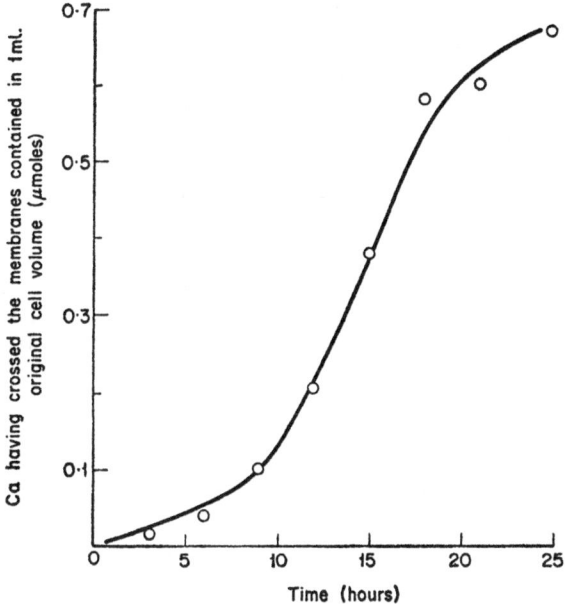

FIG. 2. Ca movement in the experiment shown in Fig. 1, after correction for the volume change.

measurements for the volume change of the cells gives the amount of calcium that crossed the surface area of 1 ml. of the original cells (Fig. 2).

Calcium extrusion in red cells

The entry of Ca into fresh cells is too slow to be measured accurately, so it is a matter of debate whether the undoubtedly low permeability is sufficient explanation for the maintenance of the intracellular calcium concentration. It is possible that during the 100-day life span of the cell even this low permeability might allow a considerable increase in cellular calcium concentration while the old cells contribute to the average calcium concentration in the cell population. If the large calcium gradient across the membrane is explained by its impermeability, there is still the unsolved problem of how the red cell in the making rids itself of calcium so effectively. The existence of a metabolically driven calcium extrusion mechanism therefore seemed probable at the outset.

We approached the problem of demonstrating a calcium pump by using cells into which we introduced calcium, magnesium and ATP by way of reversal of haemolysis. The procedure applied was similar to that proposed by Whittam (1962), except that we did not incubate the cells at 37°C after restoring isotonicity. The rapid rate of calcium exit did not allow this step. The cell suspensions were cooled in ice shortly after the haemolysis procedure. Under these conditions the cells seal to calcium, magnesium, and ATP, but they do not seal to Na and K. Haemolysis took place in a 5-fold (in some instances 10-fold) volume of water which contained 0·1–5 mM $CaCl_2$, usually 2 mM $MgCl_2$ and 0·5–4 mM ATP as Mg-, Tris- or Na-salt. Isotonicity was restored with KCl (or NaCl) and the cells were washed and suspended in a solution of 130 mM Na, 5 mM K, 2 mM Mg, 1 mM Ca, and 20 mM Tris, all as chloride at pH 7·3. The haemocrit was usually 0·3. After trichloroacetic acid precipitation of medium or cells, calcium in the supernatant was measured by EDTA titration in early experiments and in later experiments by absorption flame photometry. ATP was assayed by an indirect enzymatic method involving phosphorylation of 3-phosphoglycerate and oxidation of DPNH, ultraviolet absorption or fluorescence of the latter being measured. ATPase activity was assayed by measuring inorganic phosphate.

Cells, resealed in the way described, extruded calcium at a rapid rate against a considerable chemical gradient at 37°C or 28°C, provided that they contained ATP (Schatzmann, 1966; Schatzmann & Vincenzi, 1969; Vincenzi, 1968). Cells which were thoroughly depleted of their energy stores by incubation without glucose at 37°C for 15–17 hr did not display any measurable calcium loss during 60 min. The net outward movement of calcium from cells which contain ATP is highly temperature dependent as indicated by a Q_{10} of 3·5.

Cells with a high calcium concentrate showed a considerable increase in ATP splitting activity (Schatzmann & Vincenzi, 1969). The extra ATPase activity is induced by Ca inside the cells but not by Ca in the medium (Schatzmann & Vincenzi, 1969). By relating the extra ATP hydrolysis to the initial rate of calcium transport we found that 1·3 moles of ATP were split per 1 mole of calcium transported (Schatzmann & Vincenzi, 1969).

Omission of magnesium abolishes the rapid extrusion of calcium. The requirement for Mg is rather difficult to demonstrate, because it is not easy to obtain Mg-free cells by reversal of haemolysis and because lowering of the cellular Mg-concentration inevitably decreases the ionic Ca concentration as Ca-ATP and Mg-ATP have similar stability constants.

Figure 3 shows two experiments in which we succeeded, by trial and error, in setting the calcium, magnesium, and ATP concentrations in the haemolysing fluid so that the initial concentrations of free calcium in the cells were similar at a high and low concentration of intracellular magnesium. It can be seen that at a low concentration of intracellular magnesium the rate of calcium extrusion was considerably lower and that this decrease cannot possibly be due to a lack of free Ca^{2+}. It is necessary in calculating $[Ca^{2+}]$ from the measured total of [Ca], [Mg], and [ATP] in the cells to make the somewhat uncertain assumption that the apparent stability constants of the ATP-metal complexes at a given pH are similar for the cells and a simple solution, for example 0·1 M KCl.

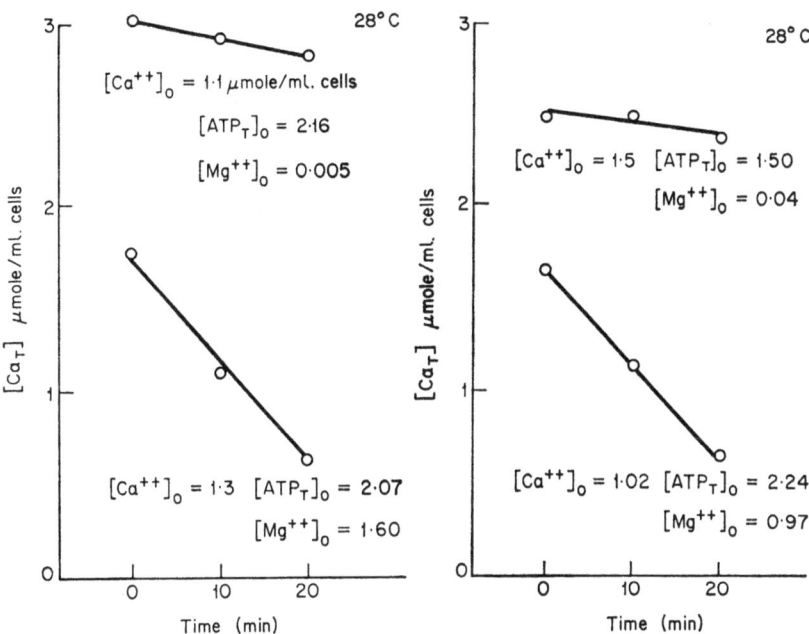

FIG. 3. Two single experiments with two different blood samples showing Mg requirement for Ca transport. Citrated blood, cells washed four times and starved for 15 hr in the absence of glucose. Haemolysis in ten-fold volume of water in the presence of Tris-ATP (pH 7·2) and Ca with or without added Mg. Total [Ca] (ordinate) measured by flame photometry in the cells after deproteinization with equal volume 10% trichloroacetic acid. Incubation at 28°C in approximately three-fold volume of medium containing (mM) 130 Na, 5 K, 1 Ca, 2 Mg, 20 Tris, 161 Cl, pH 7·3. The medium for the low Mg cells contained no Mg. Figures near the curves give values at zero time for total cellular [ATP], free [Ca^{2+}] and free [Mg^{2+}] calculated from measured total cellular concentration and stability constant of the ATP-metal complexes at pH 7·3. Notice marked decrease of Ca-extrusion rate in low-Mg cells at similar or higher cellular [Ca^{2+}].

Strontium is transported by the system in the same way as calcium (Schatzmann & Vincenzi, 1969). The strontium movement can be reduced by the simultaneous presence of calcium in the cells, most probably by competition of the two ions for the same site in the membrane. From such experiments a rough estimate of the saturation concentration for calcium can be given. It seems to be somewhere near 2 μmoles/ml. cells haemolysed in 1 mM Mg-ATP. Na is neither necessary nor inhibitory up to 130 mM because the calcium extrusion is not different in a medium in which all Na is replaced by K. Magnesium is clearly not transported under the conditions described (Schatzmann & Vincenzi, 1969).

We have preliminary evidence that calcium extrusion can also be demonstrated in intact cells. Cells starved in serum for 23 hr yield calcium against a gradient if the serum is supplemented with adenosine (and glucose) (Fuhrmann & Schatzmann, unpublished); this confirms a finding by Weed (1968).

Calcium fluxes in red cells

To confirm the results obtained by measuring net calcium movement in resealed cells we carried out experiments designed to give information about unidirectional fluxes of calcium in the presence and absence of ATP inside resealed cells. The main objective was to ensure that ATP does not increase the membrane permeability for calcium, and the procedure was the same as in the experiments with chemical determination of calcium.

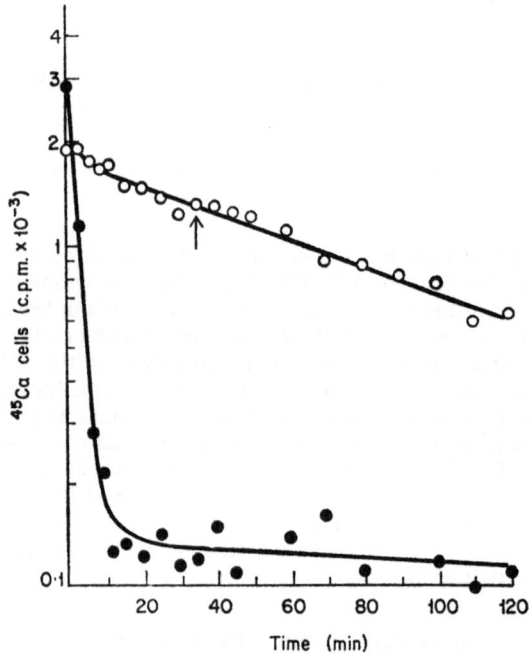

Time (min)

Fig. 4. Efflux of ^{45}Ca from resealed cells (from citrated blood, washed 4 times and starved for 15 hr at 37°C without glucose). Haemolysis in 0·75 mM Ca, 2 mM Mg and 1 mM Tris-ATP, pH 7·2 (●—●), or no ATP (○—○). Medium: 200-fold volume of solution of Fig. 3 with 1 mM Na$_2$HPO$_4$. Temperature 37°C. Ordinate: Radioactivity contained in approx. 1·5 μl. of cells on Millipore filters. The activity of external solution trapped in the filter at the end of experiment was 53 c.p.m. in ATP sample and 44 c.p.m. in ATP-free sample. Notice large increase in efflux rate in presence of cellular ATP. Addition of adenosine 5 mM and glucose 5 mM to medium at 35 min (arrow) did not accelerate efflux in ATP-free cells.

From the same batch of washed cells (from citrated blood), one sample was haemolysed in the presence of 1 mM Tris-ATP (pH 7·2), and the other in the absence of ATP. Both samples were exposed to 2 mM $MgCl_2$ and 0·75 mM $CaCl_2$ during haemolysis. For efflux experiments the calcium in the haemolysing fluid was labelled with ^{45}Ca. After restoring isotonicity the suspension was diluted at 4°C with 100-fold volume of the usual external medium containing 1 mM Ca. The cells were spun down and

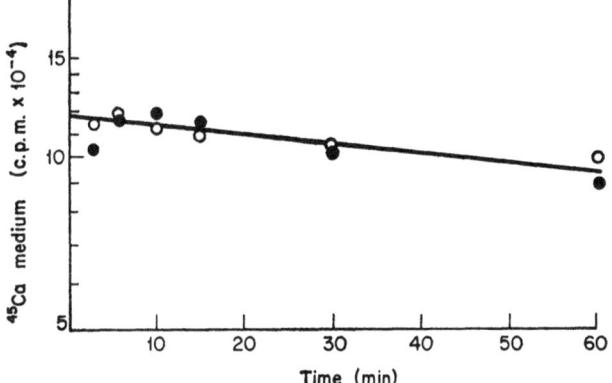

FIG. 5. Influx of ^{45}Ca into resealed cells. Cells treated as in Fig. 4. Medium as in Fig. 4 but without phosphate. Cell-to-medium volume ratio approximately 1:1. ●, Cells with 1 mM ATP; ○, cells without ATP. Ordinate: radioactivity counted in 0.1 ml. of medium. Because at equilibrium $[Ca]_{cells} \geqslant [Ca]_{medium}$, the radioactivity in the medium might fall to or below half the initial value when equilibrium is attained. There is no difference in influx rate with and without ATP.

suspended in a 200-fold volume of medium at 37°C. At intervals, 0·3 ml. of this suspension was filtered through Millipore filters (pore diameter 3 μm) and the cells on the filter were washed with 0·3 ml. of medium under light suction (-10 cm H_2O). The filters were dried at room temperature, placed at the bottom of counting vials, and covered with 10 ml. of toluene which contained the scintillation fluor (BBOT, Ciba). The vials were counted in a liquid scintillation counter (Nuclear, Chicago). For the influx experiment the calcium of the medium was labelled with ^{45}Ca; equal volumes of resealed cells and medium were mixed and brought to 37°C. At intervals samples of the medium were obtained by centrifuging 0·4 ml. of the suspension for 2 min in an ultramicro-centrifuge (Ecco-Quick) reaching 10,000 rev/min in 6 sec. A mixture of 0·1 ml. of supernatant medium and 3 ml. of methanol was counted in 10 ml. of toluene-fluor. The result of two single experiments is shown in Figs. 4 and 5. It is clear that the rate of efflux was increased very significantly by the presence of ATP in the cells whereas the influx rates of ^{45}Ca into cells free of ATP and cells containing ATP were equal within experimental error. These experiments confirm that the net calcium movement caused by ATP is induced by an increase in efflux rather than a decrease of influx. This was expected from the low influx rate and the rapid net outward movement observed in the standard experiment with resealed, ATP loaded cells. The experiments rule out the possibility that ATP in the concentration used increases the permeability of the membrane for calcium.

Conclusions

It is reasonable to assume that if ATP maintains an uphill calcium transport it does so by supplying the necessary energy and must therefore be hydrolysed in the process. Dunham & Glynn (1961) showed that isolated human red cell membranes display an ATPase activity which is stimulated by the simultaneous presence of calcium and magnesium. Wins & Schoffeniels (1966) and Vincenzi & Schatzmann (1967) confirmed that the maximal activity is reached at about 0·1 mM Ca (in the presence of 2–5 mM Mg), and that the activity declines at higher calcium concentrations The activation by calcium can be detected at concentrations as low as 10^{-7} M (Vincenzi & Schatzmann, 1967). The interaction between calcium and magnesium is complex and defies analysis. When magnesium concentrations are increased there is a shift in the concentration-effect curve for calcium to higher calcium concentrations (Dunham & Glynn, 1961). Increasing calcium concentrations shift the concentration-effect curve for magnesium to higher magnesium concentrations, but high calcium concentrations (1 or 5 mM) are not fully antagonized by magnesium, irrespective of whether total or free ionic concentrations are considered (Schatzmann & Vincenzi, 1969). Strontium can replace calcium in the activation of the ATPase (Schatzmann & Vincenzi, 1969). The Ca–Mg-activated ATPase is not affected by alterations in the Na/K ratio of the medium (Schatzmann & Vincenzi, 1969) and is insensitive towards cardiac glycosides (Dunham & Glynn, 1961). It is, however, inhibited by the organic mercurial mersalyl, as can be seen from Fig. 6, and by ethacrynic acid, the concentrations for 50% inhibition being $2·3 \times 10^{-5}$ M and $5·5 \times 10^{-5}$ M, respectively.

The calcium transport system is similar to the ATPase in that it requires Mg, accepts Sr instead of Ca, functions in the absence of Na, is not affected by cardiac glycosides but is inhibited by mersalyl and ethacrynic acid (Schatzmann & Vincenzi, 1969; Vincenzi, 1968). The latter point is somewhat ambiguous because mersalyl in concentrations necessary to cause an effect might increase the passive permeability such that a direct effect on the transport mechanism cannot be demonstrated with certainty.

The fact that the two systems are in accord in a number of points suggests that they are two aspects of the same phenomenon. It seems an attractive theory that the Ca–Mg activated ATPase plays a part in the Ca transport similar to that of the Na–K–Mg activated ATPase in Na–K transport. The differences between the two systems are evident, however:

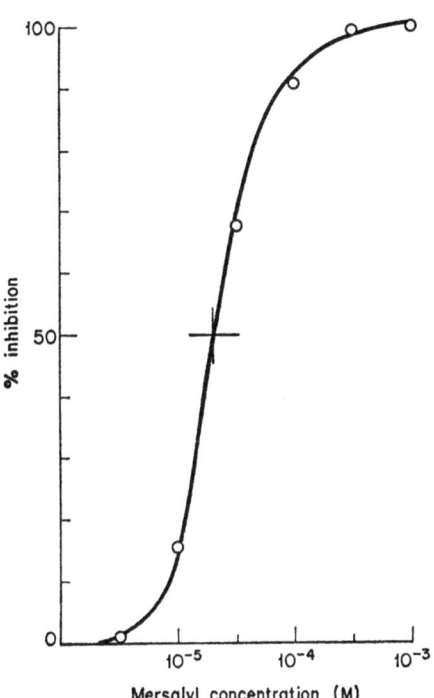

FIG. 6. Concentration-response curve for mersalyl inhibition of ATPase of isolated human red cell membranes. The Na–K stimulated ATPase is suppressed by ouabain. The measurements include the Ca–Mg-activated ATPase and a small fraction of non-specific (Mg-activated) ATPase: Ca, 0·1 mM, Mg 4 mM, ouabain 10^{-4} w/v.

(1) The Ca transport system is simpler in that no cation movement in the opposite direction seems to be involved. Mg is not transported into the cell when Ca is extruded (Schatzmann, 1967), and does not activate the ATPase when applied on the outside of resealed Ca and ATP loaded cells. Na can be ruled out because it can be omitted. K cannot strictly be ruled out from our experiments but is not a likely candidate. (2) Downhill Ca-flow in parallel channels or Ca-exchange across the transport systems are negligible compared with the ATP-dependent directional Ca-flow. (3) The efficiency of the Ca-transport system is less in terms of the stoichiometric relation, which is one Ca ion (or less) per mole of ATP-γ-pyrophosphate bond, compared with three Na ions per mole of pyrophosphate bond in the Na–K system.

The search for a more specific inhibitor has so far been unsuccessful. Chlorpromazine was reported to be effective in sarcoplasmic reticulum (Balzer, Makinose & Hasselbach, 1968). We found an incomplete

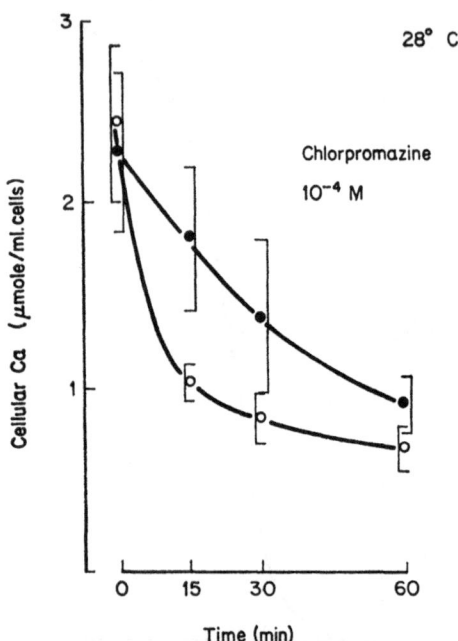

FIG. 7. Inhibitory action of chlorpromazine 10^{-4} M on net Ca extrusion from resealed cells at 28°C. Cells haemolysed in five-fold volume of water with 2 mM Mg-ATP, 0·75 mM Ca, 5 mM Tris-Cl, pH 7·3. Drug present in haemolysing fluid and medium. Medium: 130 Na, 5 K, 2 Mg, 1 Ca, 20 Tris, 161 Cl (mM). O, Control (haematocrit 0·265), ●, chlorpromazine (haematocrit 0·272). Ordinate: Ca determined in the cells. The drug treated sample approaches the same final value of [Ca]$_{cells}$, which shows that permeability is not increased.

inhibition when 10^{-4} M was applied both in the haemolysing fluid and in the medium (Fig. 7). Dinitrophenol 10^{-3} M, oligomycin 2×10^{-5} w/v (applied inside and outside), caffeine 10^{-3} w/v and Cu^{2+} 10^{-4} M were not effective.

Recently Baker, Blaustein, Hodgkin & Steinhardt (1967) reported that in the giant axon of *Loligo forbesi*, Ca-influx is accelerated by internal Na. It is probable that this observation reflects a coupled Na–Ca exchange. Such a mechanism might drive an active transport of calcium outwards by way of the Na pump which maintains a Na gradient directed inwards. Evidence for a similar Na–Ca exchange which might account for Ca outward transport in heart muscle was presented by Reuter & Seitz (1968). We have no indication that calcium movement in red cells might depend on this sort of mechanism, for it occurs in the absence of a sodium gradient and persists when sodium is omitted altogether. The fact that it is not sensitive to ouabain and oligomycin in concentrations that block the sodium pump reveals that it is not directly dependent on the sodium pump either. On the contrary, the red cell mechanism has much in common with the Ca accumulating system observed in sarcoplasmic reticulum (see article by Hasselbach, pp. 75–84).

REFERENCES

BAKER, P. F., BLAUSTEIN, M. P., HODGKIN, A. L. & STEINHARDT, R. A. (1967). *J. Physiol., Lond.*, **192**, 43P
BALZER, H., MAKINOSE, M. & HASSELBACH, W. (1968). *Arch. Pharmak. exp. Path.*, **260**, 444
DUNHAM, E. T. & GLYNN, I. M. (1961). *J. Physiol., Lond.*, **156**, 274
HARRISON, D. G. & LONG, C. (1968). *J. Physiol., Lond.*, **199**, 367
HARRISON, D. G., LONG, C. & SIDLE, A. B. (1968). *Biochem. J.*, **108**, 40P
PASSOW, H. (1963). In *Cell Interphase Reactions*, ed. BROWN, H. D. p. 57. New York: Scholars Library
PONDER, E. (1953). *J. gen. Physiol.*, **36**, 767
RUMMEL, W., SEIFEN, E. & BALDAUF, J. (1962). *Arch. exp. Path. Pharmak.*, **244**, 172
REUTER, H. & SEITZ, N. (1968). *J. Physiol., Lond.*, **195**, 451
SCHATZMANN, H. J. (1966). *Experientia*, **22**, 364
SCHATZMANN, H. J. (1967). *Protides of the Biological Fluids*, **15**, 251
SCHATZMANN, H. J. & VINCENZI, F. F. (1969). *J. Physiol., Lond.*, **201**, 369
VINCENZI, F. F. (1968). *Proc. Western Pharmac. Soc.*, **11**, 58
VINCENZI, F. F. & SCHATZMANN, H. J. (1967). *Helv. physiol. Acta*, **25**, CR 233
WEED, R. I. (1968). *1st int. Symp. on Metabolism and Membrane Permeability of Erythrocytes and Thrombocytes*, ed. DEUTSCH, E., Vienna: Gerlach
WHITTAM, R. (1962). *Biochem. J.*, **84**, 110
WINS, P. & SCHOFFENIELS, E. (1966). *Biochim. biophys. Acta*, **120**, 341

SODIUM–CALCIUM EXCHANGE ACROSS THE NERVE CELL MEMBRANE

P. F. BAKER

The Physiological Laboratory, Cambridge

If the intracellular concentration of free calcium were determined solely by the Donnan ratio, then calcium should be accumulated in most cells. In quantitative terms, with an internal potential 50–60 mV negative to the outside and external calcium concentrations of 1–10 mM, the predicted intracellular concentration of free calcium should be 0·1–1·0 M. These values are very different from those found by experiment. In the few large nerve and muscle cells where direct measurement has proved possible, the free intracellular calcium concentration is at most 10^{-5} M and may be as low as 10^{-7} M (Hodgkin & Keynes, 1955; Portzehl, Caldwell & Rüegg, 1964; Luxoro & Yanez, 1968). A discrepancy in this direction must mean that in these cells Ca ions are actively excluded from the cytoplasm. Although comparable measurements have not been possible, it seems likely that most other cells behave similarly.

More than one mechanism seems to participate in maintaining a low cytoplasmic concentration of calcium. Other papers in this symposium discuss how calcium is sequestered into cytoplasmic organelles such as mitochondria and parts of the endoplasmic reticulum, and how in red cells—which lack intracellular organelles—the hydrolysis of ATP provides energy for the ejection of calcium across the plasma membrane. The experiments described in this paper provide evidence of a further mechanism for transporting calcium ions across cell membranes. A special feature of this mechanism is that at least part of the energy required to eject Ca from the cell seems to be provided by sodium ions moving into the cell down their electrochemical gradient. Although this Na-dependent transport of calcium must ultimately require the operation of the sodium pump, the properties of this Na–Ca exchange system are quite different from those of the Na–K exchange pump (Table 1). A further difference between the mechanisms which maintain low intracellular concentrations of Na and Ca in nerve is that for Na an inwardly directed passive leak is balanced by an outwardly directed pump, whereas for Ca both the influx and efflux are dependent on the relative concentrations of Na on the two sides of the cell membrane. It is most convenient

TABLE 1. *Comparison of the Ca-dependent and K-dependent components of the Na efflux from squid axons*

Treatment	Ouabain-insensitive Na–Ca exchange	Ouabain-sensitive Na–K exchange
Ouabain	No effect at 10^{-3} M	Inhibited by 10^{-7} M
External monovalent cations	K: activates with low affinity	K: activates with high affinity
	Na: activates with low affinity up to 100 mM and then inhibits	Na: inhibits
	Li: activates with low affinity	Li: inhibits
	Choline: no effect	Choline: no effect
Ca-free media	Markedly reduced	Slightly increased
Internal Na	Approximately proportional to $[Na]_i^2$	Approximately proportional to $[Na]_i$
Lanthanum (0·1 mM)	Inhibits	No effect
Tetrodotoxin (10^{-6} g/ml.)	No effect	No effect
Cyanide (2 mM)	Inhibits	Inhibits
Dinitrophenol (0·4 mM) pH 8·0	No effect	Increases Na–Na exchange in K-free sea water
Internal EGTA (5 mM)	Inhibits	No effect

to discuss, in turn, the properties of the systems which effect Na-dependent Ca influx and efflux, and only then to consider whether these should be considered separate entities or merely two modes of operation of the same system.

Na-dependent calcium influx

The first evidence for the existence of a Na-dependent Ca influx came from a study of the Na efflux from squid axons. Hodgkin & Keynes (1955) and Caldwell, Hodgkin, Keynes & Shaw (1960) had shown that, in the presence or absence of external potassium, complete replacement of external Na by choline, lithium, or sugar caused an increase in the Na efflux. When this 'Na-free effect' was examined in more detail by Baker, Blaustein, Manil & Steinhardt (1967), it transpired that the increased efflux into Li differed from that into choline and dextrose in being insensitive to quite high doses of the cardiac glycoside ouabain. Further investigation of this ouabain-insensitive Na efflux into Li sea water revealed two very interesting properties: (1) that the efflux was dependent on the presence of external Ca ions; Sr could replace Ca in maintaining the Na efflux but Mg, although it reduced the activation due to Ca ions, could not replace Ca, and (2) that the ouabain-insensitive, Ca dependent component of the Na efflux was much more sensitive to changes in internal Na concentration than was the ouabain-sensitive Na–K exchange pump. The Ca-dependent Na efflux increased approximately as the square of the internal Na concentration whereas Na–K exchange increased linearly (Table 1).

The discovery of this Ca-dependent Na efflux posed three important questions: (1) Can a Ca-dependent Na efflux be demonstrated in the absence of ouabain? (2) Is Li essential for the appearance of a Ca-dependent Na efflux, or can the efflux be demonstrated in the presence of other external cations? (3) Is the Ca-dependent Na efflux linked to Ca influx into the axon?

The recent paper of Baker, Blaustein, Hodgkin & Steinhardt (1969) provides answers to all three questions. In the absence of ouabain, Ca-dependence of the Na efflux is readily shown in Li sea water but it is less obvious in other external solutions. This arises because removal of external Ca tends to increase the ouabain-sensitive component of the Na efflux and thus masks any small reduction in the ouabain-insensitive component. Working in the presence of ouabain, however, it can be shown that Li is not essential for Ca-dependent Na efflux. Figure 1 shows that there is a small Ca-dependent component of the Na efflux into dextrose sea water which is increased by partial replacement of the sugar by both Li and Na.

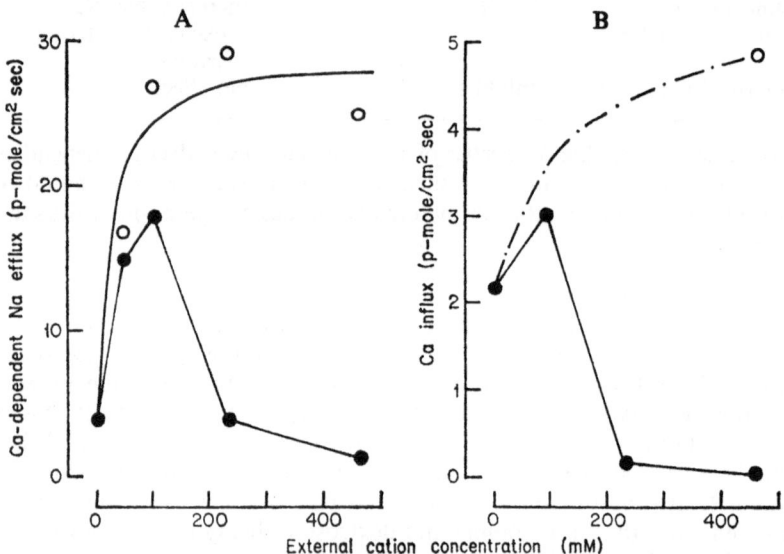

FIG. 1. Collected results showing the effects of external Na (●) and Li (○) ions on the Ca-dependent Na efflux (A) and on the Ca influx (B) in axons of *Loligo forbesi*. Isotonicity was maintained throughout by dextrose. Sea waters contained 460 mM of monovalent cation or an osmotically equivalent amount of dextrose, 55 mM $MgCl_2$, 11 mM $CaCl_2$ 2·5 mM $NaHCO_3$, and 10^{-5} M ouabain. Temperature 18°–20°C.

TABLE 2. *Comparison of the Ca-dependent Na efflux from squid axons with the Ca influx in both squid and crab nerve*

Treatment	Ca-dependent Na efflux	Ca influx in squid	Ca influx in crab
Replacement of external Na by Li, dextrose or choline	Increased	Increased	Increased
Ca concentration giving half maximal activation	In Li, 3 mM In Na, about 80–100 mM	— —	In Li, 2–3 mM In Na, about 100 mM
Dependence on internal Na	Approximately proportional to $[Na]_i^2$	Strongly dependent on $[Na]_i$	Strongly dependent on $[Na]_i$
Ouabain (up to 10^{-3} M)	No effect	No effect	No effect
Lanthanum (0·1 mM)	Inhibits	—	Reduces Ca uptake
Ca substitutes: Strontium	Can replace Ca	—	Competes with Ca and is transported
Magnesium	Competes with Ca	—	Competes with Ca. Not known whether transported
Cyanide (2 mM)	Inhibits	Reduced (but see text)	—
Tetrodotoxin (10^{-6} g/ml.)	No effect	—	No effect
Isobutanol (100 mM)	Reduces total ouabain-insensitive Na efflux	—	Reduces but does not inhibit completely

Complete replacement by Na causes inhibition, although some Ca-dependent Na efflux is normally detectable in full Na sea water. Li produces rather little inhibition. K and Cs behave like Li; choline is like dextrose.

A rather similar picture has been found for the Ca influx into squid axons. Because each measurement needs a separate axon, not as many points have been obtained as for the Na efflux; but there seems no doubt that the influx is lowest in Na sea water, is considerably increased in dextrose sea water, and is further increased by partial replacement of dextrose with Na or by complete replacement with Li (Fig. 1).

The simplest explanation of these results is that the increases in both Na efflux and Ca influx which follow replacement of external Na by dextrose are further promoted by the presence of certain monovalent cations in the external medium. Thus Li, K, and Cs, but not choline, promote Ca entry and Na loss, while the results with Na can be understood in terms of two actions. At low concentrations Na also promotes Ca entry, but at higher concentrations it displaces Ca from its binding sites with a resultant inhibition of Ca influx and Ca-dependent Na efflux.

These similarities in the properties of the Ca-dependent Na efflux and Ca influx suggest that the two fluxes may be linked, although the exact stoichiometry seems to be somewhat variable. Further evidence for the parallel behaviour of these fluxes is summarized in Table 2. The most striking feature is the marked dependence of both Ca-dependent Na efflux and Ca influx on internal sodium. In squid axons, the Ca influx from Na, Li, or dextrose sea water is increased three- to ten-fold by increasing the internal Na content four-fold. Table 2 also includes data for crab nerve (Baker & Blaustein, 1968). Because of the heterogeneity of crab nerve it is only possible to make influx measurements but, in general, the Na-dependent Ca influx is very similar to that in squid axons (see Table 3). One advantage of crab nerve is that under conditions of increased Ca influx it is possible to show a net gain in internal calcium. Although no analysis has been made of the effects of internal sodium on calcium entry into other tissues, a reduction in external Na has been shown to increase Ca entry into a variety of tissues, for example, frog skeletal muscle (Cosmos & Harris, 1961), frog heart (Niedergerke, 1963),

TABLE 3. *Dependence on internal sodium concentration of the Ca influx into leg nerves of the spider crab* Maia squinado

Pretreatment of nerve	Ca^{2+} uptake	
Soaked for 1 hr in sea water	0.385 ± 0.052	(6)
Soaked for 1 hr in K-free sea water	0.856 ± 0.050	(3)
Soaked for 1 hr in sea water + 1 mM ouabain	0.983 ± 0.139	(3)
Soaked for 30 min in Ca-free choline sea water	0.213 ± 0.031	(3)
Soaked for 30 min in Ca-free choline sea water and then stimulated in Ca-free Na sea water	0.834 ± 0.053	(3)
Stimulated (30/sec, 5 min) in sea water	1.000 ± 0.08	(6)
Stimulated (30/sec, 5 min) in Ca-free Na sea water	0.796 ± 0.096	(7)
Stimulated (30/sec, 5 min) in Ca-free Li sea water	0.192 ± 0.033	(3)

Ca uptake (mmol/kg nerve during a 7 min exposure to ^{45}Ca) is expressed as mean ± S.E. of the mean. Three crabs were used and the number of nerves exposed to each treatment is given in brackets. All influx measurements were made from Li sea water. In the same series of experiments the mean Ca uptake by four Na-loaded nerves immersed in Na sea water was 0.162 ± 0.01 mmol/kg nerve in 7 min. Temperature 16.5° C.

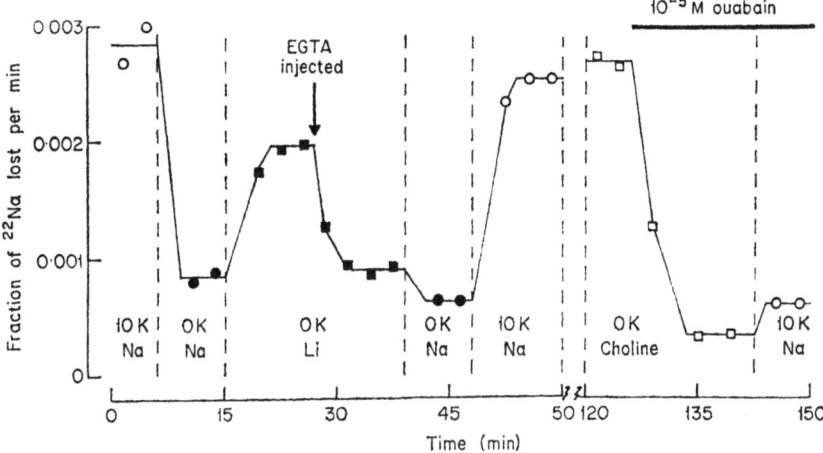

FIG. 2. Experiment illustrating the effect of injecting EGTA on the K-dependent and Li-dependent components of the Na efflux. The final EGTA concentration inside the axon was about 5 mM. Temperature 21°C. Axon diameter 650 μm.

mammalian liver (Judah & Ahmed, 1964), and smooth muscle (Goodford, 1967). It would be of interest to repeat on a mammalian nerve preparation, for example, synaptosomes or the mammalian C-fibre preparation used by Rang & Ritchie (1968), the kind of experiments performed on crab nerve by Baker & Blaustein (1968).

In view of the fact that Ca is moving into the axon down its electro-chemical gradient, the observation in Table 2 that the metabolic inhibitor cyanide reduces both the Ca influx and Ca-dependent Na efflux may appear somewhat surprising. This point still awaits clarification; but the position is certainly complex. Blaustein & Hodgkin (1969) have shown that in cyanide-poisoned squid axons the concentration of free calcium is increased in the axoplasm, and it is quite possible that it is this rise in intracellular Ca rather than the decline in energy-rich phosphate compounds which blocks the Na efflux. In an attempt to answer this point axons were injected with ethyleneglycol *bis*-(aminoethylether)-*N,N'*-tetra-acetic acid (EGTA) in order to maintain a constant low Ca concentration; but in all instances the EGTA blocked the Ca-dependent Na efflux without affecting the ouabain-sensitive and K-dependent Na efflux (Fig. 2). The explanation of this rather striking effect is not clear. It could be that some critical free Ca level is necessary in axoplasm; but attempts to rule this out by injecting Ca-EGTA buffers (Portzehl *et al.*, 1964) all gave results similar to EGTA alone. Further experiments are clearly needed, especially to find out what effect EGTA has on the Ca influx. Until the EGTA effect is fully understood, it may be very difficult to decide why the Ca-dependent Na efflux declines in cyanide. An alternative approach and a worthwhile one would be to use perfused or dialysed axons.

Na-dependent Ca efflux

The Na and Ca effluxes from squid axons differ strikingly in their responses to poisoning with cyanide. The Na efflux falls slowly—the decline parallels the fall in energy-rich phosphate compounds in the axon— and remains low for as long as the cyanide is applied (Caldwell *et al.*, 1960); whereas Rojas & Hidalgo (1968) and Blaustein & Hodgkin (1969) have shown that although the Ca efflux may decline initially, it rises after about 1–3 hr to a value 5–15 times greater than normal. This rise can be blocked by injection of EGTA and seems to coincide with a rise in the free Ca concentration in the cell. Leaving aside the source of this calcium but remembering that there is not a comparable increase in Ca influx from cyanide, it is obviously of interest to enquire what is the energy source for the large uphill movement of calcium which can take place in cyanide poisoned axons.

Part of the efflux appears to be in exchange for external calcium because the efflux is reduced to about one-third when external Ca is replaced by Mg. A further large reduction in efflux occurs, however, when external Na is replaced by either Li, dextrose, or K, and this suggests that part of the Ca efflux may be an exchange for external Na ions.

Baker, Blaustein, Hodgkin & Steinhardt (1967) and Blaustein & Hodgkin (1969) also examined the properties of the Ca efflux from unpoisoned axons. The absolute value of the efflux into sea water varied between 0·08 and 0·31 pmole cm^{-2} sec^{-1} with a mean of 0·24 pmole cm^{-2} sec^{-1}. As the mean Ca influx from sea water is 0·16 pmole cm^{-2} sec^{-1}, the two fluxes are about equal. Using these data, and assuming (1) that the calcium fluxes are entirely passive and independent of each other, (2) that the internal and external Ca concentrations are about 10 μM and 11 mM, respectively, and (3) that the internal potential is about 65 mV negative to the outside, the Ussing flux ratio predicts that the influx should be about 2×10^5 greater than the efflux (Blaustein & Hodgkin, 1969). Unless the calcium fluxes are purely Ca–Ca exchange, it follows that the outflux of calcium must require energy.

The Ca efflux has a Q_{10} of 2–3 and is unaffected by ouabain but is very dependent on the ionic composition of the external solution. The effects of external cations are essentially similar to those described for cyanide-poisoned axons: the efflux is reduced by about one-third by replacing external Ca by Mg and further reduced by replacing external NaCl by LiCl, choline chloride or dextrose. The effect of removing external Na is only fully reversible in Ca-free media (Fig. 3). This might arise because in Na-free media which contain Ca there is a large influx of cold Ca which saturates or inhibits the mechanism extruding Ca.

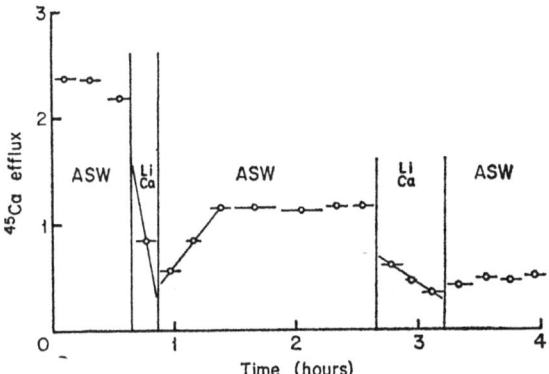

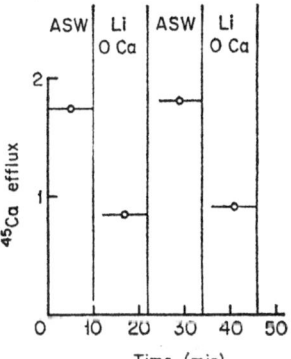

FIG. 3. Dependence of Ca efflux on external Na ions. External Na was replaced isosmotically by Li: (A) in the presence of Ca, and (B) in the absence of Ca. Mg was present in both experiments. Ordinate: fraction of ^{45}Ca lost/min (data multiplied by 10^4). Abscissa: time. (A), the first diagram, and (B), the second diagram, represent two separate experiments. Temperature 19°–21°C.

These calcium and sodium-dependent components of the Ca efflux account for about half of the resting Ca efflux. In a single experiment using Ca and Mg-free choline, the Ca efflux was reduced to about one-tenth of its normal value. It thus seems possible that—under physiological conditions—the bulk of the efflux of Ca occurs in exchange for the ions Na, Mg, and Ca.

Birks, Burstyn & Firth (1968), Reuter & Seitz (1968), and Blaustein & Hodgkin (1969) have discussed the possibility that a Na–Ca exchange could maintain the observed low level of free Ca in cytoplasm without the involvement of any energy source other than the Na and possibly the K ion gradients. Blaustein & Hodgkin (1969) conclude that a simple Na–Ca exchange in which one Ca ion moves out for two Na ions moving in could only lower the internal Ca to about one-hundredth of its value in the external medium, for squid axons that is, from 11 mM to 100 μM. This is not enough; but by making use of somewhat more complex schemes—for example, one calcium ion moving out in exchange for three sodium ions moving in—it would be possible to achieve an intracellular Ca concentration of about 1 μM.

These models are very attractive but they are still far from proved. Two particularly important pieces of evidence would be: (1) demonstration of Na–Ca exchange in the complete absence of energy-rich phosphate compounds, and (2) demonstration of an influx of Na associated with the Na-dependent Ca efflux. CN-poisoning never completely removes all the ATP, and it is possible that Na–Ca exchange may make use of the traces of ATP (about 100 μM) in these fibres. In this connection it may be relevant to point out that in many axons an early effect of CN is to produce a transient fall in the Ca efflux, which might imply that under normal conditions some of the Ca efflux requires energy-rich phosphate compounds. Perhaps the best experimental preparation for examining this problem would be the dialysed axon (Brinley & Mullins, 1968). The second piece of evidence might be difficult to obtain because of the small size of the fluxes involved. A further experiment would be to examine the influence of internal Na on the Ca efflux. If increasing internal sodium were to reduce Ca efflux as well as increase Ca influx, this would result in an even faster build up of free Ca in the cell.

Two modes of behavior of the same system?

The Ca influx seems to be linked to part of the Na efflux, and so there is a possibility that this system might operate in the opposite direction with part of the Na influx linked to Ca efflux. Thus the Na pump which normally uses energy to eject Na from the cell in exchange for K can operate in reverse, the downhill movements of Na and K through the pump synthesizing ATP (Garrahan & Glynn, 1967). At present it is not possible to say whether or not the two Na-dependent Ca fluxes represent movements in opposite directions through the same system; but if this interpretation is correct, the Ca-dependent Na efflux—which only becomes prominent when the electrochemical gradient for Na is reduced—might represent a movement in a direction opposite to that predominating in normal conditions. There is considerable scope for a comparative study of the effects of a variety of experimental manipulations on the Ca influx and Ca efflux. Information is lacking on the kinetics of the Ca efflux; it would for instance be useful to know the relative affinities of the efflux mechanism for the external cations Na, Ca, and Mg—both individually and in mixtures. Perhaps it is even more urgent to find a specific inhibitor of Na–Ca exchange because this might provide an answer as to whether or not the

Na-dependent Ca fluxes take place through the same or different systems. With the possible exception of lanthanum which has a number of other effects, no specific inhibitor has yet been found.

Implications of Na–Ca exchange

Whatever the exact relation between the two Na-dependent systems for transporting calcium, it seems likely that they have a number of roles in biological systems. The possibility has already been mentioned that the intracellular level of free Ca may ultimately be determined by the electro-chemical gradient for sodium; the observation that an increase in the internal Na concentration causes an increase in Ca influx may help to explain a number of actions of the cardiac glycosides especially those in Ca-dependent processes such as contraction and secretion. It should be emphasized that it may be misleading to generalize about a system which has, so far, only been clearly demonstrated in peripheral nerve.

The cardiac glycosides exert a number of actions on cells:

(1) they increase the force with which the heart beats (see, for example, Wilbrandt 1966);

(2) they increase the basal and stimulus-dependent secretions of a number of tissues, including the adrenal medulla (Banks, 1967), the neuromuscular junction (Birks & Cohen, 1968a,b), and the cells of the pancreas (Hales & Milner, 1968a,b); and

(3) they produce a wide variety of apparently unrelated effects on cellular metabolism (see, for example, Ho & Jeanrenaud, 1967; Bihler, 1968).

In most of these actions cardiac glycosides can be mimicked by K-free solutions, and it seems likely that their primary action is to slow the Na pump. This, in turn, may lead to a rise in intracellular Na and, by the mechanism described in this paper, to an increased Ca influx (see also Glynn, 1969; Birks & Cohen, 1968b; Baker, 1968; Baker *et al.*, 1969). There is some evidence for this chain of events in secretory tissues where the actions of ouabain require the presence of both external Ca and Na. There are strong reasons for thinking that the Na is required inside the cells (Birks & Cohen, 1968b; Banks, Biggins, Bishop, Christian & Currie, 1969; Muchnik & Venosa, 1969). Although there is less direct information for participation of Na–Ca exchange in cardiac muscle, there is no evidence against it. The simplest hypothesis is that a rise in intracellular Na would lead to an increased background level of Ca, and superimposed on this would be the Ca release triggered by the action potential; but it is not impossible that during the activation of muscle Na–Ca exchange may also play a part in the release of Ca from intracellular stores.

Perhaps the clearest evidence that the cardiac glycosides increase Ca uptake comes from the recent studies of Tower (1968) on slices of cat cerebral cortex. A number of earlier workers (for example, Ruscak & Whittam, 1967) had shown that ouabain increases the respiration of brain slices, an effect which is seen only in the presence of both external Ca and

Na ions. Bourke & Tower (1966) argued that this ouabain-induced rise in respiration may reflect the accumulation of calcium within the mitochondria and Tower (1968) has presented evidence for this. The Na-dependent Ca influx discussed in this paper would provide a link between the inhibition of the Na pump by ouabain and the rise in cell calcium. This view is supported by the finding that the ouabain-induced rise in respiration requires sodium (Ruscak & Whittam, 1967). The observations of Wollenberger (1947) suggest that cardiac glycosides may induce a similar Ca accumulation in cardiac muscle.

In conclusion, it seems likely that Na–Ca exchange may provide the cell membrane with an important mechanism for the control of the concentration of free calcium inside the cell. In normal conditions, the internal Ca concentration is probably stabilized by the Na pump which maintains fairly constant electrochemical gradients for Na and K. Anything altering the intracellular level of Na will result in a gain in intracellular Ca and may lead to profound changes in cell function and metabolism. At present we do not know whether Na–Ca exchange is sensitive to membrane potential; but should this prove to be the case, it would be necessary to consider the very interesting possibility that Na–Ca exchange may also play some part in the potential-dependent movements of Ca which are generally considered to underlie the release of nervous transmitters and the activation of muscle.

REFERENCES

BAKER, P. F. (1968). *J. gen. Physiol.*, **51**, 172s

BAKER, P. F. & BLAUSTEIN, M. P. (1968). *Biochim. biophys. Acta*, **150**, 167

BAKER, P. F., BLAUSTEIN, M. P., HODGKIN, A. L. & STEINHARDT, R. A. (1967). *J. Physiol., Lond.*, **192**, 43p

BAKER, P. F., BLAUSTEIN, M. P., HODGKIN, A. L. & STEINHARDT, R. A. (1969). *J. Physiol., Lond.*, **200**, 431

BAKER, P. F., BLAUSTEIN, M. P., MANIL, J. & STEINHARDT, R. A. (1967). *J. Physiol., Lond.*, **191**, 100p

BANKS, P. (1967). *J. Physiol., Lond.*, **193**, 631

BANKS, P., BIGGINS, R., BISHOP, R., CHRISTIAN, B. & CURRIE, N. (1969). *J. Physiol., Lond.*, **200**, 745

BIHLER, I. (1968). *Biochim. biophys. Acta*, **163**, 401

BIRKS, R. I., BURSTYN, P. G. R. & FIRTH, D. R. (1968). *J. gen. Physiol.*, **52**, 887

BIRKS, R. I. & COHEN, M. W. (1968a). *Proc. R. Soc., B*, **170**, 381

BIRKS, R. I. & COHEN, M. W. (1968b). *Proc. R. Soc., B*, **170**, 401

BLAUSTEIN, M. P. & HODGKIN, A. L. (1969). *J. Physiol., Lond.*, **200**, 497

BOURKE, R. S. & TOWER, D. B. (1966). *J. Neurochem.*, **13**, 1099

BRINLEY, F. J. & MULLINS, L. J. (1968). *J. gen. Physiol.*, **52**, 181

CALDWELL, P. D., HODGKIN, A. L., KEYNES, R. D. & SHAW, T. I. (1960). *J. Physiol., Lond.*, **152**, 561

COSMOS, E. & HARRIS, E. J. (1961). *J. gen. Physiol.*, **44**, 1121

GARRAHAN, P. J. & GLYNN, I. M. (1967). *J. Physiol., Lond.*, **192**, 237

GLYNN, I. M. (1969). *Digitalis* (University of Indiana Symposium, 1966), ed. FISCH, C. & SURAWICZ, B. New York: Grune & Stratton

GOODFORD, P. J. (1967). *J. Physiol., Lond.*, **192**, 145

HALES, C. N. & MILNER, R. D. G. (1968a). *J. Physiol., Lond.*, **194**, 725

HALES, C. N. & MILNER, R. D. G. (1968b). *J. Physiol., Lond.*, **199**, 177

HO, R. J. & JEANRENAUD, B. (1967). *Biochim. biophys. Acta*, **144**, 61

HODGKIN, A. L. & KEYNES, R. D. (1955). *J. Physiol., Lond.*, **128**, 28

JUDAH, J. D. & AHMED, K. (1964). *Biol. Rev.*, **39**, 160

LUXORO, M. & YANEZ, E. (1968). *J. gen. Physiol.*, **51**, suppl. 115

MUCHNIK, S. & VENOSA, R. A. (1969). *Nature, Lond.*, **222**, 169

NIEDERGERKE, R. (1963). *J. Physiol., Lond.*, **167**, 515

PORTZEHL, H., CALDWELL, P. C. & RÜEGG, J. C. (1964). *Biochim. biophys. Acta*, **79**, 581

RANG, H. P. & RITCHIE, J. M. (1968). *J. Physiol., Lond.*, **196**, 163

REUTER, H. & SEITZ, N. (1968). *J. Physiol., Lond.*, **195**, 451

ROJAS, E. & HIDALGO, C. (1968). *Biochim. biophys. Acta*, **163**, 550

RUSCAK, M. & WHITTAM, R. (1967). *J. Physiol., Lond.*, **190**, 595

TOWER, D. B. (1968). *Exp. Brain Res.*, **6**, 273

WILBRANDT, W. (1966). In *The Myocardial Cell: Structure, Function and Modification by Cardiac Drugs*, ed. BRILLER, S. A & CONN, H. L. Jr., p. 297. Philadelphia: Univ. of Pennsylvania Press

WOLLENBERGER, A. (1947). *J. Pharmac. exp. Ther.*, **91**, 39

ACTIVE TRANSPORT OF CALCIUM BY INTESTINE: STUDIES WITH A CALCIUM ACTIVITY ELECTRODE

D. SCHACHTER, S. KOWARSKI & PHYLLIS REID

Department of Physiology, College of Physicians and Surgeons, Columbia University, New York

Previous studies have demonstrated an intestinal mechanism which transfers calcium from the mucosa to the serosa *in vitro* (Rasmussen, 1959; Harrison & Harrison, 1960; Schachter, Dowdle & Schenker, 1960a,b; Finkelstein & Schachter, 1962; Schachter, Kowarski, Finkelstein & Ma, 1966; Schachter, Kowarski & Reid, 1967) or to the bloodstream *in vivo* (Wasserman, Kallfelz & Comar, 1961; Krawitt & Schedl, 1968) against concentration and electrical potential gradients. The transport is dependent on oxidative phosphorylation (Schachter & Rosen, 1959), and is rate-limited (Schachter *et al*, 1960a); it is relatively specific for Ca as compared with other cations (Schachter & Rosen, 1959; Schachter *et al*., 1960a) and is competitively inhibited by certain cations and actively transported hexoses (Schachter *et al*., 1960b; Schachter, Kimberg & Schenker, 1961). Moreover, it is dependent on vitamin D (Harrison and Harrison, 1960; Schachter *et al*., 1961, 1967) and varies in the rat with growth, pregnancy, and the level of dietary Ca (Schachter *et al*., 1960a; Kimberg, Schachter & Schenker, 1961). These properties suggest that the mechanism might be classified as an active transport. In earlier studies (Schachter *et al*., 1960a) a murexide technique was used to demonstrate that most of the Ca transferred to the serosal surfaces of everted gut sacs *in vitro* was ionized, and it was concluded that the mechanism is probably a cation pump. Evidence for this mechanism was provided more directly and precisely with a calcium activity electrode. The experimental results described here demonstrate net transfer of calcium ion *in vitro* against activity gradients of the ion. In addition, evidence is presented to show how the effects of vitamin D, growth, dietary Ca, and other variables influence the transport by affecting the cation pump.

The methods used in these studies have been described elsewhere and will be only briefly reviewed here. Albino male rats of the Sherman strain were used to prepare everted gut sacs. These were about 4 cm long and were incubated in Warburg vessels in a medium containing 151 mM NaCl; 4 mM KCl; 20 mM fructose; 1 mM $CaCl_2$; 4 mM Tris buffer, pH 7·4; and sufficient $^{45}CaCl_2$ to yield approximately 5000 cpm/ml. when assayed in Bray's solution (Bray, 1960) in a liquid scintillation spectrometer. The same medium was used to fill the gut sacs (serosal medium) and to bathe them (mucosal medium). Usually sacs from five to seven animals were incubated under O_2 at 37°C for 2·5 hr, and then the sacs were drained and the final mucosal and serosal media were pooled separately.

Vitamin D-depleted rats were prepared (Gordan & Schachter, 1963). Vitamin D repletion was with 20,000 i.u. vitamin D_3 given subcutaneously in ethanol/propylene glycol (1:5) 18 hr before the experiments. In the same way, male golden hamsters weighing 40–45 g were depleted of vitamin D and repleted.

Calcium was estimated chemically as previously described (Schachter *et al.*, 1960a). Activity of the calcium ion was estimated with the Orion calcium-specific electrode Model 92–20. The electrode was routinely soaked in 1·0 mM $CaCl_2$ for 2 hr before use, a procedure which reduced the tendency to drift. Calcium standards varying from 0·25 to 2·5 mM $CaCl_2$ were prepared in the standard incubation medium which did not contain ^{45}Ca or fructose, and the calcium ion potentials were measured against a calomel electrode, using a Corning No. 10 pH meter. The potentials measured in this concentration range exhibited Nernst potential behaviour: they were directly proportional to the logarithm of the calcium ion concentration and, with the instrument employed, a 2-fold increase in calcium concentration resulted in an average increment of 6·8 mV (range 6·2–7·0). Because the measured potentials of standard solutions tended to drift with time, each unknown determination was bracketed by at least two standards. All values reported are the means of triplicate estimations and were reproducible within 8%. The calcium ion concentrations of media from the gut sac experiments were calculated with reference to the standard solutions. The final calcium ion activity ratio of serosal/mucosal was calculated directly from the measured potentials of the serosal and mucosal media, using the following relationship which is derived from the Nernst equation:

$$E_s - E_m = \frac{RT}{2F} \log_{10} \left(\frac{aCa_s}{aCa_m} \right)$$

where E_s and E_m are the calcium potentials measured in the serosal and mucosal media, respectively, $RT/2F$ is the Nernst potential factor for a divalent sensing electrode (29.58 mV at 25°C) and aCa_s and aCa_m are the calcium ion activities in the serosal and mucosal media, respectively.

As a test of the measurements of calcium potential, the dissociation constant of the calcium-citrate complex was estimated in the presence of 150 mM NaCl at pH 7·4. The mean value observed for the pK (M^{-1}) in three determinations was 3·4, as compared with published values (Bjerrum, Schwarzenbach & Sillén, 1957) of 3·2 determined by other methods.

Time course of incubation

The ability of duodenal sacs to transfer calcium ions against activity gradients was demonstrated directly in time course experiments (Fig. 1). Duodenal sacs were prepared and incubated as described, groups of five

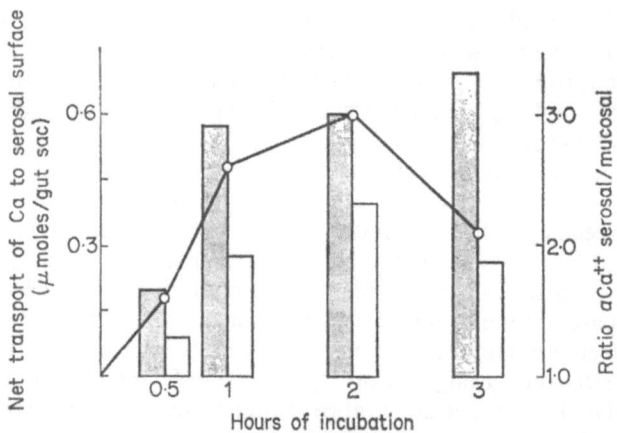

FIG. 1. Time course of transport of total ■ and ionized □ calcium by rat duodenal sacs. Duodenal, everted gut sacs were incubated, and at the indicated times groups of five were drained. Net transport of total Ca (dark bars) and of calcium ion (light bars) are indicated on the left-hand vertical axis. The final calcium ion activity ratio serosal/mucosal is indicated on the right-hand vertical axis.

were drained at various intervals, and the transfer of total calcium and of calcium ion was estimated. A net transfer of calcium ion from the mucosal medium to the serosal medium had occurred at 30 min, the earliest time point tested, and this increased until the 2 hr time point, when it accounted for approximately 60% of the net transport of total calcium. Correspondingly, the gradient of calcium ion activity serosal/mucosal increased from the value of 1·0 before incubation to a maximal value of approximately 3 at 2 hr. Between 2 and 3 hr, the Ca ion activity of the serosal medium decreased whereas the total Ca concentration increased, possibly as a result of a release of calcium-binding ligands from the intestine after prolonged incubation.

Effects of hexoses

Previous studies have demonstrated that calcium transport *in vitro* is at least partially dependent on a hexose which is metabolized in the ambient medium, and that actively transported but non-metabolized hexoses decrease the transport (Schachter *et al.*, 1961). Accordingly, the effects of a number of hexoses on transport of calcium ion were examined with duodenal sacs. Rats were fasted for 48 hr before these experiments, and the results are summarized in Table 1. The metabolic hexoses fructose, mannose,

TABLE 1. *Effects of various hexoses on transport of total and ionized Ca across duodenal gut sacs*

Hexose added	Net transfer to serosal surface (μmoles/sac per hr)		Total Ca conc. ratio serosal/ mucosal	Δ Ca^{2+} potential, serosal– mucosal (mV)	Ca^{2+} activity ratio serosal/ mucosal
	Total Ca	Ca ion			
None	+0·12	+0·02	2·6	+8·9	2·0
Fructose	+0·32	+0·18	7·1	+16·1	3·5
Mannose	+0·28	+0·10	4·8	+12·5	2·6
Glucose	+0·20	+0·07	3·5	+10·9	2·3
3-CH$_3$-D-glucose	+0·10	−0·10	2·4	+4·1	1·4

Groups of five duodenal sacs were prepared and incubated as described in the text, except that a hexose (20 mM) was added in place of fructose. After incubation at 37°C for 2·5 hr, media were pooled for determination of total Ca and ionized Ca.

and glucose all increased net transport to the serosal surface of total Ca and Ca ion as well as the final Ca activity ratios serosal/mucosal. Moreover, the increment in net transport of Ca ion was directly related to that of total Ca, with fructose most effective and glucose least. The increases in Ca ion transport observed with fructose and glucose, respectively, accounted for approximately 80% and 63% of the increments in total Ca transport. The actively transported hexose 3-O-methylglucose which cannot be metabolized decreased net transfer of Ca ion and the final Ca activity ratio serosal/mucosal.

Effects of vitamin D
Duodenal gut sacs prepared from groups of vitamin D depleted and repleted rats were used to examine the effects of vitamin D on the transport of calcium ion, and the results of four experiments are summarized in Fig. 2. As noted previously (Schachter *et al.*, 1961), vitamin D increases net transport of total Ca to the serosal surface and the final Ca concentration gradients, serosal/mucosal. The results demonstrated that increased transport of calcium ion accounted for approximately 80% of the vitamin dependent increment in net transfer. The final Ca ion activity gradients,

serosal/mucosal, averaged 2·1 for the vitamin-treated, as compared with 1·0 for the untreated, rats ($P < 0·001$). Although some net transport of calcium to the serosal surface was observed with the vitamin-depleted tissues, only approximately 30% as compared with 66% for the vitamin-repleted sacs represented calcium ion. It is reasonable to conclude that vitamin D acts primarily to increase the transport of calcium ion: that is, via a cation

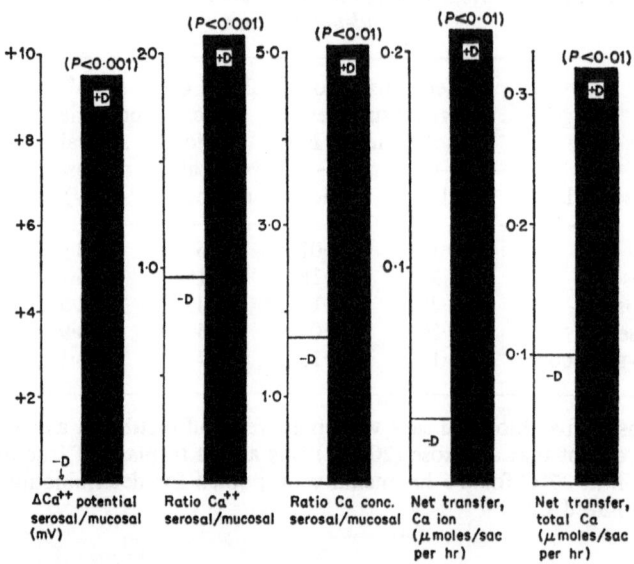

FIG. 2. Effects of vitamin D on transport of total Ca and of Ca ions by rat duodenal sacs *in vitro*. Results are mean values of four experiments in each of which between five and seven rats were used as vitamin-deficient (light bars) or -repleted groups (dark bars). *P* values for the effects of vitamin D are indicated over the dark bars.

pump. The residual net transport observed with vitamin deficient sacs, a small fraction of the activity of normal tissues, may represent mainly the transfer of calcium bound to ligands.

Effects of dietary calcium on duodenum and ileum

Previous studies have demonstrated that the calcium transport mechanism is markedly stimulated after a diet deficient in the mineral, and this adaptive response is dependent on vitamin D (Schachter *et al.*, 1960a; Kimberg *et al.*, 1961). To determine whether the effects of dietary Ca are associated with changes in the active transport of calcium ion, groups of ten weanling male rats were fed the purified diets (Kimberg *et al.*, 1961), which also contained 0·5% P and either 1·2% Ca (high Ca diet) or 0·02%

Ca (low Ca diet). Each rat was given 10 i.u. of vitamin D_3 in cottonseed oil 3 times weekly by mouth, and after 4 weeks a single duodenal and ileal gut sac was prepared from each and tested *in vitro*. The results of five separate experiments are summarized in Figs. 3 and 4. The results in Fig. 3 demonstrate that after Ca deprivation there are marked increases in the calcium ion potential differences serosal minus mucosal ($P < 0.001$) and the corresponding gradients of calcium ion activity, serosal/mucosal ($P < 0.02$). Ileal segments showed similar though less marked increases

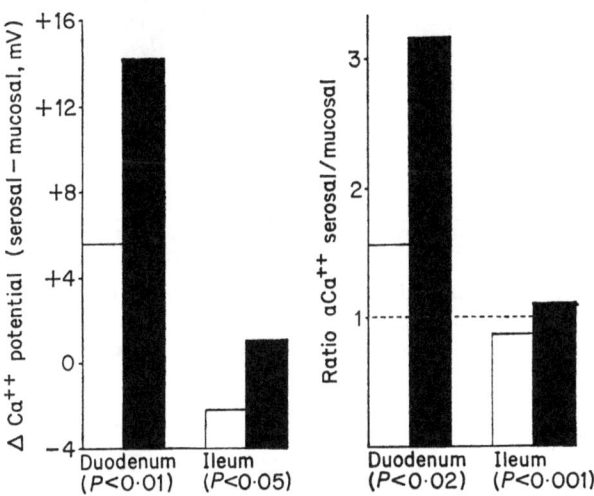

FIG. 3. Effects of diets high (light bars) or low (dark bars) in calcium on transport by duodenal and ileal sacs *in vitro*. Mean values of five experiments, as described in the text, are shown for the final calcium ion activity ratios serosal/mucosal and the final differences in calcium ion potentials serosal-minus-mucosal. P values for the differences due to dietary calcium are shown.

($P < 0.05$ and $P < 0.001$, respectively). Values for net transport of total Ca and Ca ion are shown in Fig. 4. Both parameters were approximately twice as great for duodenal segments from rats on the low as compared with the high Ca diet ($P < 0.01$), and the increment in Ca ion transport accounted for approximately 55% of the increase in total Ca transport. The mean values observed with the ileal segments varied with dietary Ca in a similar manner, but the values for net transport were low and the differences with diet not statistically significant. Transport of total calcium and of calcium ions was considerably greater across duodenal as compared with ileal sacs ($P < 0.01$) on either the low or the high Ca diet (Figs. 3 and 4). The increment in Ca ion transport accounted for approximately 59% of the increase in total Ca transport observed with duodenal as compared to ileal sacs.

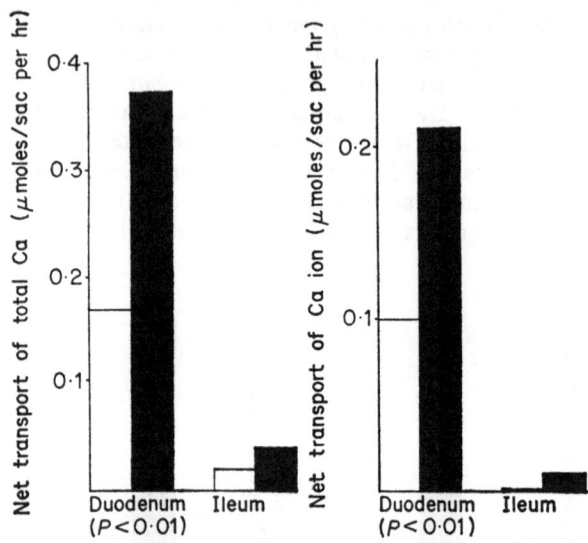

FIG. 4. Effects of dietary calcium (high Ca, light bars; low Ca, dark bars) on net transport of total calcium and of calcium ion by duodenal and ileal sacs *in vitro*. Mean values shown are from the same experiments illustrated in Fig. 3.

Effects of age

Duodenal sacs from young growing rats transport Ca *in vitro* more readily than do sacs from old animals (Schachter *et al.*, 1960a). Table 2 summarizes the results of two experiments in each of which three groups of rats

TABLE 2. *Effects of age on transport of total and ionized Ca across duodenal gut sacs*

Exp.	Group mean weight (g)	Net transfer to serosal surface (μmoles/sac per hr) Total Ca	Ca ion	Total Ca conc. ratio serosal/ mucosal	Δ Ca^{2+} potential serosal– mucosal (mV)	Ca^{2+} activity ratio serosal/ mucosal
1	49	+0·32	+0·16	6·5	+10·0	2·2
	158	+0·23	+0·19	4·3	+8·7	2·0
	531	+0·02	0	1·4	−1·3	0·9
2	58	+0·31	+0·12	6·9	+15·5	3·4
	209	+0·15	+0·04	3·5	+10·2	2·2
	522	+0·02	−0·06	1·4	−1·6	0·9

Groups of five rats of each weight were used to prepare and test duodenal sacs.

of various ages were used to prepare and test duodenal sacs for transport of calcium ion and of total calcium. The mean Ca ion activity ratios serosal/mucosal were greatest in the youngest age groups, which weighed 49 g and 58 g, with values of 2·2 and 3·4, respectively. The ratios were less in the intermediate groups, and in the oldest rats, which weighed over 500 g, the final mean ratios were 0·9, which gave no indication of active cation transport to the serosa. Values for net transport to the serosal surfaces varied correspondingly, with no detectable net transport of Ca ion in the oldest age group. Net transport of total Ca in the youngest groups (49 g and 58 g) was approximately 16 times that in the oldest groups (522 g and 531 g); increased transport of calcium ion by the youngest tissues accounted for approximately 58% of the increment in Ca transferred.

Effects of pregnancy
Duodenal sacs from female rats in the third week of pregnancy transfer calcium more readily than do sacs from non-pregnant female controls (Schachter *et al.*, 1960a). Table 3 summarizes three experiments in which duodenal sacs from pregnant and control rats were studied for calcium ion transport *in vitro*. The final potential differences for calcium ion, serosal minus mucosal, were +4·0 to +5·5 mV in the controls and +8·7 to +13·1 mV in the pregnant rats ($P < 0.02$), and the activity ratios serosal/mucosal

TABLE 3. *Effects of pregnancy on transport of total and ionized Ca across duodenal gut sacs*

Exp.	Group	Net transfer to serosal surface (μmoles/sac per hr)		Total Ca conc. ratio serosal/mucosal	Δ Ca²⁺ potential serosal–mucosal (mV)	Ca²⁺ activity ratio serosal/mucosal
		Total Ca	Ca ion			
1	Non-preg.	+0·22	+0·06	3·6	+4·0	1·4
	Preg.	+0·27	+0·13	5·7	+8·7	2·0
2	Non-preg.	+0·11	+0·02	2·4	+5·5	1·5
	Preg.	+0·20	+0·03	5·3	+10·1	2·2
3	Non-preg.	+0·08	0	2·2	+5·0	1·5
	Preg.	+0·23	+0·03	6·0	+13·1	2·8

Pregnant female rats were in the third week of pregnancy and were compared with control females of the same non-pregnant weight. Five rats were used in each group in experiments 2 and 3, and ten rats in each group in experiment 1. The effects of pregnancy on the calcium ion activity ratios serosal/mucosal and differences in calcium ion potential serosal–mucosal were significant at the $P < 0.02$ level.

D. Schachter, S. Kowarski & Phyllis Reid

varied correspondingly. Net transport to the serosal surfaces of both total Ca and Ca ion was increased by pregnancy in each experiment. Values for the increments in net transport, particularly of Ca ion, however, tended to be small and variable. The results suggest that only approximately 31% (mean of three experiments) of the increment in total Ca transferred to the serosa as a result of pregnancy could be accounted for by an increase in Ca ion transport, but further experiments are required to confirm this apparently low value.

Effects of replacing NaCl

Because a number of active transport mechanisms require ambient NaCl (Crane, 1960), it was of interest to examine with the divalent, calcium-sensing electrode the possible influence of NaCl on the transport of Ca ion. Groups of five duodenal sacs were incubated in the usual medium, except that a fraction of the NaCl was replaced isosmotically with manni-tol. The results, illustrated by a representative experiment in Fig. 5, demonstrate that replacement of approximately 30% of the NaCl increases the Ca^{2+} activity gradients, serosal/mucosal, the net transport of total calcium and of calcium ion to the serosal surfaces (the increment in Ca ion

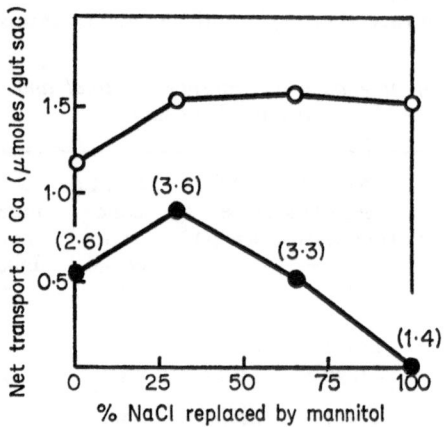

FIG. 5. Effects of replacing NaCl by isosmotic mannitol on calcium trans-port *in vitro*. Groups of five duodenal gut sacs prepared from rats were incubated as usual except that for each group a fraction of the ambient NaCl (mucosal and serosal media) was replaced isosmotically by mannitol. Sacs were incubated for 2·5 hr at 37°C under O_2. Net uptake of total calcium at the mucosal surface (○—○) and net transport to the serosal surface (●—●) are plotted. The values in parenthesis are the final calcium ion activity gradients serosal/mucosal estimated with the calcium sensing electrode. For estimations with the electrode standard solutions of calcium were prepared to contain the same NaCl and mannitol concentrations used in the incubation media. The experiment shown is representative of three similar experiments.

transport accounted for 51% of the increase in total Ca transferred), and the net uptake of Ca at the mucosal surface. Further replacement of NaCl progressively decreased net transport to the serosa and the activity gradients, serosal/mucosal, whereas the mucosal uptake of Ca remained unchanged. The results indicate that NaCl is required specifically for the transfer of Ca ion from the mucosa to or toward the serosa. They provide additional evidence that mucosal uptake and transfer to the serosa can be dissociated and identified as separate steps. A possible two-step transfer across the mucosa—*entry* at the mucosal surface versus *exit* from the opposite face of the mucosal cell—has been described, with supporting evidence, previously (Finkelstein & Schachter, 1962; Schachter, 1967; Schachter *et al.*, 1961, 1966, 1967). It is noteworthy that previous observations of the effects of metabolic inhibitors and the requirements for metabolizable hexoses indicate that the exit step, transfer to or toward the serosa, is dependent on cellular metabolism.

The replacement of NaCl defines two steps in Ca transport *in vitro*, with NaCl required for the second step dependent on cellular metabolism. It was therefore of interest to examine the influence of NaCl on ^{47}Ca concentrations in intestinal tissue compartments and the surrounding media after incubation. The procedure has been described (Schachter *et al.*, 1966) and involves scraping the mucosa from the underlying intestinal coats after incubation of sacs. The published observations (Schachter *et al.*, 1966) show two 'uphill' concentration differences, (*a*) higher concentrations in the mucosa as compared with the mucosal medium, and (*b*) higher concentrations in the underlying intestinal coats and serosal medium as compared with the mucosal tissue. The former, (*a*), results, in part at least, from mucosal binding of radioactive Ca and is relatively insensitive to changes in cellular metabolism. The latter, (*b*), is observed only when sacs are capable of active transport across the entire wall. The evidence, in conjunction with measurements of net transport, indicates that the uphill concentration difference between the serosal medium and the mucosal tissue results from the second, exit step of calcium transport, and this step requires cellular energy (Schachter *et al.*, 1966).

Figure 6 illustrates one of three experiments in which NaCl was completely replaced with isosmotic mannitol. For comparison, results obtained on the effects of incubation under N_2 as compared with O_2 (Schachter *et al.*, 1966) are also shown. In the presence of NaCl the usual profile of concentration differences were observed with duodenal sacs, and they are generally similar to the oxygenated condition (Fig. 6). In the absence of NaCl, the most striking change was a reversal of the normally uphill concentration difference between serosal medium and mucosal tissue, an effect also seen with sacs incubated under N_2. The replacement of NaCl differed from incubation under N_2 however, in that the concentration difference between mucosal tissue and mucosal medium increased considerably in the absence of NaCl but decreased somewhat under N_2.

Studies with the golden hamster

In many mammalian species, transport of Ca *in vitro* is maximal with segments of duodenum, whereas maximal transport is observed with the ileum of the golden hamster (Schachter *et al.*, 1960a), frog (Schachter, 1967), and in newborn rats (Batt & Schachter, 1969). The question arises whether the ileal transport is fundamentally different from that in the

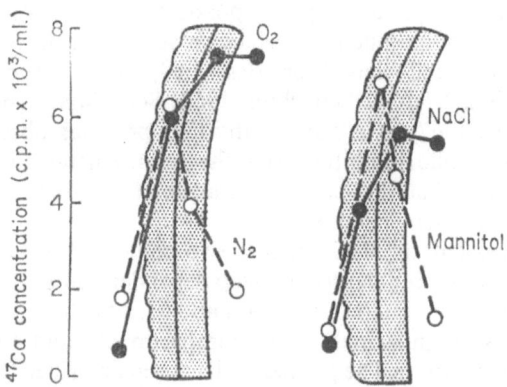

FIG. 6. Effects of replacing NaCl by isosmotic mannitol on ^{47}Ca concentrations in media and duodenal tissue compartments. Groups of five rat duodenal sacs were incubated in the usual medium or with the NaCl completely replaced isosmotically by mannitol. ^{47}Ca was used to facilitate estimation of tissue radioactivity. Sacs were incubated 2·5 hr at 37°C under O_2, and the final ^{47}Ca concentrations in media and tissue compartments estimated as described in the text. Values for tissue compartments are in terms of tissue water content. The mean values are illustrated on a cross-section of the intestinal wall and represent the following four compartments (from left to right): mucosa medium, mucosal tissue, underlying intestinal coats and serosal medium. The experiment shown is representative of three similar experiments. Values for incubation under O_2 versus N_2 are taken from an earlier paper (Schachter *et al.*, 1966).

duodenum, and experiments were therefore designed to test whether hamster ileal transport is vitamin D-dependent or capable of transferring Ca ion against activity gradients. Male golden hamsters weighing 40–45 g were maintained on the vitamin D-depletion regimen for 3 weeks. Groups of five animals either deficient or repleted were used, one duodenal and ileal sac prepared from each, and the sacs tested as described above. Figure 7 summarizes the results of five separate experiments. In the vitamin D-repleted groups, net transport of Ca ion against activity gradients was observed with both duodenum and ileum. Moreover, the final activity gradients for Ca^{2+}, serosal/mucosal, were significantly greater with ileum as compared with duodenum ($P < 0·02$). Vitamin D increased the net transport of Ca ion in duodenum ($P < 0.05$) and ileum ($P < 0.01$), with corresponding increases in the activity gradients, serosal/mucosal ($P < 0.02$ and $P < 0.01$, respectively). The results demonstrate, therefore, that the ileal transport in the hamster is fundamentally similar to that in the duodenum of the hamster or rat; in other words, it is a calcium cation pump dependent on vitamin D.

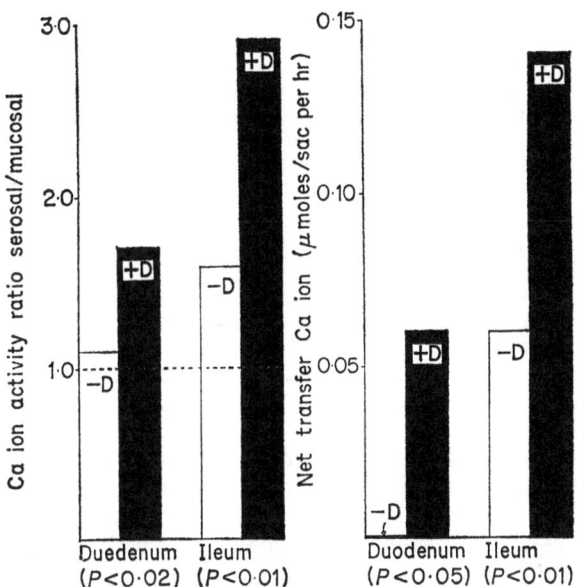

FIG. 7. Effects of vitamin D on transport of calcium ion by sacs of hamster duodenum and ileum. Methods are identical to those described for rats (text) except that the incubation media contained 0·5 mM $CaCl_2$ initially. Results shown are means for five separate experiments in each of which five vitamin D-deficient and five repleted animals were used. P values for the effects of vitamin D are shown.

The various factors which have been shown to increase Ca transport *in vitro* generally increase the net transport of Ca ion. Table 4 lists for each variable studied the value of the fraction, increment in Ca ion transport/ increment in total Ca transport. This fraction was greatest for the influence of vitamin D in both rat (0·82) and hamster (0·73 and 0·88). In rats the next greatest values were for the effects of hexose *in vitro* (0·65), age, the low Ca diet and duodenum versus ileum (0·55–0·59), and the influence of NaCl (0·51). Finally, as a preliminary observation, pregnancy in the rat seemed to yield the lowest value (0·31). It seems that primary increments in Ca ion transport could account for the influence of many factors which increase Ca transport *in vitro*.

TABLE 4. *Effects of various factors on the ratio (increment in calcium ion transport to the serosal surface)/(increment in total calcium transport to the serosal surface)*

Species	Factor	Ratio (increment in calcium ion transport)/ (increment in total calcium transport)
Rat	Vitamin D repletion	0·82
	Duodenum versus ileum	0·59
	Low calcium diet	0·55
	Young versus old rats	0·58
	Metabolizable hexose in incubation medium	0·65
	NaCl versus mannitol in medium	0·51
	Pregnancy	0·31
Hamster	Vitamin D repletion:	
	Duodenum	0·88
	Ileum	0·73

Conclusions

It has been established (Finkelstein & Schachter, 1962; Schachter, 1963) that the electrical potential differences between serosal and mucosal surfaces of rat duodenal sacs incubated *in vitro* under the conditions described here are positive in sign. In conjunction with the present results, therefore, rat duodenum is capable of net transport of calcium ion against electrochemical potential gradients; in other words, the transfer involves an active cation pump. As active transfer of calcium ion accounted for most of the increments in total Ca transport secondary to a number of variables—segment of intestine, metabolizable hexose in the medium, presence of NaCl, young versus old, non-growing rats, dietary calcium, and vitamin D—it is reasonable to suggest that adaptive and other variations in calcium transport depend on changes in the cation pump. This conclusion is particularly justified with respect to the mechanism of action of vitamin D; these results for rat and hamster demonstrate that *in vitro* approximately 70–90% of the total effect of the vitamin on Ca transport results from increments in Ca ion transfer. Recent studies implicate a mucosal calcium-binding protein in the action of the sterol vitamin in chickens (Wasserman & Taylor, 1966) and rats (Schachter *et al.*, 1967). Although the precise role of the protein is not yet defined, its function is of considerable interest as it may underlie the molecular basis of calcium transport by representing wholly or in part a transport carrier. In view of the present results the protein may be a structural element of a cation pump.

The transport of Ca across intestinal mucosa *in vitro* has been characterized as a two-step process in previous reports (Schachter *et al.*, 1966) and an hypothesis has been described which characterizes the steps as entry into and exit from the mucosal cells. Although both steps seem to be rate-limited, the exit step is apparently the active process which requires cellular energy. The present observations of replacement of extracellular NaCl with isosmotic mannitol provide additional evidence in favour of the entry–exit model. In the absence of NaCl the two steps are completely dissociated through the impairment of the exit process. Extracellular NaCl thus appears essential for the energy-coupled step, a finding which, superficially at least, resembles the requirement for Na^+ for active transport of hexoses (Crane, 1960; Csáky, 1961) and amino acids (Csáky, 1961) by the small intestine. Further experiments are required to determine whether the requirement for Ca transfer is specific for the sodium ion. Replacement of 30% of the extra-cellular NaCl increased both steps of the Ca transport, effects which may result from competition between Na^+ and Ca^{2+} for essential cellular sites. Such competition has already been suggested for nerve (Hodgkin, Huxley & Katz, 1949) and kidney (Walser & Robinson, 1963) cells.

The experiments with hamster intestine indicate that the ileal mechanism for Ca transport in this species is also an active cation pump dependent on vitamin D. The evolutionary or functional significance of this ileal site, which differs from the usual duodenal localization in other mammals, remains unknown, but some general significance is indicated by a similar localization to the distal small intestine of the frog (Schachter, 1967) and the newborn rat (Batt & Schachter, 1969). In the case of the hamster, the ileal mechanism may relate to the peculiarly efficient Ca-absorbing mechanism observed in the native American pack rat (*Neotoma* sp.) inasmuch as hamsters and pack rats belong to the same family of rodents (Cricetidae) (Schmidt-Nielsen, 1964). Schmidt-Nielsen (1964) noted that pack rats absorb appreciable quantities of calcium fed as either calcium carbonate or calcium oxalate, compounds whose Ca cannot be absorbed by most mammals. The pack rats so fed produced a creamy urine, thick with a precipitate of calcium carbonate, and many observers have noted that laboratory hamsters may also excrete a thick, creamy urine.

The Ca transport noted in the ileum of the hamster is significant with respect to the function of the soluble, vitamin D-dependent calcium-binding protein (Schachter *et al.*, 1967; Wassermann & Taylor, 1966). As reported previously (Schachter *et al.*, 1967), the activity of the calcium-binding protein noted in the supernatant of mucosal homogenates correlates closely with calcium transport in the rat, but in the hamster the calcium-binding activity is maximal in duodenum and very much less in ileum; transport, however, varies in the opposite direction. Studies are in progress to determine whether this apparent divergence results from changes in the intracellular localization or integration of the calcium-binding protein.

Finally, the effects of age on Ca transport should again be emphasized, in view of the report of Helback, Forte, and Saltman (1966), who studied Ca fluxes *in vitro* with segments from rats weighing 400 g. They found no evidence for active transport of calcium, a finding to be expected in view of the effects of age. As noted in previous studies (Schachter *et al.*, 1960a,b) and again demonstrated clearly in the present experiments, active transport of Ca is minimal or absent with segments from old, non-growing rats on casual laboratory diets.

This research was supported by United States Public Health Service Grant AM-01483. One of us (DS) was a Career Scientist of the Health Research Council of the City of New York under Contract I-183.

REFERENCES

BATT, E. R. & SCHACHTER, D. (1969). *Am. J. Physiol.*, **216**, 1064

BJERRUM, J., SCHWARZENBACH, G. & SILLÉN, L. G. (1957). *Stability Constants* Part I: *Organic Ligands*, p. 39. London: The Chemical Society

BRAY, G. A. (1960). *Analyt. Biochem.*, **1**, 279

CRANE, R. K. (1960). *Physiol. Rev.*, **40**, 789

CSÁKY, T. A. (1961). *Am. J. Physiol.*, **201**, 999

FINKELSTEIN, J. D. & SCHACHTER, D. (1962). *Am. J. Physiol.*, **203**, 873

GORDAN, G. S. & SCHACHTER, D. (1963). *Proc. Soc. exp. Biol. Med.*, **113**, 760

HARRISON, H. E. & HARRISON, H. C. (1960). *Am. J. Physiol.*, **199**, 265

HELBOCK, H. J., FORTE, J. G. & SALTMAN, P. (1966). *Biochem. biophys. Acta*, **126**, 81

HODGKIN, A. L., HUXLEY, A. F. & KATZ, B. (1949). *Arch. sci. physiol.*, **3**, 129

KIMBERG, D. V., SCHACHTER, D. & SCHENKER, H. (1961). *Am. J. Physiol.*, **200**, 1256

KRAWITT, E. L. & SCHEDL, H. P. (1968). *Am. J. Physiol.*, **214**, 232

RASMUSSEN, H. (1959). *Endocrinology*, **65**, 517

SCHACHTER, D. (1963). In *The Transfer of Calcium and Strontium Across Biological Membranes*, ed. WASSERMAN, R. H., p. 197. New York: Academic Press

SCHACHTER, D. (1967). In *L'Osteomalacie*, ed. HIOCO, D. J., p. 199. Paris: Masson & Cie.

SCHACHTER, D., KIMBERG, D. V. & SCHENKER, H. (1961). *Am. J. Physiol.*, **200**, 1263

SCHACHTER, D., DOWDLE, E. B. & SCHENKER, H. (1960a). *Am. J. Physiol.*, **198**, 263

SCHACHTER, D., DOWDLE, E. B. & SCHENKER, H. (1960b). *Am. J. Physiol.*, **198**, 275

SCHACHTER, D., KOWARSKI, S., FINKELSTEIN, J. D. & MA, R. W. (1966). *Am. J. Physiol.*, **211**, 1131

SCHACHTER, D., KOWARSKI, S. & REID, P. (1967). *J. clin. Invest.*, **46**, 1113

SCHACHTER, D. & ROSEN, S. M. (1959). *Am. J. Physiol.*, **196**, 357

SCHMIDT-NIELSEN, K. (1964). *Desert Animals*, p. 146. Oxford: Clarendon Press

WALSER, M. & ROBINSON, B. H. B. (1963). In *The Transfer of Calcium and Strontium Across Biological Membranes*, ed. WASSERMAN, R. H., p. 310. New York: Academic Press

WASSERMAN, R. H. & TAYLOR, A. N. (1966). *Science, N.Y.*, **152**, 791

WASSERMAN, R. H., KALLFELZ, F. A. & COMAR, C. L. (1961). *Science, N.Y.*, **133**, 883

DISCUSSION TO SESSION III

Reuter (Bern)
I would like to comment on the Na–Ca exchange mechanism discussed by Dr. Baker. In earlier experiments we found that more than 80% of calcium efflux from mammalian cardiac preparations is due to an exchange of external Ca^{2+} and Na^+ against internal Ca^{2+}. Continuing these experiments Dr. Glitsch, Dr. Scholz and I found that changes of internal sodium concentration had appreciable effects on Ca influx. Thus an increase in internal sodium concentration increased Ca influx, while a decrease of internal sodium decreased the influx. This effect of internal sodium on Ca influx was obtained irrespective of whether Ca influx was measured from sodium-poor or sodium-rich solution. Ca influx from sodium-poor solution, however, corresponds to an about equal net uptake of Ca while this was not the case with normal external sodium concentration. Correspondingly, Ca efflux from preparations loaded with sodium was inhibited in sodium-poor solution and increased in sodium-rich solution. These results support further the previous assumption of a Na–Ca exchange and show clearly the similarities of the systems in squid axon and mammalian cardiac muscle.

Loewenstein (New York)
Dr. Baker, does a rise in intracellular calcium concentration affect the sodium efflux?

Baker (Cambridge)
We do not know.

Goodford (London)
I would like to ask Dr. Hasselbach if there is an electrical potential difference between sarcoplasmic vesicles or mitochondria and their surroundings, which might affect the calcium distribution.

Hasselbach (Heidelberg)
As far as the energetics of the accumulation of the calcium oxalate is concerned, it is not affected by the existence of such a potential. As far as the passive calcium movement is concerned it does not change when the normal medium of low ionic strength (mainly KCl) is replaced by 1 M NH_4Cl or 1 M sodium acetate. This is in agreement with light scattering measurements which indicate that the membranes are permeable to Na^+, K^+, NH_4^+, Cl^- ions.

Birks (Montreal)

Does Dr. Schatzmann have any reason to believe, or not to believe, that the active extrusion of Ca^{2+} may suffer from competition with intracellular sodium? A comparison of efflux rates of Ca^{2+} from cold-stored erythrocytes which have accumulated Ca^{2+}, and either been allowed or not allowed to accumulate sodium, might provide some useful information on this question. Have you any information at this time?

Schatzmann (Bern)

Ca-ATPase is not sensitive to replacement of K by Na. Ca-extrusion proceeds both from cells with 130 mM Na or K (inside and outside). It was faster from K-cells but neither am I certain whether the difference was real nor do I know what it might mean. Replacement of Na by Iris did not accelerate the exit of Ca.

Woodin (Oxford)

Can I ask Dr. Hasselbach what is the anion which is deposited with calcium during *in vivo* accumulation in the sarcoplasmic reticulum?

The hydrolysis of ATP suggests that deposition of calcium phosphate may occur. However, it seems to me that deposition of solid calcium phosphate might be an embarrassment for the living cell in being irreversible under physiological conditions.

Hasselbach (Heidelberg)

As long as the phosphate precipitates are small they remain soluble, especially as the free calcium concentration in the myoplasm is extremely low.

Baker (Cambridge)

I should like to ask Dr. Hasselbach how many calcium transporting sites there are per μ^2 of sarcoplasmic reticulum. Is the density of ferritin binding sites uniform along the longitudinal reticulum? Or are there regions lacking pumps which might, perhaps, be more concerned in calcium release?

Hasselbach (Heidelberg)

We have reason to assume that the distance between transporting sites is < 100 Å. The arguments which I gave for this figure do not exclude small regions in the sarcoplasmic membranes having no transporting sites. On the other hand there is no argument against the possibility that pumping sites and releasing sites are not the same.

Hodgkin (Cambridge)

Dr. Schatzmann, how many moles of calcium were transported out of red cells per mole of ATP?

Schatzmann (Bern)
The ratio was 0·75 mole Ca/1 mole ATP.

Caldwell (Bristol)
By my calculation there is probably enough free energy available from the splitting of one molecule of ATP for the transport of one calcium ion from red cells. This is because Garrahan & Glynn (1967) have calculated the free energy normally available to be about 13,000 cal./mole of ATP.

Norman (California)
I would like to ask Dr. Schachter three questions. (1) Is the calcium binding protein really found in the supernatant or is it possible that it arises on breaking up of the cell and might it not therefore be membrane bound *in vivo*? (2) Have you demonstrated a *de novo* synthesis of calcium binding protein in response to vitamin D? (3) Is the rate of appearance of the calcium binding protein in response to physiological doses of vitamin D consistent with the known delayed action of the vitamin?

Schachter (New York)
The protein is found in the supernatant after homogenization of the mucosa in isotonic media. Of course, it is possible that the procedure of homogenization could release a protein normally bound to membrane *in vivo*. We have demonstrated *de novo* synthesis of calcium binding protein in response to vitamin D. Using ^{14}C-labelled amino acids as precursors, we demonstrated an increased rate of biosynthesis as well as an increased pool size in the mucosa following vitamin D. The rate of appearance of the protein has been consistent with the time course of action of the vitamin in our experiments thus far.

Rasmussen (Philadelphia)
Does the calcium binding protein have any ATPase or phosphatase activity?

Schachter (New York)
It does not have ATP'ase activity, but we have not tested various other phoshate esters.

Robinson (London)
What evidence has Dr. Schachter that the exit step in calcium transport across the intestine is sodium dependent? In his experiment in which he compared the Ca^{2+} fluxes in the presence of saline and mannitol to derive this conclusion, it would appear he has omitted to consider the possibility of differential water flow. In *in vivo* studies with loops of rat duodenum and jejunum we have observed different rates of fluid transfer when isotonic solutions of saline and mannitol were used to study the absorption of calcium.

Dr. Schachter showed that the majority of the calcium binding protein was located in the duodenum and this was associated with the active transport of calcium in that segment. What is the physiological significance of the calcium binding protein when most of the calcium absorption *in vivo* appears to occur in the ileum (Marcus & Lengemann, 1962; Cramer, 1965) where only small amounts of calcium binding protein are found and where Dr. Schachter does not find active transport in the normal animal? The inference would be that it represents a homeostatic mechanism which must be controlled by external factors such as the level of plasma calcium or parathyroid activity. This would explain the raised amounts of calcium binding protein that Dr. Wasserman's group finds in conditions of calcium deficiency or pregnancy. Both these, incidentally, are characterized by parathyroid hyperactivity.

Schachter (*New York*)
In the *in vitro* studies with everted sacs mucosal and serosal media were of the same composition and no difference in net water flow was observed. Active transport of calcium can be demonstrated in ileum following calcium deprivation; it probably operates at a reduced level in the normal animal but it is not detectable.

REFERENCES

CRAMER, C. F. (1965). *Can. J. Physiol. Pharm.*, **43**, 75
GARRAHAN, P. J. & GLYNN, I. M. (1967). *J. Physiol., Lond.*, **192**, 237
MARCUS, C. S. & LENGEMANN, F. W. (1962). *J. Nutr.*, **77**, 155
WASSERMAN, R. H., TAYLOR, A. N., CORRADINO, R. A. & KALLFELZ, F. A. (1968). In *Les Tissues Calcifiés* (5th Eur. Symp.), ed. MILHAND, G., OWEN, M. & BLACKWOOD, H. J. J., p. 253. Paris: Soc. d'Edition d'enseignement Sup.

Session IV

Calcium and Cellular Function

Chairmen:

N. HALES
Department of Biochemistry, University of Cambridge

P. F. BAKER
Physiological Laboratory, University of Cambridge

ROLE OF CALCIUM IONS IN NEUROMUSCULAR TRANSMISSION

R. RAHAMIMOFF

Department of Physiology, Hebrew University,
Hadassah Medical School, Jerusalem, Israel

The important role of calcium ions in neuromuscular transmission was recognized three-quarters of a century ago (Locke, 1894). If a neuro-muscular preparation of the frog is immersed in normal Ringer solution, an adequate stimulus to the nerve produces a twitch. When the calcium concentration in the medium is reduced to, say, one-eighth of the normal, a similar stimulus produces no mechanical response. If the effect of partial or total calcium deprivation is examined on the various steps leading from the nerve stimulus to muscle contraction, almost every stage, depending on the experimental conditions and the time of exposure, is found to be affected to some extent: the initiation of the action potential, its conduction along the nerve (Frankenhauser & Hodgkin, 1957; Frankenhauser, 1957), the release of the transmitter from the motor nerve terminal (Katz & Miledi, 1965a), the reaction of the transmitter with the postsynaptic membrane (Takeuchi & Takeuchi, 1962; Nastuk & Liu, 1966), the genera-tion of the action potential in the muscle, and finally the contraction of the muscle (see Sandow, 1965). It appears, however, that the link which is most sensitive to reduction of external calcium is the nerve terminal.

The sensitivity of the nerve terminal to calcium is illustrated by an experiment performed by Katz & Miledi (1965c) in which the nerve muscle preparation was immersed in a solution without calcium. The electrical activity was recorded extracellularly by a $CaCl_2$ filled micropipette. The left half of Fig. 1 shows superimposed traces and the right half auto-matically obtained averages. When a 'braking current' is applied to the micropipette and no calcium is allowed to diffuse out (Fig. 1a), stimulation of the nerve produces an action potential in the terminals, an action potential which is unable to induce transmitter release. Transmitter is released, however, when the electric current on the pipette is adjusted in such a way that Ca ions are applied electrophoretically. (Fig. 1b and 1c). When Ca efflux from the micropipette is stopped again and enough time is allowed for the previously applied calcium to diffuse away (Fig. 1d), no transmitter is liberated. The postsynaptic sensitivity is not greatly affected under these conditions.

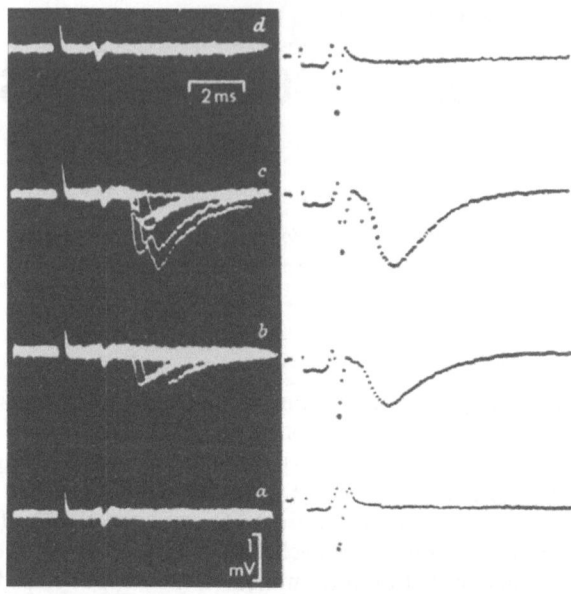

FIG. 1. Effect of calcium on neuromuscular transmission. Focal external recording from a junctional spot of frog sartorius. Muscle was immersed in Ca-free solution containing 0·84 mM Mg. A 0·5 M $CaCl_2$ pipette was used for recording; efflux of Ca was controlled electrophoretically. In the four records, a to d, Ca efflux was stopped initially by applying sufficient negative voltage to the pipette. The bias was then reduced in two steps and finally reapplied. On the left, several superimposed traces are shown at each stage; on the right, automatically obtained average of 600 traces at each stage by a computer for averaging transients. Averaging step of approximately 80 μsec. Time calibration applies to records on the left only. (From Katz & Miledi, 1965c.)

These and other experiments show (Katz, 1969) that in the absence of calcium the action potential still invades the terminals but fails to induce release of transmitter. The main role of calcium is to mediate between the presynaptic depolarization and transmitter liberation. In this respect the neuromuscular junction resembles other synapses as well as neurosecretory endings (Hutter & Kostial, 1954; Takeuchi & Takeuchi, 1962; Miledi & Slater, 1966; Douglas & Rubin, 1961; Douglas & Poisner, 1964).

In the rest of this paper I shall consider only this particular process—the effect of calcium and other ions on the voltage dependent release of transmitter. I shall not deal at all with the effect of divalent ions on the spontaneous release of transmitter or on post synaptic sensitivity. Out of the various aspects of Ca action, I would like to mention four in detail: the timing of calcium action; the quantitative relation between calcium and release of transmitter; the effect of sodium ions; and the effect of other

Group IIa divalent ions on release of transmitter. Two parameters will be taken to measure transmitter release: the mean number of quanta released per nerve impulse (quantal content, m) and the mean amplitude of the end plate potential (e.p.p.) (Del Castillo & Katz, 1954b). The latter parameter is valid only when the mean size of the spontaneous miniature e.p.p. remains constant (see Katz, 1962, 1966).

Timing of calcium action

Acetylcholine is not released immediately upon the arrival of the action potential at the nerve terminal. There is a delay of one-half to several milliseconds between these two events (Katz & Miledi, 1965b). The question arises, when are Ca ions necessary? Are they essential during the actual release of transmitter or during the depolarization of the terminal by the action potential? To answer this question Katz & Miledi (1967c)

(a) (b) (c)

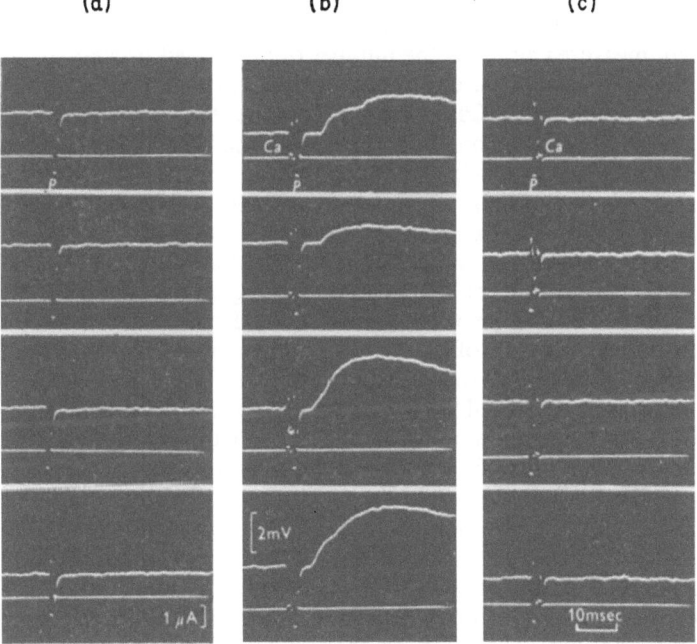

FIG. 2. Effect of ionophoretic pulses of calcium (Ca) on end-plate response. Depolarizing pulses (*P*) and calcium were applied from a twin-barrel micropipette to a small part of a neuromuscular junction in the frog sartorius. Intracellular recording from the end-plate region of the muscle fibre. Bottom traces show current pulses through the pipette. Temperature 4°C. Column (a), Depolarizing pulse alone. Column (b), Calcium pulse precedes depolarizing pulse. Column (c), Depolarization precedes calcium pulse. (From Katz & Miledi, 1967d.)

performed the following experiment (Fig. 2). A neuromuscular preparation was immersed in a solution containing low [Ca] (less than 0·1 mM) 1 mM Mg and tetrodotoxin. An intracellular micropipette was used for recording. A double-barrelled extracellular micropipette was employed, one side being filled with NaCl for depolarization of the terminals and the other side with $CaCl_2$. Figure 2a shows that depolarization alone is not sufficient to induce release. When the depolarization is preceded by a Ca pulse, transmitter is released (Fig. 2b). Applying Ca immediately after the depolarizing pulse does not produce any release of transmitter (Fig. 2c). Such experiments show that Ca action is very rapid on a biological time scale, and that it must be present either before or during the depolarization. It is ineffective if applied after the depolarization, even before the onset of transmitter liberation, that is, during the period of synaptic delay.

Quantitative relation between calcium and acetylcholine output

The action of calcium on the liberation of transmitter at the neuromuscular junction is not an all-or-none phenomenon. The amount of transmitter released, measured either as the number of preformed packets (or quanta) or as the amplitude of the end plate potential, depends on the extracellular calcium concentration, $[Ca]_o$. Figure 3 illustrates that an increase of 25% in $[Ca]_o$ has a large effect on e.p.p. amplitude; the latter increases about two fold. In this range of $[Ca]_o$, the release can be augmented nearly 16 times if $[Ca]_o$ is doubled (Dodge & Rahamimoff, 1967). This highly nonlinear relation is also obtained when magnesium is present in the bathing medium, Mg being a competitive antagonist to Ca (Jenkinson, 1957) (see Fig. 4a).

Replotting the results of Fig. 4a on double logarithmic coordinates yields straight lines with a slope of nearly 4 (Fig. 4b). In this range of low calcium concentrations the slope of the log Ca − log e.p.p. relation was quite consistent in different fibres, the average value being 3·78 ± 0·2 S.D. in twenty-eight experiments. This slope was, however, by no means constant when the $[Ca]_o$ was extended to a higher range. Near the physiological $[Ca]_o$ of 1·8 mM the slope was only between one and two.

When these results were obtained (Dodge & Rahamimoff, 1967), we chose to interpret them along a hypothesis suggested by Del Castillo & Katz (Del Castillo and Katz, 1954a). This hypothesis assumes that the nerve terminal possesses specific sites for calcium, and these are appropriately named X. Calcium ions combine with these sites to form a reversible CaX complex which is somehow responsible for transmitter release.

$$Ca + X \overset{K_1}{\rightleftharpoons} CaX \qquad (1)$$

K_1 is the dissociation constant ($K_1 = [Ca][X]/[CaX]$). Mg ions can combine with the same site, but the resulting MgX complex is ineffective in release:

$$Mg + X \overset{K_2}{\rightleftharpoons} MgX \qquad (2)$$

where K_2 is the dissociation constant for the MgX complex.

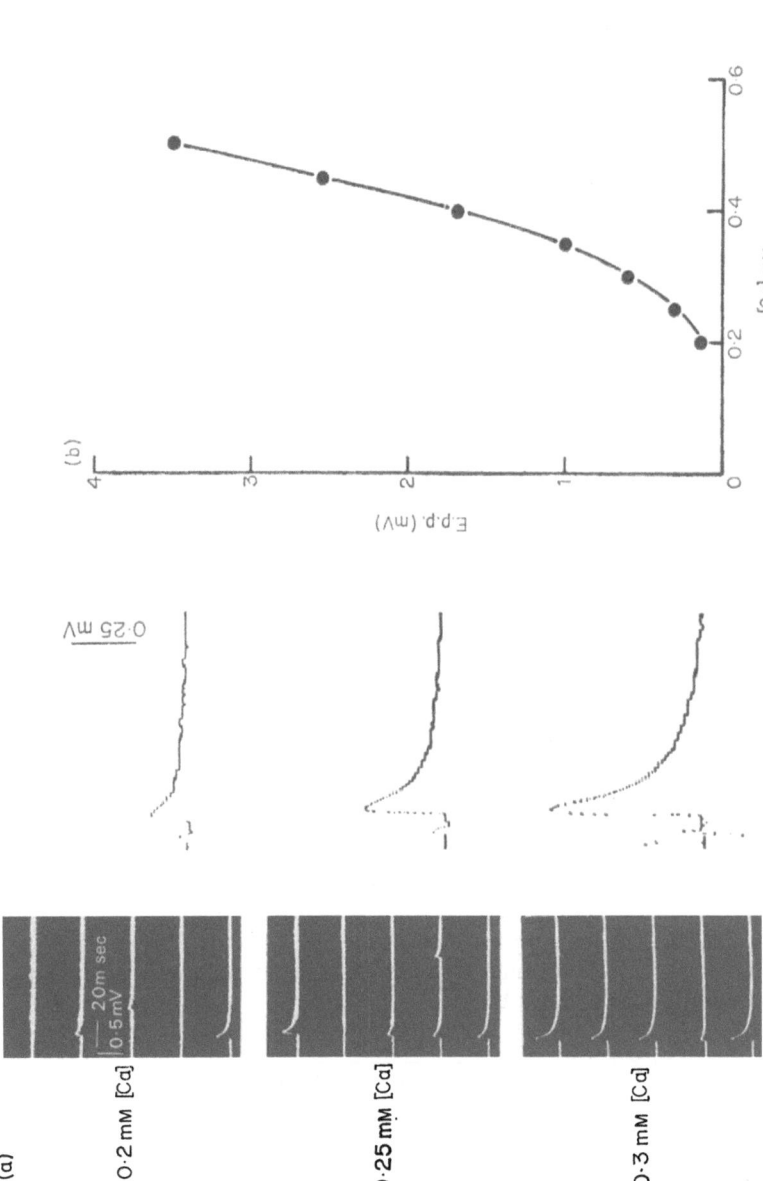

FIG. 3. Dependence of transmitter release on external calcium concentration. (a) Sample records of e.p.p.s. (left) and automatically averaged response to 256 stimuli (right). Averaging step, 160 μsec. Three different calcium concentrations as shown in figure. The same calibrations apply to all concentrations. (b) From the same end-plate. Relation between e.p.p. amplitude and calcium concentration on linear co-ordinates. The results include the three concentrations of calcium used in (a). 1 mM Mg present throughout the experiment. The e.p.p. amplitude is a reliable measure of the quantal release, since the variation in minature e.p.p. amplitude was small. (From Dodge & Rahamimoff, 1967.)

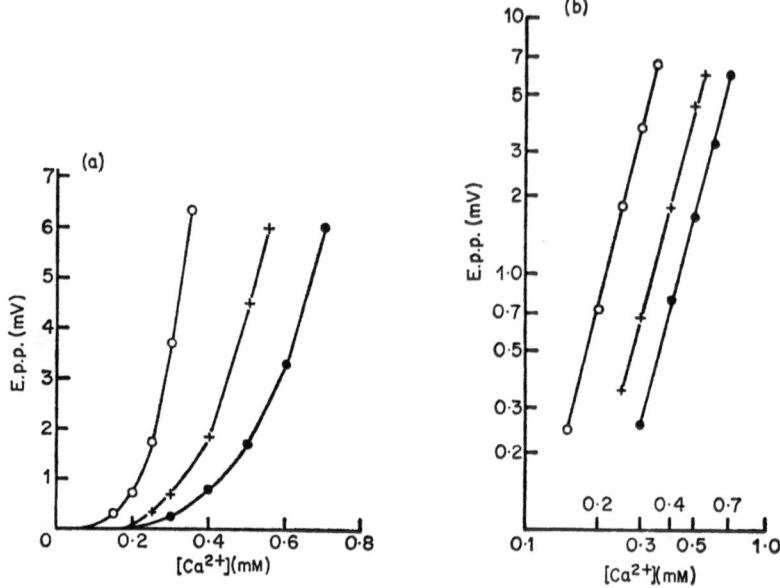

FIG. 4. Relation between Ca^{2+} concentration and amplitude of e.p.p. (a) Linear plot at three different Mg^{2+} concentrations. $\bigcirc$, 0·5 mM Mg^{2+}; $+$, 2·0 mM Mg^{2+}; $\bullet$, 4·0 mM Mg^{2+}. Each value represents the average amplitude of 128 or 256 e.p.p.s. (b) Same as (a) on double-logarithmic co-ordinates. (From Dodge & Rahamimoff, 1967.)

In the presence of both Ca and Mg ions the fraction of sites in the form of CaX will be given according to the mass action law by

$$Ca X = \frac{W[\text{Ca}]}{1 + \dfrac{[\text{Ca}]}{K_1} + \dfrac{[\text{Mg}]}{K_2}} \tag{3}$$

W is a constant, and in the simplest case W has the value of $1/K_1$.

A difficulty arises if one assumes that the e.p.p. or the quantal content m is directly proportional to CaX:

$$m \propto Ca X = \frac{W[\text{Ca}]}{1 + [\text{Ca}]/K_1 + [\text{Mg}]/K_2} \tag{4}$$

The graphical representation of equation (4) is a rectangular hyperbola with a linear start in the low range of calcium concentrations. The experimental results (Fig. 4a) show a marked initial non-linearity curving upwards.

This difficulty can be overcome if one assumes that for transmitter release a cooperative action of several Ca ions or CaX complexes is necessary. Experimentally, the initial slope of the log Ca − log e.p.p. relation was close to 4, therefore the minimal number of cooperating calciums would be 4. The simplest model is that four CaX must act simultaneously for the release of one quantum of transmitter. Such an assumption leads to the following expression:

$$m \propto (CaX)^4 = \left(\frac{W[Ca]}{1 + [Ca]/K_1 + [Mg]/K_2}\right)^4 \tag{5}$$

or, omitting the proportionality sign,

$$m = k(CaX)^4 = k\left(\frac{W[Ca]}{1 + [Ca]/K_1 + [Mg]/K_2}\right)^4 \tag{6}$$

k is the maximal number of quanta the release mechanism is able to liberate at 'infinite' concentration of Ca.

The expression in equation (6) can be tested in several ways:

(1) Taking a fourth root of both sides of equation (6) yields a relation similar to Michaelis & Menten's. Application of the Lineweaver & Burk (1934) double reciprocal transformation must give a straight line if the experimental results are properly described by equation (6). That this is the case can be seen from Fig. 5a. The action of the competitive inhibitor Mg is to change the slope of the double reciprocal relation without changing the intercept significantly (Fig. 5b). Some of the constants of equation (6) can be calculated from the slope and the intercept (see Dodge & Rahamimoff, 1967).

(2) When the constants of equation (6) have been calculated, there is still a further test. If the $[Ca]_o$ is kept constant, the $[Mg]_o$ can be increased. In terms of equation (6) only the denominator changes. In such conditions a straight line is expected when one plots the logarithm of the the e.p.p. or m as function of the logarithm of the denominator of equation (6). Such a relation is shown in Fig. 6b. The line has a slope of approximately −4. Hence, the e.p.p. amplitude changes as the fourth power of both the numerator and the denominator of equation (6).

(3) The above two tests were performed at relatively low calcium concentrations. If the range of $[Ca]_o$ is extended upwards, the e.p.p. exceeds threshold, and a twitch appears after nerve stimulation. This can be avoided by reducing the postsynaptic sensitivity with (+)-tubocurarine. When under these conditions the experiment is repeated, the constants estimated, and a curve is drawn according to equation (6), then one can see that there is fair agreement between the theoretical prediction and the experimental results (Fig. 7).

To summarize this part, the quantitative relation between $[Ca]_o$ and transmitter release shows a sigmoidal curve. The analysis of this curve is consistent with the idea of calcium binding 'sites' on the terminal. At least four adjacent sites have to be occupied simultaneously to give a release

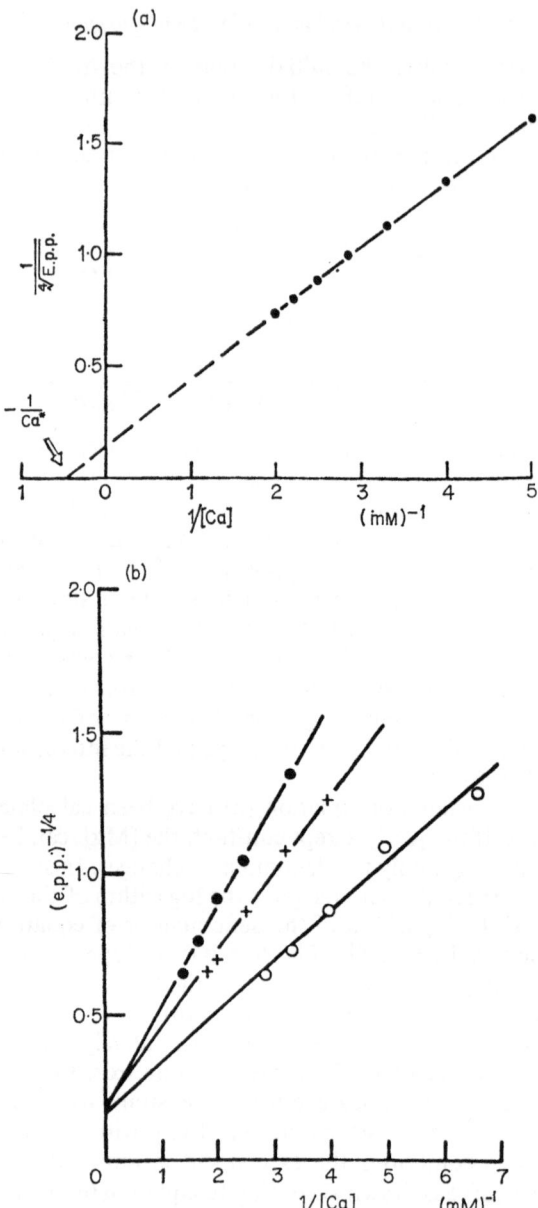

FIG. 5. Lineweaver–Burk plot for the relationship between $(CaX)^4$ and response. Ordinate: reciprocal of fourth root of e.p.p. Abscissa: reciprocal of [Ca]. (a) Same end-plate as in Fig. 3b. The intersection with the abscissa gives $-1/Ca$ and from this value the dissociation constant K_1 can be calculated. (b) Same end-plate as in Fig. 4. The straight lines are at different Mg concentrations and have approximately the same intercept. (From Dodge & Rahamimoff, 1967.)

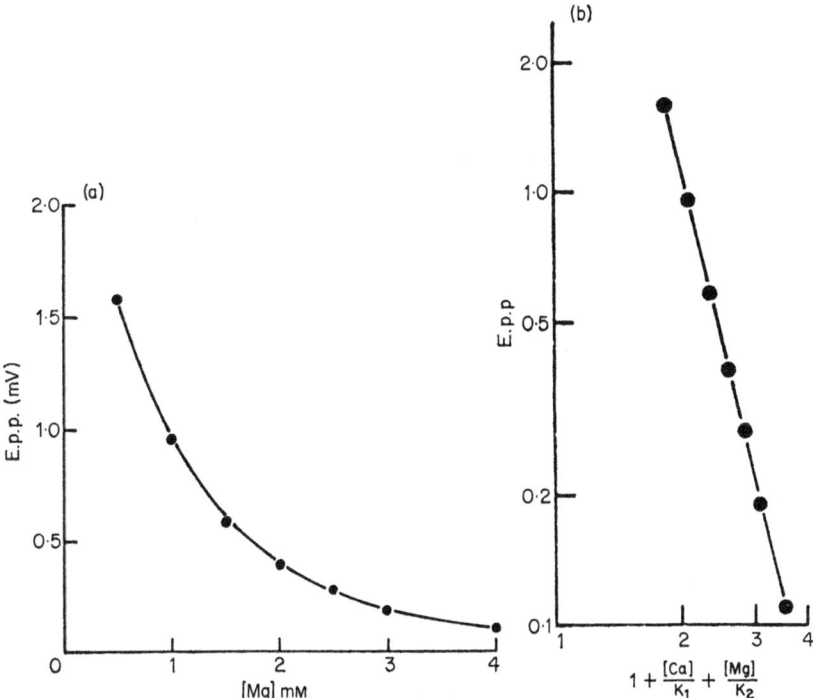

FIG. 6. Relation between e.p.p. amplitude and Mg^{2+} concentration at constant extracellular [Ca]. (a) Linear co-ordinates. Each point represents the average of 256 responses (except the first, 128) 0·4 mM Ca present throughout the experiment. (b) Double logarithmic plot of e.p.p. amplitude against $(1 + Ca/K_1 + Mg/K_2)$. $K_1 = 0·7$ mM; $K_2 = 2·0$ mM; Ca = 0·4 mM. (From Dodge & Rahamimoff, 1967.)

of one quantum of transmitter. The nature of these sites is unknown. They might be channels in the membrane, opened by depolarization, through which calcium ions move into the terminal. Alternatively, they might be molecules with which calcium ions bind to create conditions for release of a quantum. The kinetic evidence presented here cannot distinguish between such possibilities. A more direct approach will probably be necessary. In any case, however, it seems that at some stage of the release process a cooperative action of at least four calcium ions is required.

Effects of sodium ions on transmitter release

Let us now turn to the question of whether other constituents of the normal Ringer solution affect the release of transmitter from the nerve endings. Recent experiments of Katz & Miledi (1967a) showed that sodium ions are not necessary at all for transmitter liberation. Figure 8 illustrates

R. *Rahamimoff*

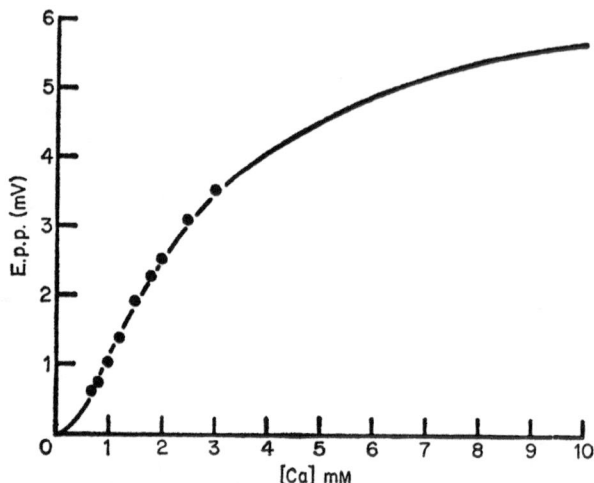

FIG. 7. Effect of calcium concentration on e.p.p. amplitude in a curarized preparation. 5×10^{-6} g/ml (+)-tubocurarine present throughout the experiment. Linear co-ordinates. The first three points are averages of 128 responses. The other points are averages of 64 responses. The sigmoidal curve follows the equation

$$\text{e.p.p.} = \left[\frac{2 \cdot 7[\text{Ca}]}{1 + [\text{Ca}]/0 \cdot 6} \right]^4$$

and the points are the experimental results. (From Dodge & Rahamimoff, 1967.)

an experiment performed on the frog neuromuscular junction, which was bathed in a medium containing no sodium ions. Sodium was replaced by calcium. Under these conditions transmitter is still liberated spontaneously (Fig. 8a), and extra release is evoked by depolarization of the terminals (Fig. 8b).

Although sodium ions are not necessary for release, their concentration affects the number of quanta liberated by a nerve impulse. Some years ago Birks & Cohen (1965) and Kelly (1965) demonstrated that, under certain conditions (when [Ca]$_o$ is low), reduction in external Na concentration increases the amplitude of the e.p.p. and the quantal output of transmitter (Fig. 9). At first glance this seemed an unexpected finding. Reduction in [Na]$_o$ diminishes the postsynaptic sensitivity to acetylcholine (Takeuchi & Takeuchi, 1960) and even if the same amount of transmitter is liberated, the postsynaptic effect must be smaller. Furthermore, reduction in [Na]$_o$ decreases the amplitude of the action potential (Hodgin & Katz, 1949) and as the number of quanta liberated depends on the presynaptic action potential (Katz & Miledi, 1967b,c), transmitter output would be expected to be smaller. This apparent paradox can be resolved, however, in line with the suggestion of Niedergerke & Lüttgau (1957) for cardiac muscle if

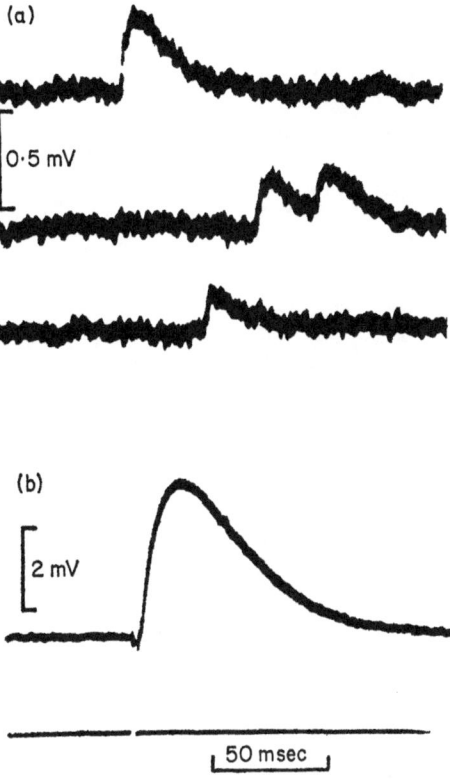

FIG. 8. Release of transmitter in a sodium-free medium containing 83 mM CaCl₂, 2 mM KCl, and $2 \cdot 10^{-6}$ g/ml. neostigmine methylsulphate. Intracellular recording at 6·5°C. Frog neuromuscular preparation. (a) Spontaneous miniature end-plate potentials. (b) End-plate potential (upper trace) evoked by a depolarizing pulse of about 2 msec (lower trace). (From Katz & Miledi, 1967b.)

Ca and Na compete, presumably for a common site. Reduction in $[Na]_o$ will make more sites available for Ca, which means that Ca becomes more effective and transmitter output is increased.

This hypothesis was examined (Colomo & Rahamimoff, 1968) by measuring the Ca-release relation in solutions containing 60% and 100% of the normal sodium concentration (Fig. 10). In the range of $[Ca]_o$ examined, the effect of reducing sodium is to produce an almost parallel shift of the log Ca − log release relation. This result suggests that Ca and Na compete on a common site in the process of transmitter release. In terms of equation (6), however, this is not final proof, for Na might alter one of the proportionality constants. The question of whether $[Na]_o$ and $[Ca]_o$ compete at the same site was further examined by testing the triple

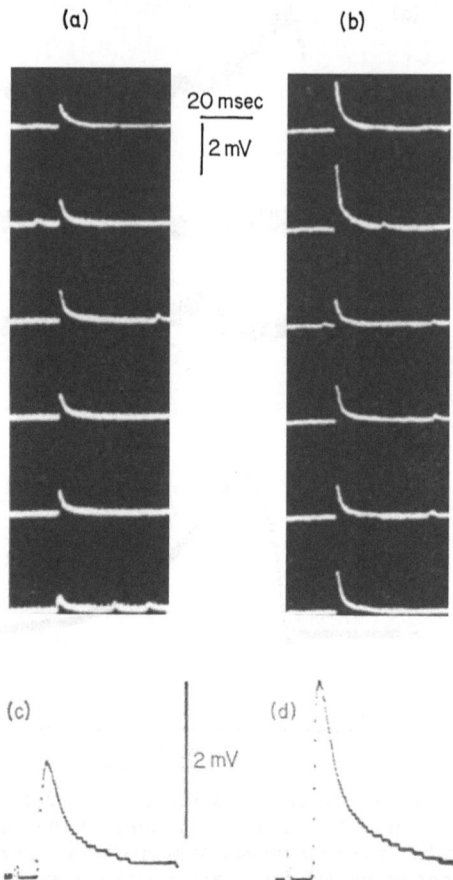

FIG. 9. Inhibitory effect of sodium ions on transmitter release. All results from the same frog sartorius neuromuscular junction. (a) Sample of e.p.p.s obtained at 116 mM Na. (b) Sample of e.p.p.s obtained at 70 mM Na. (c) Automatically obtained average of 256 responses at 116 mM Na. Averaging step 80 μsec. (d) Same as (c) at 70 mM Na. Average amplitude of miniature e.p.p. in 116 mM Na was 254 μV ($n = 190$) and in 70 mM was 197 μV ($n = 196$). Quantal content in 116 mM Na was 5·63 and in 70 mM 13·60. 0·4 mM calcium and 2·0 mM magnesium present throughout the experiment. Room temperature of approximately 20°C. (From Colomo & Rahamimoff, 1968.)

interaction of Ca, Mg, and Na. If Na and Ca compete on a common site, then reduction in [Na]$_o$ makes more sites available for Ca. When Mg is also present in the medium, part of the vacated sites will be occupied by Mg and be ineffective in release. Therefore, the higher the Mg concentration, the smaller would be the effect of Na withdrawal on this hypothesis.

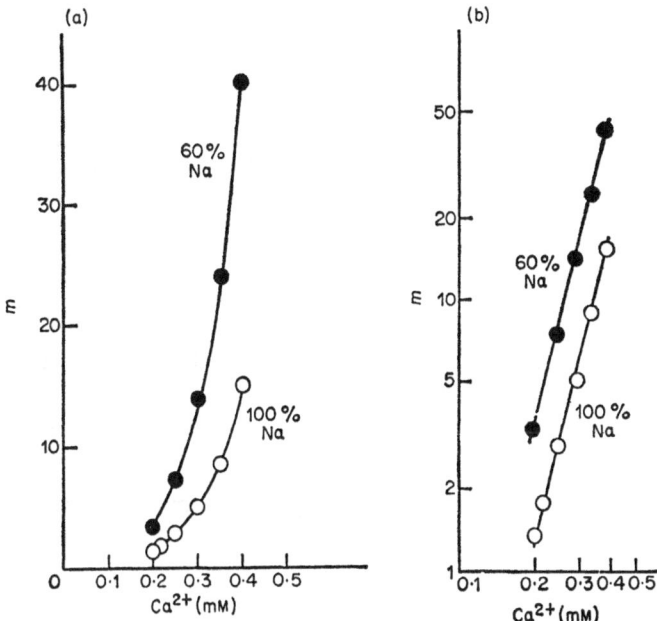

FIG. 10. Relation between calcium concentration and quantal content at two different sodium concentrations. (○) −116 mM and (●) −70 mM. All results from the same frog sartorius neuromuscular junction. Each point represents average response to between 64 and 256 stimuli. Room temperature of about 20°C. (a) Linear plot. (b) Same on double logarithmic co-ordinates. Magnesium, 1 mM, was present throughout the experiment. (From Rahamimoff & Colomo, 1967.)

Table 1 shows that this is so. In the presence of 8 mM Mg the effect of Na withdrawal is a 24% increase of transmitter, while in the presence of 1 mM Mg the effect is much larger—149% increase.

These results indicate that in normal ionic environment, the release mechanism is partially inactivated by sodium ions.

While extracellular sodium ions inhibit the release of transmitter, intracellular sodium has an opposite effect. Birks & Cohen (1968a,b) have shown that agents which increase the concentration of sodium inside the nerve terminals augment the amount of transmitter liberated. For example cardiac glycosides are known to inhibit the Na-pump and hence elevate $[Na]_i$. Their effect is to increase the quantal content. It is of interest to note that the effect is not immediate. It takes approximately 15 min to see the increase in m. It seems therefore that it is not the inhibition of the pump mechanism itself which is the relevant factor, but the accumulation of Na inside the nerve terminal. We have seen in this symposium that movements of Ca through the nerve membrane are dependent on sodium

TABLE 1. *Effect of reducing sodium concentration on the quantal
content (m)*

	1 mM Mg	8 mM Mg
Average m at normal [Na] ($m_{1.0}$)	11·08 (64)	0·64 (256)
Average m at low [Na] ($m_{0.6}$)	27·62 (64)	0·79 (256)
$m_{0.6}/m_{1.0}$	2·49	1·24

In parentheses: number of responses in each average. Calcium concentration
was kept constant throughout the experiment: 0·4 mM (Rahamimoff &
Colomo, 1967).

(Baker, this volume, Baker, Blaustein, Hodgkin & Steinhardt, 1969). It is
therefore quite possible that the release promoting effect of intracellular
Na is due to an increase of the Ca influx into the nerve terminals.

Effect of various divalent ions on transmitter release
The next point to discuss is the specificity of calcium ions in the process of
transmitter release. It has been shown repeatedly that the preceding element
to calcium in Group IIa of the periodic table—magnesium—not only
cannot replace calcium in transmitter liberation, but inhibits the release by
competing with calcium ions on a postulated common site (Del Castillo &
Engback, 1954; del Castillo & Katz, 1954a; Jenkinson, 1957; Dodge &
Rahamimoff, 1967; Hubbard, Jones & Landau, 1968). It is of interest to
see what influence the other members of Group IIa have on the release of
acetylcholine at the neuromuscular junction. Blioch, Glagoleva, Liberman
& Nenashev (1968) found that beryllium, which is the first element in
Group IIa, inhibits transmitter liberation.

(a)   (b)   (c)

FIG. 11. Non-linear summation of the effects of calcium and strontium on
e.p.p. amplitude and transmitter release. (a) Average response to 140 stimuli
in a medium containing 0·4 mM Ca. (b) Average response to the same number
of stimuli in a medium containing 2·0 mM Sr. (c) The combined effect of calcium and strontium on e.p.p. amplitude. The individual averaged response
is 0·25 mV in Ca and 0·44 mV in Sr. The response to the same number of
stimuli in a medium containing both ions together is 1·25 mV instead of 0·69,
about 80% more. 1·0 mM Mg present throughout the experiment. (From
Meiri & Rahamimoff, unpublished results.)

The heavier ions in this group, strontium and barium, are able to replace Ca in release of acetylcholine (Miledi, 1966). The release in Sr solutions is quantal in nature and is in general similar to that in Ca solutions (Dodge, Miledi & Rahamimoff, 1969).

Calcium and strontium exert their effect on the release by acting presumably on the identical releasing sites. Some evidence on this point is presented in Figs. 11 and 12 (from unpublished work of Meiri & Rahamimoff). If calcium and strontium were activating the release process by two independent mechanisms, then it might be expected that during the combined actions of calcium and strontium, the release should be the algebraic sum of the individual releases. This is not so. In the lower range of the release versus divalent ion relation the combined effect exceeds the sum of the individual responses. For example, when the activating divalent ion is calcium (Fig. 11a) the average e.p.p. amplitude was 0.25 mV, and when it

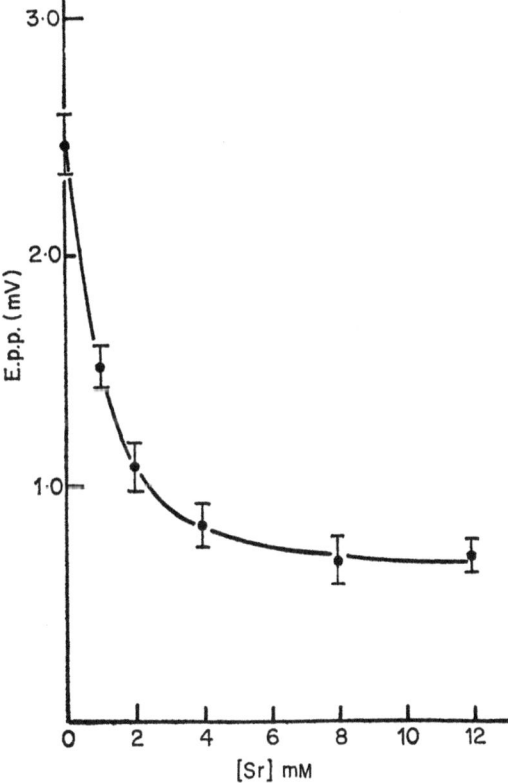

FIG. 12. Inhibitory effect of strontium ions on transmitter release. Frog sartorius neuromuscular preparation. Each point is the average response to 60 stimuli. Ca = 1·8 mM and (+)-tubocurarine 2×10^{-6} g/ml. present throughout the experiment. (From Meiri & Rahamimoff, unpublished results.)

is strontium, 0·44 mV. When calcium and strontium were both present, instead of an e.p.p. of 0·69 mV the response was almost double this value —1·25 mV. This is not surprising, however, if Sr and Ca act on the same sites, and account is taken of the sigmoidal relation between Ca (or Sr) and release (Fig. 7 here, and Fig. 8 in Dodge, Miledi & Rahamimoff, 1969); for doubling the concentration can then give much more than twice the response. It appears therefore that Sr combines with the same sites as calcium to give SrX:

$$Sr + X \rightleftharpoons SrX \tag{7}$$

and it seems that Ca and Sr can co-operate in acetylcholine liberation.

Strontium, however, is less effective than calcium in transmitter release. This might formally be due either to lower affinity of Sr for the X site, or to a lower effectiveness of SrX in the release process. Of course both possibilities might exist together. Figure 12 illustrates the lower effectiveness of SrX; addition of Sr, to a preparation whose release mechanism is already substantially activated by 1·8 mM calcium, reduces the average amplitude of the e.p.p. (and the release, because the changes in miniature e.p.p. amplitude are relatively small). These experiments indicate that calcium and strontium can compete and therefore presumably act on the same site.

Thus it appears that ionic charge is not enough to induce release of transmitter. There are enormous qualitative as well as quantitative differences among the various alkali earth metal ions in their influence on release.

The findings reported here may be summarized as follows: Calcium ions play a crucial part in the coupling between nerve terminal depolarization and release of transmitter at the neuromuscular junction. They are necessary apparently during the depolarization itself and not during the actual release of acetylcholine.

The kinetic analysis of calcium action suggests that calcium ions combine with specific sites on the nerve terminal and that cooperative action of several calcium ions (at least four) occurs at some stage of the transmitter release process. The physico-chemical nature of these postulated sites remains unknown.

Several other ions act on the calcium activated process of transmitter liberation. Under normal conditions extracellular sodium ions partially inactivate the release mechanism, probably by competing with calcium. Among the Group IIa metal ions, two different patterns of action exist. Beryllium and magnesium are not able to replace calcium and inhibit the release of acetylcholine in the presence of calcium. Strontium and barium can replace calcium (in the release process) but are much less effective.

I am grateful to Prof. B. Katz for reading the manuscript. Part of the work presented here was supported by a grant from the Research Fund of the Hebrew University—Hadassah Medical School, Jerusalem.

REFERENCES

BAKER, P. F., BLAUSTEIN, M. P., HODGKIN, A. L. & STEINHARDT, R. A. (1969). *J. Physiol., Lond.,* **200**, 431

BIRKS, R. I. & COHEN, M. W. (1965). In *Muscle,* ed. PAUL, W. M., DANIEL, E. D., KAY, C. M. & MONCKTON, G., p. 403. Oxford: Pergamon Press

BIRK, R. I. & COHEN, M. W. (1968a). *Proc. R. Soc., B,* **170**, 381

BIRKS, R. I. & COHEN, M. W. (1968b). *Proc. R. Soc., B,* **170**, 401

BLIOCH, Z. L., GLAGOLEVA, I. M., LIBERMAN, E. A. & NENASHEV, V. A. (1968). *J. Physiol., Lond.,* **199**, 11

COLOMO, F. & RAHAMIMOFF, R. (1968). *J. Physiol., Lond.,* **198**, 203

DEL CASTILLO, J. & ENGBACK, L. (1954). *J. Physiol., Lond.,* **124**, 370

DEL CASTILLO, J. & KATZ, B. (1954a). *J. Physiol., Lond.,* **124**, 553

DEL CASTILLO, J. & KATZ, B. (1954b). *J. Physiol., Lond.,* **124**, 560

DODGE, F. A. & RAHAMIMOFF, R. (1967). *J. Physiol., Lond.,* **193**, 419

DODGE, F. A., MILEDI, R. & RAHAMIMOFF, R. (1969). *J. Physiol., Lond.,* **200**, 267

DOUGLAS, W. W. & RUBIN, R. P. (1961). *J. Physiol., Lond.,* **159**, 40

DOUGLAS, W. W. & POISNER, A. M. (1964). *J. Physiol., Lond.,* **172**, 1

FRANKENHAUSER, B. (1957). *J. Physiol., Lond.,* **137**, 245

FRANKENHAUSER, B. & HODGKIN, A. L. (1957). *J. Physiol., Lond.,* **137**, 218

HODGKIN, A. L. & KATZ, B. (1949). *J. Physiol., Lond.,* **108**, 37

HUBBARD, J. I., JONES, S. F. & LANDAU, E. M. (1968). *J. Physiol., Lond.,* **196**, 75

HUTTER, O. F. & KOSTIAL, K. (1954). *J. Physiol., Lond.,* **124**, 234

JENKINSON, D. H. (1957). *J. Physiol., Lond.,* **138**, 434

KATZ, B. (1962). The Croonian Lecture. *Proc. R. Soc., B,* **155**, 455

KATZ, B. (1966). *Nerve, Muscle and Synapse,* p. 193. New York: McGraw-Hill

KATZ, B. (1969). The release of neural transmitter substances. *The Sherrington Lectures, No. X.* Liverpool: Liverpool University Press

KATZ, B. & MILEDI, R. (1965a). *Proc. R. Soc. B,* **161**, 453

KATZ, B. & MILEDI, R. (1965b). *Proc. R. Soc., B,* **161**, 483

KATZ, B. & MILEDI, R. (1965c). *Proc. R. Soc., B,* **161**, 496

KATZ, B. & MILEDI, R. (1967a). *Nature, Lond.,* **215**, 651

KATZ, B. & MILEDI, R. (1967b). *Proc. R. Soc., B,* **167**, 1

KATZ, B. & MILEDI, R. (1967c). *Proc. R. Soc., B,* **167**, 23

KATZ, B. & MILEDI, R. (1967d). *J. Physiol., Lond.,* **189**, 535

KELLY, J. S. (1965). *Nature, Lond.,* **205**, 296

LINEWEAVER, H. & BURK, D. (1934). *J. Am. chem. Soc.,* **56**, 658

LOCKE, F. S. (1894). *Zbl. Physiol.,* **8**, 166

MILEDI, R. (1966). *Nature, Lond.,* **212**, 1233

MILEDI, R. & SLATER, C. R. (1966). *J. Physiol., Lond.,* **184**, 473

NASTUK, W. L. & LIU, JANE H. (1966). *Science, N.Y.,* **154**, 266

NIEDERGERKE, R. & LÜTTGAU, H. C. (1957). *Nature, Lond.,* **179**, 1066

RAHAMIMOFF, R. & COLOMO, F. (1967). *Nature, Lond.,* **215**, 1174

SANDOW, A. (1965). *Pharmac. Rev.,* **17**, 265

TAKEUCHI, A. & TAKEUCHI, N. (1960). *J. Physiol., Lond.,* **154**, 52

TAKEUCHI, A. & TAKEUCHI, N. (1962). *J. gen. Physiol.,* **45**, 1118

INVOLVEMENT OF CALCIUM IN THE SECRETION OF CATECHOLAMINES

P. BANKS

Department of Biochemistry, University of Sheffield

The role of calcium as the stimulus-secretion coupling agent in the adrenal medulla has been soundly established in the papers which Douglas and his colleagues have published since 1961 (Douglas, 1968). However, three major questions concerning the calcium-dependent secretion of catecholamines remain unanswered. These are, first, how is calcium entry into chromaffin cells regulated; second, what is the nature of the process activated by calcium; third, how is calcium removed from the cells at the end of the secretory response? Evidence touching upon these three questions is discussed here.

Calcium influx and sodium

Douglas & Rubin (1961) were the first to show that the secretory response, normally evoked from the adrenal medulla by acetylcholine (ACh) fails in calcium-free Locke solution and is enhanced in calcium-rich solutions (see Figs. 1 and 5). The list of adrenal medullary secretagogues whose effectiveness is known to be dependent on the presence of Ca^{2+} ions in the perfusing medium is now extensive (Table 1). It includes sympathomimetic compounds, such as phenylethylamine, which were once thought to evoke catecholamine secretion in the absence of calcium. During periods of secretion induced by ACh there is a marked increase in the rate of uptake of $^{45}Ca^{2+}$ from solutions containing radioactive calcium (Douglas & Poisner, 1962). Reintroduction of calcium after a period of perfusion with a Ca^{2+}-free solution causes a substantial discharge of catecholamines in the absence of any secretagogue. This is probably due to an influx of calcium into the chromaffin cells through cell membranes which become abnormally permeable in the Ca^{2+}-free solution (see Douglas, 1968).

During their early investigations, Douglas & Rubin (1961) reported that cat adrenals could still respond to ACh by secreting catecholamines when perfused with an iso-osmotic solution of sucrose which contained Ca^{2+} as the only cation. They also found that omitting K^+ from Locke's solution enhanced the secretory response. On the other hand the substitution of sucrose for NaCl had little immediate effect on the evoked response although it caused a transitory rise in the spontaneous release of catecholamines. When several periods of perfusion with NaCl-free solutions were interpolated between periods of perfusion with Locke's solution, catecholamine secretion in response to ACh was diminished even when given in Locke's solution. In a later paper (Douglas & Rubin, 1963), it was

TABLE 1. *Agents which evoke a* Ca^{2+}*-dependent secretion of catecholamines from the adrenal medulla*

Agent	Reference
Acetylcholine	Douglas & Rubin (1961)
Angiotensin	Poisner & Douglas (1966)
Amphetamine	Rubin & Jaanus (1966)
Bradykinin	Poisner & Douglas (1966)
Carbamylcholine	Banks (1967)
Histamine	Poisner & Douglas (1966)
5-Hydroxytryptamine	Poisner & Douglas (1966)
Methacholine	Poisner & Douglas (1966)
Muscarine	Poisner & Douglas (1966)
Pilocarpine	Poisner & Douglas (1966)
Phenylethylamine	Rubin & Jaanus (1966)
Potassium (high concs.)	Douglas & Rubin (1961)
Tyramine	Rubin & Jaanus (1966)

observed that once the period of increased spontaneous secretion had died down, the secretory response to ACh was increased above the control value after 14 min of perfusion with a solution having less than 10% of the normal Na^+ content.

Despite these effects of Na^+ and K^+ omission, the early observations clearly pointed to a predominant role for Ca^{2+} ions in the secretory response. This inference was supported by later studies of changes in the membrane potentials and ion currents across the membranes of chromaffin cells exposed to various secretagogues (Douglas, Kanno & Sampson, 1967a,b; Douglas & Kanno, 1967). The resting potential across the membranes of gerbil chromaffin cells in tissue culture was 29.3 ± 0.2 mV in Locke solution; as the external potassium concentration was raised from 10 to 100 mM the potential fell by about 20 mV.

At rest, therefore, the potential is defined largely by the transmembrane concentration gradient of K^+ ions. Na^+ ions must make some contribution to the resting potential, however, for the cells were slightly hyperpolarized in Na^+-free media. In the presence of 10^{-4} g ACh/ml. the potential fell to about 11 mV in Locke solution; the depolarization induced by ACh increased linearly with the logarithm of the external Na^+ concentration.

In the absence of Na^+ but in the presence of normal amounts of Ca^{2+}, ACh caused a much smaller depolarization than in normal, Na^+-rich media. In Na^+-free solutions, the depolarization induced by ACh increased as the Ca^{2+} concentration was raised from 1 to 117 mM, and even in solutions with a normal content of Na^+, the depolarization on exposure to ACh was increased by raising the Ca^{2+} concentration above 10 mM. In Ca^{2+}-free solutions the resting potentials were a little lower than usual but ACh or 50 mM K^+ still depolarized the cells, thus demonstrating that depolarization and secretion are not coupled in the absence of calcium.

The local anaesthetic amethocaine (tetracaine USP) was shown to inhibit the uptake of $^{45}Ca^{2+}$ by perfused cat adrenals and to block the secretion of catecholamines obtained when calcium was readmitted following perfusion with Ca^{2+}-free, or with Ca^{2+}-free K^+-rich, solutions. The secretory response to calcium readmission in the presence of amethocaine could be restored to some extent by increasing the Ca^{2+}-content of the solution. Amethocaine also inhibited ACh-induced secretion at lower concentrations than those needed to prevent the Ca^{2+}-induced response (Rubin, Feinstein, Jaanus & Paimre, 1967; Jaanus, Miele & Rubin, 1967). These and other studies (Meile & Rubin, 1968) suggested that amethocaine had two actions on the adrenal medulla, one of which prevented Ca^{2+}-influx and another which was antagonistic to ACh. The relative potency of these two actions varied between anaesthetics and it was concluded that there was no correlation between the ability to inhibit Ca^{2+}-evoked secretion and the ability to inhibit ACh-evoked, Ca^{2+}-dependent, secretion. Possibly amethocaine acts at two different sites.

Perfusion of gerbil adrenals confirmed that amethocaine could inhibit the secretory response evoked by ACh (Douglas & Kanno, 1967) and also showed that amethocaine only partially blocked the depolarizing action of ACh on gerbil chromaffin cells bathed in a normal Locke solution in tissue culture. In Na^+-free media containing Ca^{2+}, amethocaine completely abolished the Ca^{2+}-dependent depolarization normally triggered by ACh. In other words, amethocaine prevented the inward current carried by Ca^{2+} ions but not that carried by Na^+ ions. Thus amethocaine abolished the contribution of calcium influx to the ACh-induced depolarization.

These findings indicate that catecholamine secretion by chromaffin cells is dependent on the influx of calcium ions from the external medium. In the absence of calcium or in the presence of agents which prevent calcium influx, secretion fails. The inward movement of Na^+ and the marked depolarization it causes seem to be much less important than Ca^{2+}-influx which activates the secretory apparatus but which causes only a small depolarization. It would appear, therefore, that extracellular Na^+ is of little importance in the secretory response although the increased response observed shortly after the removal of Na^+ suggests that Na^+ may have a slight inhibitory action (Douglas & Rubin, 1963). Furthermore prolonged exposure to Na^+-free solutions leads to a deterioration of the response (Douglas & Rubin, 1961). Although extracellular sodium seems to be of little importance, evidence is accumulating that the secretory response is dependent on the intracellular concentration of sodium.

When adrenal glands were perfused with ouabain (Fig. 2a; see also Fig. 1), in order to inhibit the sodium pump and thereby increase the intracellular concentration of Na^+ at the expense of intracellular K^+, it

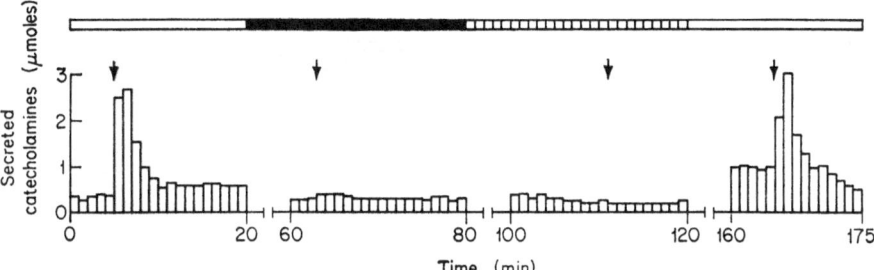

FIG. 1. Dependence of catecholamine secretion upon calcium and the ineffectiveness of ouabain upon secretion in the absence of calcium. Bovine adrenal gland perfused at about 20 ml./min in a retrograde manner at 37°C. ⊏⊐, Perfusion with Tyrode solution gassed with 95% O_2 + 5% CO_2. ▬▬▬, Perfusion with a Ca^{2+}-free Tyrode solution gassed with 95% O_2 + 5% CO_2. ⊓⊓⊓⊓⊓, Perfusion with a Ca^{2+}-free Tyrode solution containing 10^{-4}M ouabain. Solution gassed with 95% O_2 + 5% CO_2. The arrows indicate the injection of 0·5 ml. 10^{-2} M carbachol. Composition of Tyrode solution: NaCl 137 mM; KCl 2·7 mM; $MgCl_2$ 0·1 mM; $CaCl_2$ 1·8 mM; $NaHCO_3$ 11·9 mM; NaH_2PO_4 0·4 mM; glucose 5·6 mM.

was found that both the Ca^{2+}-dependent, spontaneous secretion of catecholamines and the Ca^{2+}-dependent secretion evoked by ACh were higher than during the control period (Banks, 1967). Further experiments confirmed that perfusion with K^+-free solutions, which would also impair the functioning of the K^+-dependent sodium pump, enhanced both the spontaneous and evoked secretion of catecholamines (Banks, Biggins, Bishop, Christian & Currie, 1969). These results suggest that the ability of Ca^{2+} ions to act as the stimulus-secretion coupling agent can be improved by either an increased intracellular concentration of sodium or by a decreased intracellular concentration of potassium or by both of these factors.

The work of Baker and his colleagues (Baker & Blaustein, 1968; Baker, Blaustein, Hodgkin & Steinhardt, 1967, 1969) has indicated that the influx of Ca^{2+} into the axons of *Loligo* and *Maia* increases as the internal concentration of Na^+ is raised and decreases when it is lowered. Furthermore,

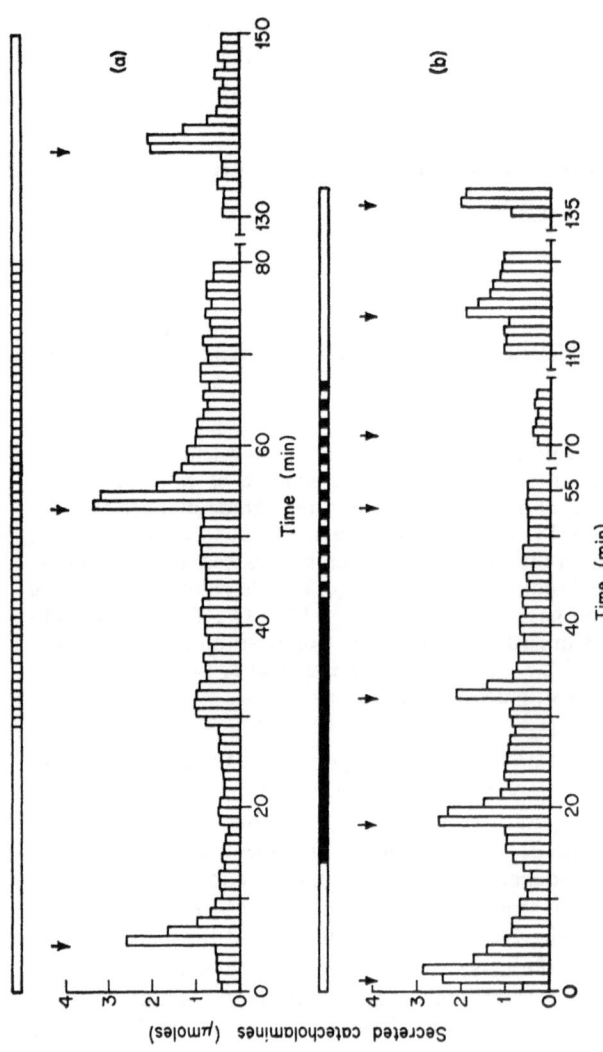

Fig. 2. (a) Facilitating effect of ouabain on the secretion of catecholamines by a bovine adrenal gland perfused in a retrograde manner with Tyrode solution at 37°C. ▭▭▭, Perfusion with Tyrode solution gassed with 95% O_2 + 5% CO_2. ▨▨▨, Perfusion with Tyrode solution containing 10^{-4} M ouabain. Solution gassed with 95% O_2 + 5% CO_2. The arrows indicate the injection of 0·5 ml. 10^{-2} M carbachol. Perfusion rate was about 20 ml./min. (b) Ineffectiveness of ouabain on secretion in a sodium-deficient medium. Bovine adrenal gland perfused at 37°C in a retrograde manner at about 20 ml./min. ▭▭▭, Perfusion with Tyrode solution gassed with 95% O_2 + 5% CO_2. ▭▭▭, Perfusion with a Tyrode solution in which all the NaCl was replaced by LiCl. Solution gassed with 95% O_2 + 5% CO_2. ▬▬▬, Perfusion with a Tyrode solution in which all the NaCl was replaced by LiCl and which contained 10^{-4} M ouabain. The arrows indicate the injection of 0.5 ml. 10^{-2} M carbachol.

they showed that extracellular Na^+ had an inhibitory action on Ca^{2+}-influx. It seemed possible, therefore, that catecholamine release was increased by ouabain because the increased intracellular concentration of Na^+ encouraged Ca^{2+}-influx both at rest and following stimulation. If this view is correct, catecholamine secretion should be impaired under conditions which lower the intracellular concentration of sodium.

Experiments were therefore carried out using Na^+-free and Na^+-deficient solutions in the hope that prolonged perfusion with such solutions would reduce the intracellular concentration of sodium (Banks *et al.*, 1969). Perfusion with Na^+-free or Na^+-deficient (12 mM Na^+) solutions based on lithium chloride or choline chloride gradually diminishes and eventually abolishes the secretory response to carbamylcholine although it can be restored to some extent by returning to perfusion with Tyrode solution. As soon as Na^+-free or Na^+-deficient solutions reach the gland, the basal secretion of catecholamines rises and then declines to about the control level. If a gland is stimulated after a LiCl-based, Na^+-deficient, medium has entered, the response does not differ significantly from the control value for about 15 min; thereafter it begins to decline (Fig. 3). According to Douglas & Rubin (1963) the response is enhanced shortly after reducing the external sodium concentration below 10 mM in sucrose-based solutions. The difference in response observed during the first few minutes of perfusion in Li^+ or sucrose-based solutions may be related to the observation (Banks *et al.*, 1969) that the decline of the response is slower in Na^+-deficient solutions based on sucrose than it is in similar solutions based on LiCl or choline chloride. Because the secretory response of the bovine adrenal medulla does not begin to fall until some 15 min after the Na^+-deficient solution has begun to pass through the gland at

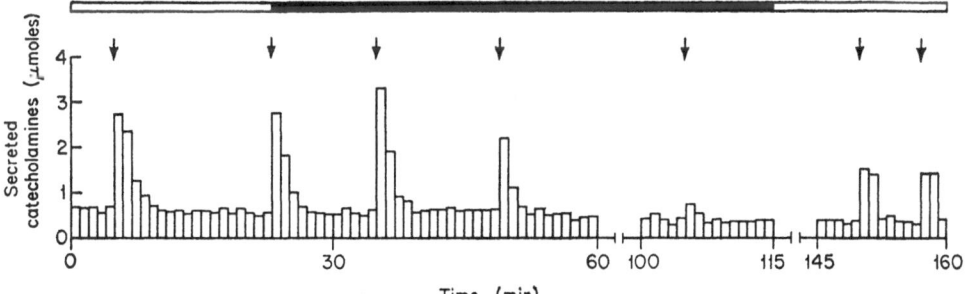

FIG. 3. Inhibitory effect of sodium deficiency on the secretion of catecholamines by a bovine adrenal gland perfused in a retrograde manner at 37°C. ▭▬▭, Perfusion with Tyrode solution gassed with 95% O_2 + 5% CO_2. ▬▬▬, Perfusion with a Tyrode solution in which all the NaCl was replaced by LiCl. Solution gassed with 95% O_2 + 5% CO_2. The arrows indicate the injection of 0·5 ml. 10^{-2} M carbachol. Perfusion rate was about 20 ml./min.

20 ml./min (Fig. 3), the decline probably results from a progressive fall in the internal concentration of Na^+ rather than from a low extracellular concentration of Na^+ *per se.*

Ouabain has no effect upon secretion in Na^+-free solutions (Fig. 2b) and high concentrations of K^+ fail to evoke the release of catecholamines after sodium has been omitted from the perfusing medium for about an hour (Banks *et al.*, 1969).

The results obtained from experiments involving ouabain and Na^+-free or deficient solutions could be explained if the Ca^{2+}-dependent secretion of catecholamines is facilitated by elevated intracellular concentrations of sodium and is diminished by lowered concentrations. Such behaviour would be expected if Ca^{2+}-entry into chromaffin cells were dependent on the intracellular concentration of sodium in a manner similar to that described for squid and crab nerves (Baker & Blaustein, 1968; Baker *et al.*, 1967, 1969).

The eventual failure of the secretory response in Na^+-free or deficient solutions could also be explained if calcium tended to accumulate in the cells under such conditions and thereby inhibited the secretory response in some unspecified way. (Ca^{2+}-influx into various tissues increases when the external Na^+ concentration is reduced (Baker *et al.*, 1969) and the calcium content of squid axons immersed in Na^+-free solutions is raised above the control level (Blaustein & Hodgkin, 1969).) Such a possibility can be tested in two ways. First, perfusion with Na^+-deficient, Ca^{2+}-free solutions should reduce the intracellular Na^+ concentration without the possibility of Ca^{2+}-accumulation. Then immediately on readmitting calcium, it should be possible to demonstrate a secretory response if Ca^{2+}-accumulation is normally responsible for inhibiting the response in Na^+-deficient media. We have failed to demonstrate either an evoked response or an increased spontaneous release of catecholamines under these conditions. This suggests that Ca^{2+}-accumulation is not responsible for the decline of the response in Na^+-deficient solutions containing the normal concentration of Ca^{2+} ions.

A second way to test the hypothesis that Ca^{2+}-accumulation in Na^+-deficient solutions blocks the secretory response is to investigate the effect of increasing the Ca^{2+} content of the medium. Figure 4 shows that when the Na^+-deficient solution entering the gland contains 7·8 mM Ca^{2+}, a very substantial secretion of catecholamines can still be evoked at a time when the response would have been abolished in a Na^+-deficient medium containing the normal amount of calcium (1·8 mM). If intracellular accumulation of calcium inhibited the secretory response in Na^+-free solutions, it might be expected to exert an even greater inhibitory effect in the experiment shown in Fig. 4. Clearly this is not so. The introduction of excess Ca^{2+} during a period of perfusion with a Na^+-deficient solution containing 1·8 mM Ca^{2+} also permits a marked secretion to be evoked at a time when the response is otherwise very small (Fig. 5). In percentage terms the extra calcium was no less effective in increasing the response 140 min after beginning perfusion with a Na^+-deficient solution than it was during the initial control period in Tyrode solution.

These experiments demonstrate that a secretory response can be evoked after long periods of perfusion in Na^+-deficient solutions if calcium entry is facilitated by increasing the extracellular concentration of calcium. It is possible, therefore, that calcium, at a given extracellular concentration, is

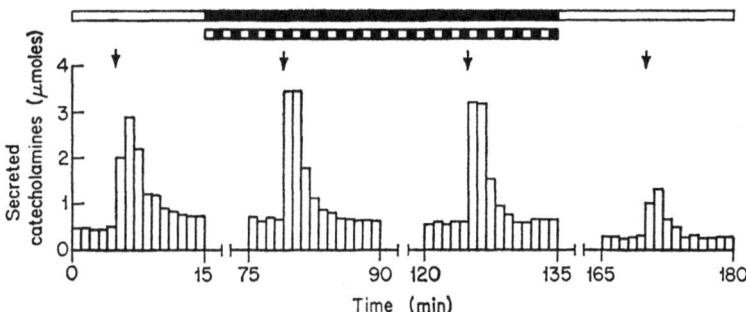

FIG. 4. Maintenance of catecholamine secretion in a sodium-deficient solution containing excess calcium. Bovine adrenal gland perfused in a retrograde manner at 37°C. ⊏▭▭⊐, Perfusion with Tyrode solution gassed with 95% O_2 + 5% CO_2. ▬▬▬▬, Perfusion with a Tyrode solution in which all the NaCl was replaced by LiCl. Solution gassed with 95% O_2 + 5% CO_2. ▪▫▪▫▪▫, Infusion of a solution containing 42 mM $CaCl_2$ and 137 mM LiCl into the perfusion medium immediately before it entered the gland to give a final calcium concentration in the perfusing solution of 7·8 mM. The arrows indicate the injection of 0·5 ml. 10^{-2} M carbachol. Perfusion rate was about 20 ml./min.

able to act as the stimulus-secretion coupling agent only if the intracellular concentration of sodium is high enough to permit an influx of calcium which is sufficient to activate the secretory apparatus.

The possibility that the internal concentration of sodium is an important factor in controlling Ca^{2+}-entry to secretory cells does not appear to be confined to chromaffin cells. For example, Hales & Milner (1968a,b) have found that insulin secretion by slices of pancreas is increased by ouabain or by incubation in K^+-free solutions and is dependent upon the external concentration of sodium. Furthermore, Birks & Cohen (1968a,b) have argued for some years that the Ca^{2+}-dependent release of acetylcholine at neuromuscular junctions is dependent upon the intraneuronal concentration of sodium. This is because treatment with ouabain or with K^+-free solutions increases the size of the endplate potential whilst no such effect is observed in Na^+-deficient solutions. Furthermore ouabain increases the frequency of miniature endplate potentials.

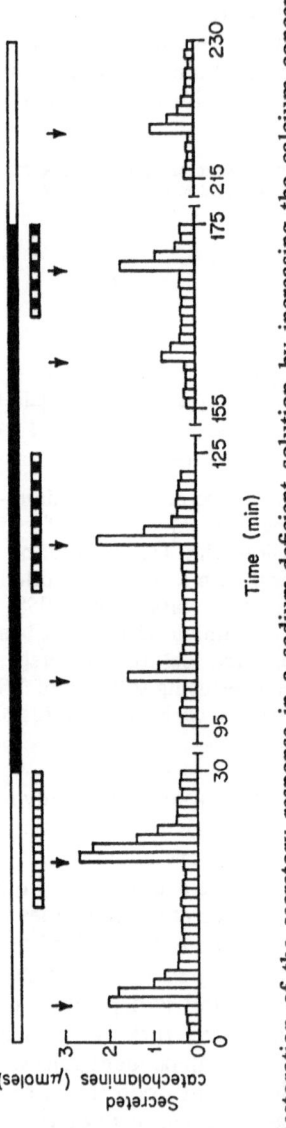

FIG. 5. Restoration of the secretory response in a sodium-deficient solution by increasing the calcium concentration. Bovine adrenal gland perfused in a retrograde manner at 37°C. ☐, Perfusion with Tyrode solution gassed with 95% O_2 + 5% CO_2. ■, Perfusion with a Tyrode solution in which all the NaCl was replaced by LiCl. Solution gassed with 95% O_2 + 5% CO_2. ▨, Infusion of a solution containing 36 mM $CaCl_2$ and 137 mM NaCl into the perfusion medium immediately before it entered the gland to give a final calcium concentration in the perfusing solution of 5·8 mM. ▨ Infusion of a solution containing 36 mM $CaCl_2$ and 137 mM LiCl into the perfusion medium immediately before it entered the gland to give a final calcium concentration in the perfusing solution of 5·8 mM. The arrows indicate the injection of 0·5 ml. 10^{-2} M carbachol.

Calcium and noradrenaline release from sympathetic nerves

There is now good evidence that the release of noradrenaline from post-ganglionic sympathetic nerves is also dependent upon the concentration of calcium in the external medium (Burn & Gibbons, 1964, 1965; Boullin, 1967; Hukovic & Muscholl, 1962; Kirpekar & Misu, 1967). Little information is available concerning the requirements for monovalent cations although there is some indication in the paper by Kirpekar & Misu (1967) that release may be enhanced in K^+-free solutions and that it can be abolished by reducing the Na^+ concentration. The latter observation is ascribed by the authors to a blockage in conduction rather than to a direct effect on the secretory mechanism. In a subsequent paper, sodium deficiency was found to enhance the release of noradrenaline evoked from splenic nerve endings by KCl (Kirpekar & Wakade, 1968). This may be a consequence of an impaired re-uptake of noradrenaline in the absence of sodium rather than of a direct effect of sodium lack on the secretory mechanism. Sodium deficiency, however, also enhanced release in animals treated with phenoxybenzamine, an inhibitor of noradrenaline uptake, so the situation remains rather confused. Clearly the existence of a sodium-dependent uptake of noradrenaline by nerve endings will complicate studies on the sodium dependence of noradrenaline release (see also Loffelholz, 1967).

The mechanism of secretion

Within chromaffin cells, catecholamines are stored in subcellular organelles termed chromaffin granules (Smith, 1968) which can be isolated from tissue homogenates by density gradient centrifugation or by filtration. Disruption of their lipoprotein membrane by osmotic shock or by treatment with detergents releases catecholamines, ATP and a group of soluble proteins, the chromogranins, into solution; the molar ratio of catecholamines to ATP is about 4 and there are about 4 μmoles of catecholamine for every milligram of soluble protein. In recent years evidence from electron microscopic and biochemical studies has helped to establish the role of chromaffin granules in the secretory response. Electron micrographs have been obtained which show the membranes of chromaffin granules fused to chromaffin cell membranes and the contents of the granules being expelled from the cells by exocytosis or reverse pinocytosis (De Roberts & Vaz Ferreira, 1957; Coupland, 1965; Diner, 1967). However, images of this type are seen only rarely and the most compelling evidence for exocytosis as the means of secretion has come from biochemical studies. During periods of stimulation, catecholamines, catabolites of ATP (Douglas, Poisner & Rubin, 1965; Banks, 1966a) and chromogranins (Banks & Helle, 1965; Kirshner, Sage, Smith & Kirshner, 1966; Blaschko, Comline, Schneider, Silver & Smith, 1967; Schneider, Smith & Winkler, 1967) appear in the perfusate in about the same ratio as they occur in chromaffin granules. There is no evidence that the lipoprotein membranes of the granules are secreted as insignificant quantities of cholesterol and phospholipids are released during periods of secretion (Schneider *et al.*, 1967; Trifaró, Poisner & Douglas, 1967). Following secretion, analysis of the adrenal medulla reveals a fall in the catecholamine, ATP (Carlsson & Hillarp, 1956; Schümann, 1958) and protein (Poisner, Trifaró & Douglas, 1967) contents but no change in the phospholipid or cholesterol contents (Poisner *et al.*, 1967). Lactic dehydrogenase, a cytoplasmic enzyme, is not released during periods of secretion, although it should pass through cell membranes more readily than chromogranin A (Schneider *et al.*, 1967). These findings are precisely what would be expected if the water soluble contents of the granules were released by exocytosis and the lipoprotein membrane remained behind in the cells. The problem now is to discover how an influx of calcium into the cells can initiate secretion by exocytosis.

Evidence has been presented that Ca^{2+} ions can release catecholamines and ATP but not protein from isolated chromaffin and nerve granules suspended in isotonic sucrose or salt solutions (Schümann & Philippu, 1963, 1968; Oka, Ohuchi, Yoshida & Imaizumi, 1965). It seems unlikely, however, that this behaviour is related to the secretory response of intact cells because the rate of release in the presence of physiological concentrations of calcium is far too low to account for the very rapid secretory response. Furthermore, calcium does not release protein from isolated medullary granules although it is released during the secretory response of the intact gland.

Chromaffin granules possess an ATPase activity (Hillarp, 1958a,b; Banks, 1965; Kirshner, Kirshner & Kamin, 1966) but it requires Mg^{2+} ions rather than Ca^{2+} ions as a co-factor so that it is unlikely to be the site of action of calcium. Indeed, no conclusive evidence exists that the granule ATPase is involved in the secretory response although Poisner & Trifaro (1967) have cited experiments in which secretion was prevented by ATPase inhibitors. The presence of ATP-catabolites in the perfusate from stimulated glands cannot be taken as evidence that the granule ATP is broken down during the secretion process since ATP perfused through resting glands is similarly degraded (Banks, 1966a). Recently it has been found that low concentrations of Mg-ATP can cause the release of catecholamines, ATP and protein but not of cholesterol from isolated chromaffin granules suspended in isotonic salt solutions (Oka *et al.*, 1965, 1967; Poisner & Trifaró, 1967; Trifaró & Poisner, 1967). In iso-osmotic sucrose Mg-ATP has no effect on isolated granules. Calcium could not substitute for magnesium and release was inhibited by N-ethylmaleimide which is known to inhibit the granule Mg-ATPase (Kirshner *et al.*, 1966). Nevertheless the role of the Mg-ATPase remains far from clear because chromaffin granules exhibit an Mg-ATP-dependent, reserpine-sensitive uptake of catecholamines which is inhibited by N-ethylmaleimide (Carlsson, Hillarp & Waldeck, 1962, 1963; Kirshner, 1962). Thus the ATPase may be related to amine uptake rather than to secretion. However, if the catecholamine release from isolated chromaffin granules induced by Mg-ATP is of physiological significance, the question of how the granules are protected from the depleting action of intracellular Mg-ATP must be answered.

Chromaffin granules, as might be expected, possess a net negative charge which would probably prevent them from closely approaching the negatively charged inner surface of the cell membrane. As Ca^{2+} ions alter the surface charge of isolated granules and cause them to aggregate, it is conceivable that in intact cells Ca^{2+} ions form bivalent bridges which enable the cell and granule membranes to establish contact (Banks, 1966b). Direct evidence supporting such a role for calcium in the mechanism of secretion has been provided by observations of the secretory granules in leucocidin-treated leucocytes suspended in normal and Ca^{2+}-free solutions (Woodin, French & Marchesi, 1963).

A model for the secretory process has been developed by Poisner & Trifaró (1967) in which stimulation frees calcium from an association with ATP in the cell membrane and thus allows it to form a bridge between the cell and granule membranes. The membrane-bound ATP is then hydrolysed by the Mg^{2+}-ATPase of the chromaffin granule and thus provides energy for the molecular rearrangements which complete the fusion of the cell and granule membranes and permit the egress of the granule contents.

Calcium dependence of catecholamine release by sympathomimetic compounds
The involvement of Ca^{2+} ions in the release of catecholamines from the adrenal medulla and from noradrenergic nerve endings by indirectly acting sympathomimetic compounds remains obscure. There is much evidence that compounds such as tyramine and phenylethylamine are able to cause a dose-dependent release of catecholamines but not of ATP from isolated chromaffin granules; the displaced amines are replaced in the storage complex by the sympathomimetic agent (Schümann & Philippu, 1962a,b). Furthermore, it has been reported that phenylethylamine and tyramine can release catecholamines from perfused adrenal glands (Schümann & Philippu, 1962c; Philippu & Schümann, 1966) and from adrenergic nerve endings (Burn & Gibbons, 1965) in the absence of calcium. However, Rubin & Jaanus (1966, 1967) have reported that phenylethylamine, tyramine, and amphetamine require the presence of calcium in order to release catecholamines from perfused adrenal glands. Moreover the release of amines by amphetamine and phenylethylamine is accompanied by a release of AMP. The reason for these apparently contradictory findings is not understood.

Removal of calcium from chromaffin cells
Termination of the secretory response and return to the resting state must be associated with the removal of the Ca^{2+} ions which entered during the secretory phase. At present virtually nothing is known about the processes

which are involved in pumping calcium out of the cell. Schatzmann (1966) has described an ATP-dependent system which is located in the membranes of red cells and is responsible for a ouabain-insensitive efflux of calcium. This system may be related to the ouabain-insensitive Mg^{2+}-ATPase described by Dunham & Glynn (1961) which is activated by low concentrations of calcium. An ATPase activity which is stimulated in the presence of calcium has been found in the microsomal fraction obtained from adrenal medullae (Banks, 1965; Goz, 1967). At present the function of this ATPase is unknown but the possibility that it is concerned with Ca^{2+} expulsion may be worth considering. There is no evidence that any particulate system exists in chromaffin cells for sequestering calcium as does the sarcoplasmic reticulum of skeletal muscle (Banks, 1964; Goz, 1967). Blaustein & Hodgkin (1969) have suggested that calcium may be removed from squid nerves by exchanging with external sodium. Thus, in view of the evidence presented above that calcium influx may be regulated by internal sodium, the possibility must also exist that calcium efflux may be driven by the influx of extracellular sodium.

Calcium in chromaffin granules
In addition to its role in secretion, calcium may play some part in the storage of catecholamines, for it accounts for 0·16% of the dry weight of chromaffin granules from which it is not readily removed by EDTA. The molar ratio of catecholamines to Ca^{2+} in the granules is between 20 and 30 (Borowitz, Fuwa & Weiner, 1965; Borowitz, 1967) and calcium is lost from isolated granules together with ATP and catecholamines during incubation at 37°C or on osmotic shock (Philippu & Schümann, 1966). Tyramine displaces catecholamines from isolated granules but is without effect on their calcium content (Philippu & Schumann, 1966). The binding of calcium to fragments of chromaffin granule membranes does not differ significantly from calcium binding to fragmented mitochondria (Borowitz, 1967).

Summary
The findings reported here can be summarized as follows. Calcium influx into chromaffin cells couples the stimulus with the secretory response and appears to be dependent upon the intracellular concentration of sodium. The role of calcium within the cell is unknown although it has been speculated that it promotes the process of exocytosis by allowing the cell and chromaffin granule membranes to make contact. Nothing is known about the processes which remove calcium from chromaffin cells. Calcium is concentrated in chromaffin granules and may be involved in the storage of catecholamines.

REFERENCES

BAKER, P. F. & BLAUSTEIN, M. P. (1968). *Biochim. biophys. Acta*, **150**, 167
BAKER, P. F., BLAUSTEIN, M. P., HODGKIN, A. L. & STEINHARDT, R. A. (1967). *J. Physiol., Lond.*, **192**, 43P

BAKER, P. F., BLAUSTEIN, M. P. HODGKIN, A. L. & STEINHARDT, R. A. (1969). *J. Physiol., Lond.*, **200**, 431

BANKS, P. (1964). D. Phil. Thesis, Oxford

BANKS, P. (1965). *Biochem. J.*, **95**, 490

BANKS, P. (1966a). *Biochem. J.*, **101**, 536

BANKS, P. (1966b). *Biochem. J.*, **101**, 18C

BANKS, P. (1967). *J. Physiol., Lond.*, **193**, 631

BANKS, P., BIGGINS, R., BISHOP, R., CHRISTIAN, B. & CURRIE, N. (1969). *J. Physiol., Lond.*, **200**, 797

BANKS, P. & HELLE, K. (1965). *Biochem. J.*, **97**, 40c

BIRKS, R. I. & COHEN, M. W. (1968a). *Proc. R. Soc., B*, **170**, 381

BIRKS, R. I. & COHEN, M. W. (1968b). *Proc. R. Soc., B*, **170**, 401

BLASCHKO, H., COMLINE, R. S., SCHNEIDER, F. H. & SMITH, A. D. (1967). *Nature, Lond.*, **215**, 58

BLAUSTEIN, M. P. & HODGKIN, A. L. (1969). *J. Physiol., Lond.*, **200**, 497

BOROWITZ, J. L. (1967). *J. cell Physiol.*, **69**, 311

BOROWITZ, J. L., FUWA, K. & WEINER, N. (1965). *Nature, Lond.*, **205**, 42

BOULLIN, D. R. (1967). *J. Physiol., Lond.*, **189**, 85

BURN, J. H. & GIBBONS, W. R. (1964). *Br. J. Pharmac. Chemother.*, **22**, 540

BURN, J. H. & GIBBONS, W. R. (1965). *J. Physiol., Lond.*, **181**, 214

CARLSSON, A. & HILLARP, N.-Å. (1956). *Acta. physiol. scand.*, **37**, 235

CARLSSON, A., HILLARP, N.-Å. & WALDECK, B. (1962). *Medina. exp.*, **6**, 47

CARLSSON, A., HILLARP, N.-Å. & WALDECK, B. (1963). *Acta physiol. scand.*, **59**, suppl. 215, p. 5

COUPLAND, R. E. (1965). *J. Anat.*, **99**, 231

DE ROBERTIS, E. D. P. & VAZ FERREIRA, A. (1957). *Exp. Cell Res.*, **12**, 568

DINER, O. (1967). *C.r. hebd. Seanc. Acad. Sci., Paris*, **265**, 616

DOUGLAS, W. W. (1968). *Br. J. Pharmac.*, **34**, 451

DOUGLAS, W. W. & KANNO, T. (1967). *Br. J. Pharmac. Chemother.*, **30**, 612

DOUGLAS, W. W., KANNO, T. & SAMPSON, S. R. (1967a). *J. Physiol., Lond.*, **188**, 107

DOUGLAS, W. W., KANNO, T. & SAMPSON, S. R. (1967b). *J. Physiol., Lond.*, **191**, 107

DOUGLAS, W. W. & POISNER, A. M. (1962). *J. Physiol., Lond.*, **162**, 385

DOUGLAS, W. W., POISNER, A. M. & RUBIN, R. P. (1965). *J. Physiol., Lond.*, **179**, 130

DOUGLAS, W. W. & RUBIN, R. P. (1961). *J. Physiol., Lond.*, **159**, 40

DOUGLAS, W. W. & RUBIN, R. P. (1963). *J. Physiol., Lond.*, **167**, 288

DUNHAM, E. T. & GLYN, I. M. (1961). *J. Physiol., Lond.*, **156**, 274

GOZ, B. (1967). *Biochem. Pharmac.*, **16**, 596

HALES, C. N. & MILNER, R. D. G. (1968a). *J. Physiol., Lond.*, **194**, 725

HALES, C. N. & MILNER, R. D. G. (1968b). *J. Physiol., Lond.*, **199**, 177

HILLARP, N.-Å. (1958a). *Acta physiol. scand.*, **42**, 144

HILLARP, N.-Å. (1958b). *Acta physiol. scand.*, **43**, 82

HUKOVIC, S. & MUSCHOLL, E. (1962). *Arch. exp. Path. Pharmak.*, **244**, 81

JAANUS, S. D., MIELE, E. & RUBIN, R. P. (1967). *Br. J. Pharmac. Chemother.*, **31**, 319

KIRPEKAR, S. M. & MISU, Y. (1967). *J. Physiol., Lond.*, **188**, 219

KIRPEKAR, S. M. & WAKADE, A. R. (1968). *J. Physiol., Lond.*, **94**, 595

KIRSHNER, N. (1962). *J. biol. Chem.*, **237**, 2311

KIRSHNER, N., KIRSHNER, A. G. & KAMIN, D. L. (1966). *Biochim. biophys. Acta*, **113**, 332

KIRSHNER, N., SAGE, H. J., SMITH, W. J. & KIRSHNER, A. G. (1966). *Science*, *N.Y.*, **154**, 529

LOFFELHOLZ, K. (1967). *Arch. exp. Path. Pharmak.*, **258**, 108

MIELE, E. & RUBIN, R. P. (1968). *J. Pharmac. exp. Ther.*, **161**, 296

OKA, M., OHUCHI, T., YOSHIDA, H. & IMAIZUMI, R. (1965). *Biochim. biophys. Acta*, **97**, 170

OKA, M., OHUCHI, T., YOSHIDA, H. & IMAIZUMI, R. (1967). *Jap. J. Pharmac.*, **17**, 199

PHILIPPU, A. & SCHÜMANN, H. J. (1966). *Arch. exp. Path. Pharmak.*, **252**, 339.

POISNER, A. M. & DOUGLAS, W. W. (1966). *Proc. Soc. exp. Biol. Med.*, **132**, 62

POISNER, A. M. & TRÍFARÓ, J. M. (1967). *Mol. Pharmac.*, **3**, 561

POISNER, A. M., TRÍFARÓ, J. M. & DOUGLAS, W. W. (1967). *Biochem. Pharmac.*, **16**, 2101

RUBIN, R. P., FEINSTEIN, M. B., JAANUS, S. D. & PAIMRE, M. (1967). *J. Pharmac. exp. Ther.*, **155**, 463

RUBIN, R. P. & JAANUS, S. D. (1966). *Arch. exp. Path. Pharmak.*, **254**, 125

RUBIN, R. P. & JAANUS, S. D. (1967). *Biochem. Pharmac.*, **16**, 1007

SCHATZMANN, H. J. (1966). *Experientia*, **22**, 364

SCHNEIDER, F. H., SMITH, A. D. & WINKLER, H. (1967). *Br. J. Pharmac. Chemother.*, **31**, 94

SCHÜMANN, H. J. (1958). *Arch. exp. Path. Pharmak.*, **233**, 237

SCHÜMANN, H. J. & PHILIPPU, A. (1962a). *Int. J. Neuropharmac.*, **1**, 179

SCHÜMANN, H. J. & PHILIPPU, A. (1962b). *Nature, Lond.*, **193**, 890

SCHÜMANN, H. J. & PHILIPPU, A. (1962c). *Experientia*, **18**, 138

SCHÜMANN, H. J. & PHILIPPU, A. (1963). *Arch. exp. Path. Pharmak.*, **244**, 466

SCHÜMANN, H. J. & PHILIPPU, A. (1968). *Arch. exp. Path. Pharmak.*, **259**, 193

SMITH, A. D. (1968). In *The Interaction of Drugs and Subcellular Components on Animal Cells*, ed. CAMPBELL, P. N., p. 239. London: J. & A. Churchill

TRÍFARÓ, J. M. & POISNER, A. M. (1967). *Mol. Pharmac.*, **3**, 572

TRÍFARÓ, J. M., POISNER, A. M. & DOUGLAS, W. W. (1967). *Biochem. Pharmac.*, **16**, 2095

WOODIN, A. M., FRENCH, J. E. & MARCHESI, V. T. (1963). *Biochem. J.*, **87**, 567

CALCIUM AND HORMONE RELEASE

E. K. MATTHEWS

Department of Pharmacology, University of Cambridge

The calcium ion plays an essential part in the liberation of transmitter at chemically operated synapses. Yet its involvement in the more general process of hormone release may be no less critical. Thus the synaptic release of acetylcholine and noradrenaline is calcium dependent (Harvey & MacIntosh, 1940; del Castillo & Stark, 1952; Huković & Muscholl, 1962; Kirpekar & Misu, 1967); but so, too, is the release of many hormones (Table 1). Furthermore, stimulation of the neurohypophysis and adrenal medulla causes a five- to eight-fold increase in the rate of $^{45}Ca^{2+}$ uptake by these tissues (Douglas & Poisner, 1962, 1964).

TABLE 1. *Calcium-dependent hormone release*

Site	Hormone	References
Pancreatic β-cell	Insulin	Grodsky & Bennett (1966) Milner & Hales (1967)
Adenohypophysis	Thyroid-stimulating hormone Luteinizing hormone	Vale, Burgus & Guillemin (1967) Vale & Guillemin (1967) Samli & Geschwind (1968)
Neurohypophysis	Vasopressin Oxytocin	Douglas & Poisner (1964) Dicker (1966); Ishida (1968)
Adrenal medulla	Adrenaline Noradrenaline	Douglas & Rubin (1961)

It is therefore apparent that calcium is of functional significance not only in neurotransmitter release mechanisms but that an obligatory requirement for this divalent cation exists also in the wider field of hormone release involving endocrine and neuroendocrine cells. In fact, the operational characteristics of hormone and neurotransmitter release mechanisms show many basic similarities and while the multistep secretory process is in either case undeniably complicated the features of major importance can now be discerned. Central to the whole theme stands the ubiquitous calcium ion.

Calcium-dependent hormone release in endocrine cells

Pancreatic islets. It has proved possible to investigate the bioelectrical aspects of islet cell function by inserting glass ultra-microelectrodes into single islet cells (Dean & Matthews, 1968). A segment of pancreas removed from a mouse is maintained at 37°C with superfusion of oxygenated Krebs–Henseleit solution containing glucose 2·8 mM. Under these conditions a membrane potential of about −20 mV is recorded from islet cells compared with a value of −41 mV from surrounding acinar cells.

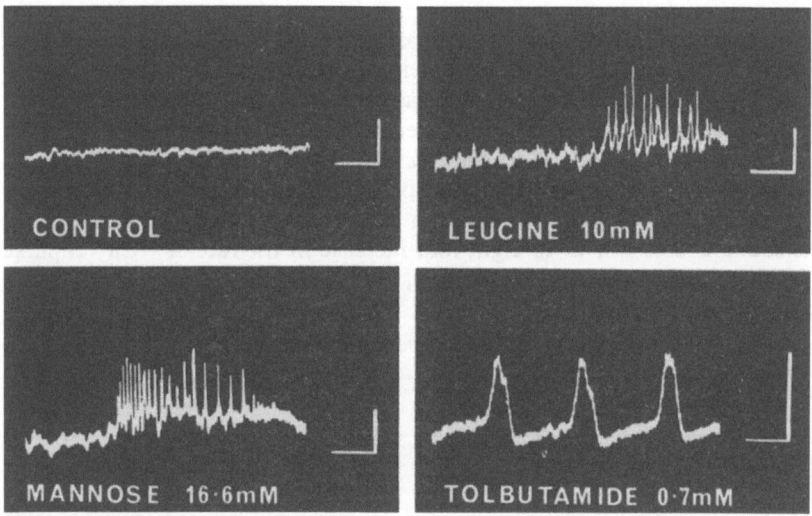

FIG. 1. Action potential discharge in islet cells elicited by L-leucine 10 mM, D-mannose 16·6 mM, and tolbutamide sodium 0·7 mM. Control recording in the presence of glucose 2·8 mM. Vertical calibration, 2 mV; horizontal calibration, 0·5 sec. (From Matthews & Dean, 1969.)

In an environment of 2·8 mM D-glucose, the islet cells were electrically quiescent but increasing the glucose concentration sixfold to 16·6 mM, or exposure to mannose 16·6 mM, leucine 10 mM or tolbutamide sodium 0·7 mM caused the appearance of small 'action potentials' superimposed on the membrane potential (Fig. 1; see also Dean & Matthews, 1968). The appearance of electrical activity in islet cells was dependent on the D-glucose concentration (Fig. 2) and it seems likely that action potential incidence and insulin release are causally related (Matthews & Dean, 1969).

One possible scheme by which action potential discharge might be correlated with insulin secretion is illustrated in Fig. 3. This model takes into account the important ionic requirements for insulin release which were demonstrated by Hales & Milner (1968).

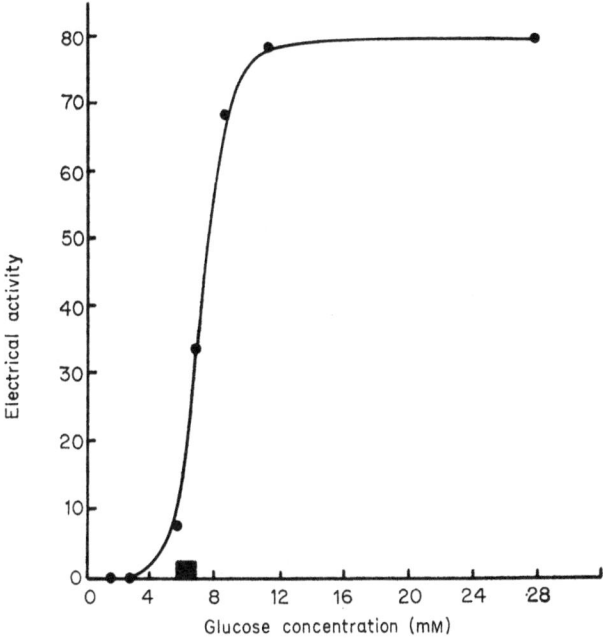

FIG. 2. Relationship between glucose concentration and electrical activity of islet cells. Electrical activity expressed as the percentage of impaled cells which show action potential discharge. The black bar indicates the normal range of blood glucose concentration in the mouse. (Redrawn from Matthews & Dean, 1969.)

The maintenance of a normal membrane potential is dependent on the calcium ion; in its absence the islet cells depolarize. It is also evident that in a calcium-free medium the release of insulin is inhibited (Fig. 4). On the other hand the action potentials elicited in islet cells by D-glucose were abolished when the calcium concentration was increased tenfold (Fig. 5). Slow wave activity was seen initially and this gradually disappeared. Milner & Hales (1967) found that increasing the calcium concentration to only twice its normal level caused a diminished insulin release (Fig. 6) from rabbit pancreas *in vitro*.

Omission of magnesium (1·13 mM) from normal Krebs solution was without effect on either the membrane potential of islet cells or the action potentials induced by glucose. Insulin secretion was likewise unaffected (Fig. 7). Conversely, increasing the magnesium concentration tenfold to 11 mM did not alter the membrane potential or prevent the appearance of action potentials in islet cells (Fig. 8). However, as in other secretory systems, a raised magnesium concentration does inhibit insulin release (Fig. 9). It may therefore be possible to uncouple electrical activity from

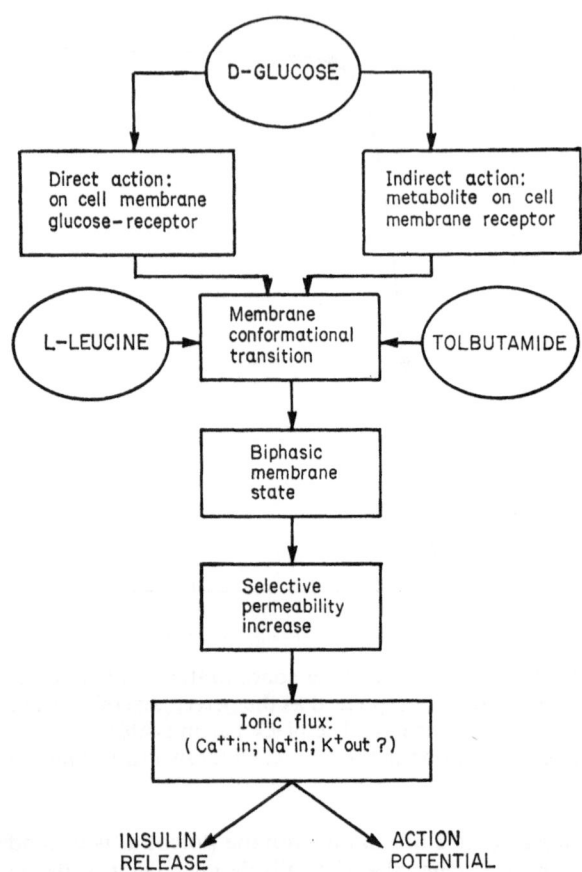

FIG. 3. Scheme by which action potential incidence might be correlated with insulin secretion. (From Matthews & Dean, 1969.)

secretory activity. Finally, exposure of the islet cells to manganese 2 mM (in the presence of a normal calcium concentration of 2·56 mM) led within a short time to complete inhibition of electrical activity (Fig. 10). This has some bearing on the origin of islet cell action potentials because, in other tissues at least, manganese appears to block calcium dependent action potentials (Hagiwara & Nakajima, 1966); more experimental evidence is clearly needed on this particular point.

Adenohypophyseal cells. The importance of calcium in neurosecretory mechanisms of the posterior pituitary has been emphasized by Douglas & Poisner (1964) (see also Dicker, 1966; Ishida, 1968). Recently others have extended these studies to include the adenohypophysis. Here the stimula-

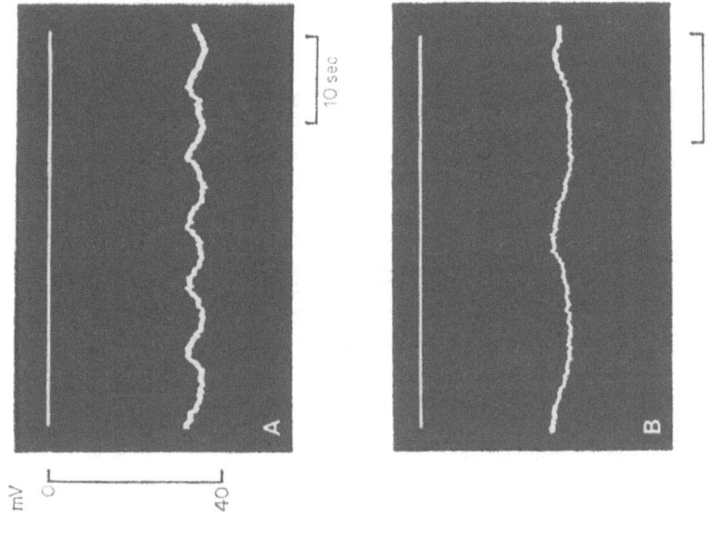

FIG. 5. Electrical activity in mouse islet cells; effect of a ten-fold increase in calcium concentration (to 25·6 mM). Records obtained 28 min after exposure to 16·6 mM glucose and 25·6 mM calcium. Tris-HCl Krebs.

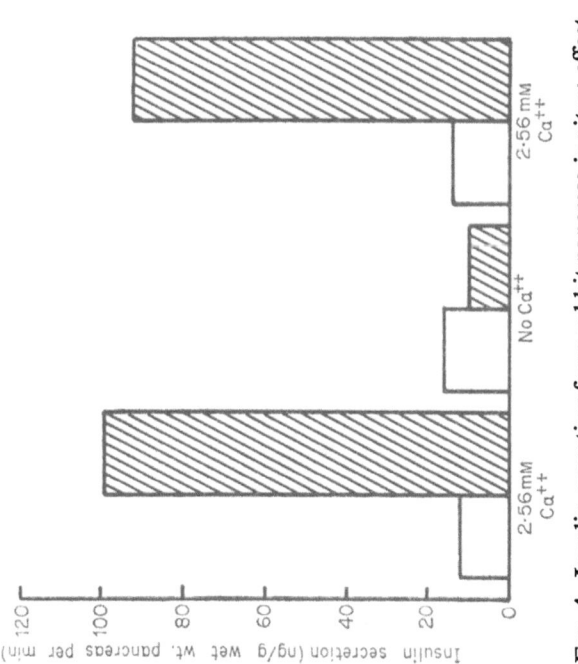

FIG. 4. Insulin secretion from rabbit pancreas *in vitro*: effect of calcium omission. Basal secretion (open columns) in 3·3 mM glucose; stimulated secretion (hatched columns) in 16·6 mM glucose. 30 min incubation periods. (From data of Hales & Milner, 1968.)

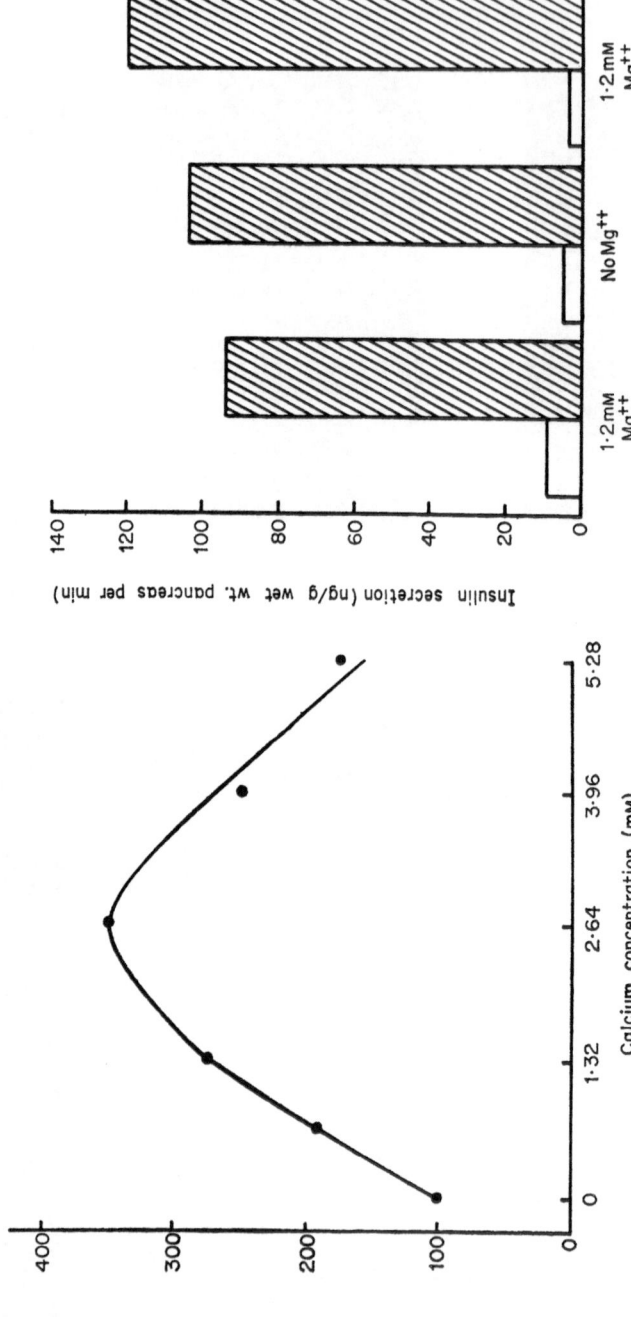

Fig. 7. Insulin secretion from rabbit pancreas in presence and absence of magnesium. Details as Fig. 4. (From data of Hales & Milner, 1968.)

Fig. 6. Insulin secretion from rabbit pancreas; effect of calcium. Stimulation of insulin secretion by 33 mM glucose. The calcium concentration was increased in each successive 30 min incubation period. (Redrawn from Hales & Milner, 1968.)

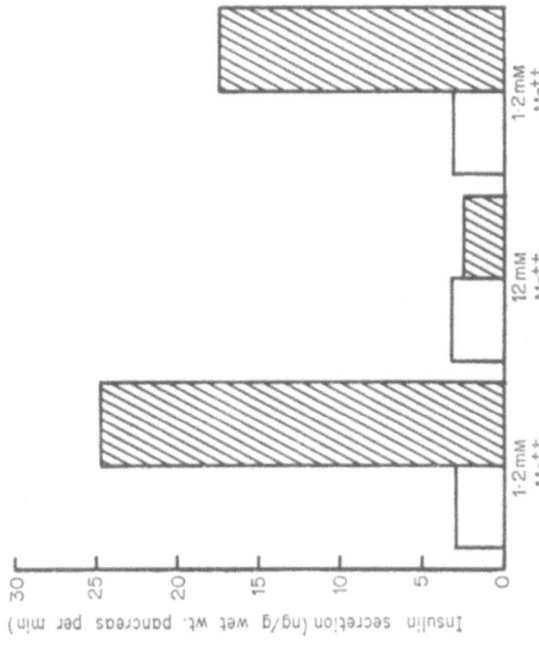

FIG. 9. Insulin secretion from rabbit pancreas: effect of magnesium concentration, increased 10-fold in test periods. Details as Fig. 4. (From data of Hales & Milner, 1968.)

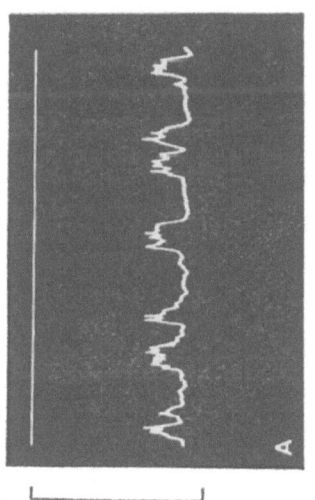

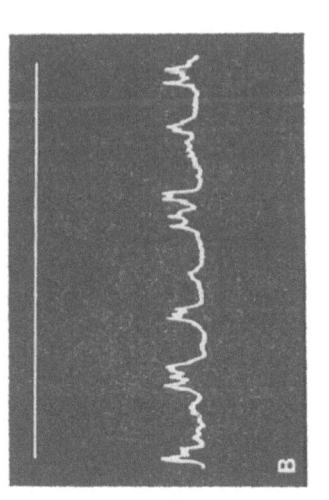

FIG. 8. Electrical activity in mouse islet cells: effect of a 10-fold increase in magnesium concentration (to 11·3 mM). Record A, 32 min, and B, 52 min after exposure to glucose 16·6 mM in presence of magnesium 11·3 mM.

tory signal for hormone release is blood-borne (as it is also in the pancreas, that is, glucose on the β-cell) rather than neuronal as in the posterior lobe.

Six major polypeptide hormones are located in the adenohypophysis—namely, growth hormone (GH); follicle-stimulating hormone (FSH); luteinizing hormone (LH); thyroid-stimulating hormone (TSH); adrenocorticotrophic hormone (ACTH); luteotrophic hormone (LTH)—and each

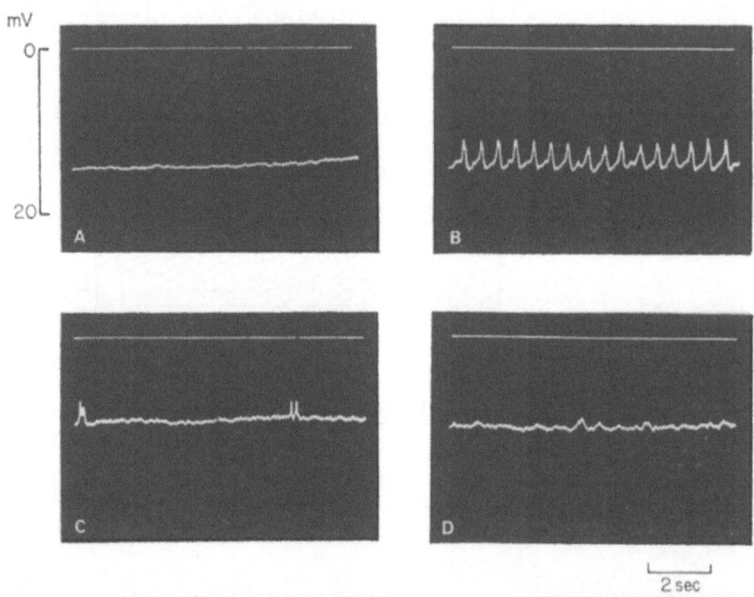

FIG. 10. Electrical activity in mouse islet cells; effect of manganese. Record A in presence of glucose 2·8 mM, B to D in presence of glucose 16·6 mM. C, 11 min, and D, 15 min, after exposure to manganese 2 mM. Tris-HCl Krebs solution.

is stored in granular form in a characteristic cell type (see Table 2) until released by its appropriate hypothalamic releasing factor.

Samli & Geschwind (1968), using a hypothalamic extract or a tenfold increase in extracellular K^+ (from 5·9 to 59 mM), showed that LH release *in vitro* is increased in the presence of calcium but not in its absence (Fig. 11). Similarly, Vale, Burgus & Guillemin (1967) and Vale & Guillemin (1967) demonstrated that the specific TSH-releasing factor or a raised K^+ concentration (25 mM) evoked the release of TSH from adenohypophyseal tissue *in vitro*; this process also was calcium dependent (Fig. 12). Both groups proposed that to stimulate hormone release the appropriate releasing factor may depolarize its adenohypophyseal target cell. They also suggested that calcium may be involved in the stimulus-secretion coupling process.

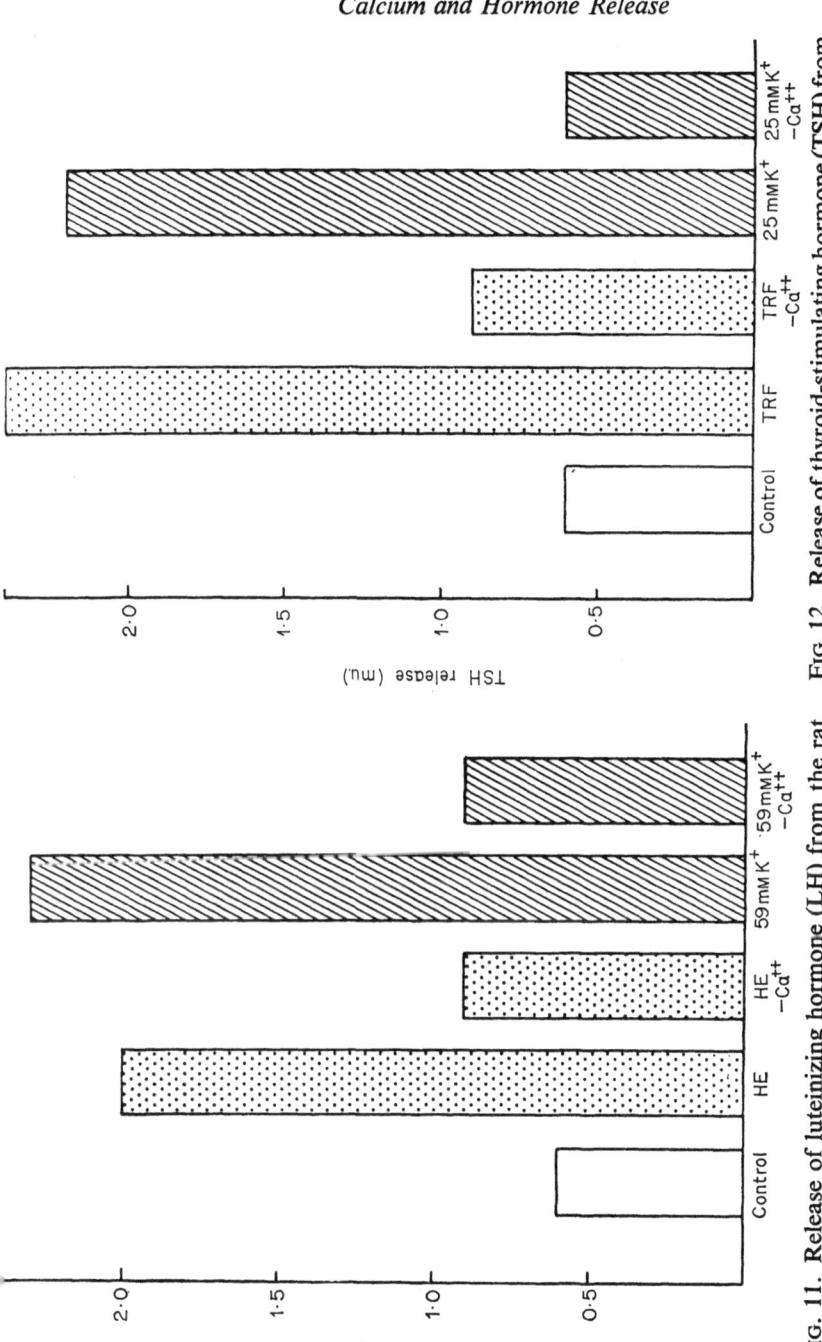

FIG. 11. Release of luteinizing hormone (LH) from the rat pituitary gland *in vitro*. Effect of altering medium Ca²⁺ or K⁺ on the release of LH. Hypothalamic extract indicated by HE. In each column LH release is expressed as µg/mg pituitary per 4 hr incubation. (From data of Samli & Geschwind, 1968.)

FIG. 12. Release of thyroid-stimulating hormone (TSH) from the rat pituitary gland *in vitro*. Effect of altering medium Ca²⁺ or K⁺ on the release of TSH. TRF indicates TSH-releasing factor. Open columns show TSH release expressed as mu./20 min incubation period and the hatched columns as mu./15 min incubation period. (Drawn from pooled data of Vale, Burgus & Guillemin, 1967; and Vale & Guillemin, 1967.)

Hormone storage and release

It seems therefore that the calcium ion is widely concerned in the secretory process. Indeed, it would appear to be implicated wherever the secretory product is stored as an exportable pre-wrapped package awaiting delivery to the cell surface. The distribution and location of these storage 'granules' and 'vesicles' is of considerable interest. Evidence from electron microscopic and subcellular fractionation studies now firmly supports the notion that the granules detected in many hormone producing cells represent the site of hormone storage.

TABLE 2. *Storage of Hormones*

Cell	Hormone	Granule diam. (mμ)	References
Adenohypophysis:			
Acidophil	GH	350	Hymer & McShan (1963)
Basophil	FSH LH	200	Perdue & McShan (1966)
Basophil	TSH	140	Siperstein & Allison (1965)
Chromophobe	ACTH	200	Siperstein (1963)
Neurohypophysis:			
Neurone	Oxytocin	100–300	Bindler, LaBella & Sanwal (1967)
	Vasopressin		Sachs (1967)
Pancreatic islet:			
α-cell	Glucagon	300	Meyer & Bencosme (1965)
β-cell	Insulin	200–300	Lazarus, Shapiro & Volk (1967)
Chromaffin	Catecholamine	50–350	Coupland (1965)
Adrenergic Neurone	Noradrenaline	50–100	Hökfelt (1968)
Cholinergic Neurone	Acetylcholine	47*	Whittaker (1965)

The granule diameters are upper limits unless otherwise indicated. * = mean.

In view of the chemical diversity of the stored hormone, from simple amine (adrenaline) to complex polypeptide (insulin:ACTH) the granule size displays a quite surprising similarity (Table 2). The average diameter is about 150 mμ for most hormone granules, somewhat less for neurotransmitter vesicles; the values in Table 2 are upper limits unless otherwise stated.

There have been several mechanisms proposed by which a hormone or neurotransmitter released from its granule or vesicular storage site eventually gains the extracellular space. All are regarded, in various ways, as calcium-dependent. A particular difficulty encountered in any postulated theory of hormone release is how a storage particle from the interior of the cell can reach the periphery for release. Another problem concerns the apparent inability to demonstrate a significant decrease in the granule or vesicle population after procedures which would evoke maximal stimulation of release. What may be involved, however, is a localized decrease of a specific rather than a general vesicle population (Hubbard & Kwanbunbumpen, 1968). These difficulties have led Whittaker (1968) and others (Lacy, Howell, Young & Fink, 1968) to suggest a microtubular system of vesicle and granule discharge. Of the various mechanisms proposed for hormone and neurotransmitter release, for both processes appear essentially similar, three perhaps merit close consideration:

1. Intracellular disruption of storage granules, followed by outward diffusion of 'free' hormone across the cell membrane.
2. Exocytosis (reverse pinocytosis); granules attach to the cell membrane and eject contents into the extracellular space.
3. Microtubular discharge; granules constrained within a tubular structure reach the cell surface by mechano-chemical activity of the tubule.

The first two theories depend on a random and the third on a preferential pathway for granule approach to the cell membrane. An increased calcium concentration is required (1) for disruption of granules, (2) for attachment of granules to cell membrane, or (3) for activation of the tubular system.

The first theory seems definitely inapplicable to certain types of secretory cell. For example, in stimulated adrenal glands the presence in the perfusate of granule protein, adenine nucleotides, and catecholamines in relative amounts similar to those present in intact storage granules indicates that intracellular disruption of storage granules followed by outward diffusion is unlikely (Douglas & Poisner, 1966; Kirshner, Sage & Smith, 1967). Such a process would demand a marked increase in the permeability of the cell membrane to large molecular weight granule proteins. Cytoplasmic proteins might then also be secreted but Schneider, Smith & Winkler (1967) found no evidence for this, and Kirshner *et al.* (1967) noted that the cytoplasmic enzyme phenethanolamine-N-methyltransferase was not secreted on stimulation. Extrusion of the entire granule through the cell membrane is unlikely because Schneider *et al.* (1967) observed in the adrenal perfusate only minute amounts of phospholipids and cholesterol, major components of the chromaffin granule lipid, and a complete absence of the characteristic granule phospholipid lysolecithin. They concluded that exocytosis (see (2) above) was the only mechanism allowing secretion of both high and low molecular weight material specifically from chromaffin granules. Furthermore, Malamed, Poisner, Trifaró & Douglas (1968) have recently shown that depleted granules remain within the cell after stimulation.

In the neurohypophysis, Thorn (1965) claims that calcium ions cause the liberation of vasopressin and oxytocin from neurophysin within the neurosecretory nerve ending, the peptide hormones becoming free to diffuse from the intracellular to the extracellular phase. Similarly, the failure of calcium ions to initiate release of vasopressin from isolated neurosecretory granules (Daniel & Lederis, 1963) while actually inhibiting vasopressin-neurophysin association *in vitro* led Ginsburg & Ireland (1966) to postulate intracellular dissociation of an extragranular hormone-neurophysin complex by calcium and an alteration in permeability of the neuronal membrane. More recently Sach & Haller (1968) have succeeded in labelling the vasopressin storage depots by isotope incorporation. Although they could demonstrate a readily releasable pool of vasopressin, no information was obtainable as to its precise cellular location. A combined chemical and morphological approach, exemplified in the study of catecholamine release from the adrenal medulla, which was beset with similar difficulties (Douglas, 1968), may be expected to yield important clues as to the actual secretory mechanism involved. In the meantime, an important role of exocytosis in neurohypophyseal hormone release cannot be excluded.

The existence of microtubules in a wide variety of cells seems beyond dispute (see Porter, 1966) but there is little morphological evidence for a highly organized microtubular cytoskeleton which might be responsible specifically for translocation of storage granules to the cell membrane by a predetermined pathway. Indeed if such a system does exist it may be much less well developed in some secretory cells than in others.

It seems worthwhile therefore to consider in quantitative kinetic terms whether a stochastic or random process with an inherent probability pattern capable of modification by, for example, calcium ions, can adequately account for the arrival of a granule at the cell membrane with subsequent fusion and secretion by exocytosis. Qualitative expression has been given to some of these ideas by Douglas (1968) and Banks (1966).

Consider first a storage granule 50–400 mμ in diameter (Table 2) under no particular mechanical restraint and free to move within the cellular cytoplasm. Such a particle will be displaced by the kinetic energy of translational or Brownian motion since its dimensions are well below 5 μm.

Treating Brownian motion as a three-dimensional random walk, $\bar{\Delta}$, the mean displacement of a particle from its original position along a given axis is expressed by Einstein's equation:

$$\bar{\Delta} = \sqrt{\frac{kTt}{3\pi a\eta}} \tag{1}$$

where k is Boltzmann's constant, T the absolute temperature, a the radius of the particle (cm), η the coefficient of viscosity (poises), and t the duration (sec) of random Brownian motion. If $\bar{\Delta}$ is the mean displacement then the distribution function of absolute displacements $|x|$ (see below) in a given direction is

$$F(|x|) = \frac{1}{\sqrt{2\pi}} \cdot \int_{-x/\bar{\Delta}}^{x/\bar{\Delta}} e^{-y^2/2} \, dy \qquad (2)$$

It is evident from equation (1) that the displacement of a granule is inversely proportional to the square root of its radius and also, importantly, to the square root of its environmental (cytoplasmic) viscosity.

There is, unfortunately, a paucity of information concerning cytoplasmic viscosity in different cell types, but according to Heilbrunn (1956) cellular cytoplasm is of relatively low viscosity, with the cortical regions often more viscous than the interior. Cytoplasm of nerve (lobster) has a viscosity of 0·06 poises and muscle (frog) a viscosity of 0·3 poises (Rieser, 1949). These values are some 6 to 30 times greater than the absolute viscosity of water which is 0·01 poises at 20°.

Using equations (1) and (2) it is possible to calculate the theoretical distribution functions of absolute displacements in a given direction by Brownian motion (Shea & Karnovsky, 1966). For example, illustrated in Fig. 13 are the absolute displacements in an interval of 1 sec for particles of 60 mμ and 400 mμ diameter in media of viscosity 0·06, 0·3, 0·5, and 10·0 poises. It can be seen that the lower the viscosity the smaller is the probability of finding a given particle close to the origin after 1 sec: that is, it is more likely to be displaced to points remote from the origin.

If, as seems not unreasonable, the viscosity of a secretory cell approximates to that of nerve (or even approaches that of muscle), Brownian motion could play an effective role in allowing the migration of a granule from the interior of the cell to the surface membrane at a rate compatible with the rapid process of secretion. For example, it can be calculated from equation (1) that an insulin storage granule of 200 mμ diameter situated 5 μ from the centre of an islet β-cell (about 20 μ diameter) could cross the remaining 5 μ to the periphery in 33 sec, assuming a cytoplasmic viscosity of 0·06 poises. If the granule were more superficially and strategically located, say within 10 nm of the inner surface of the cell membrane, this transit time would reduce to only 0·13 msec.

Although the evidence for an action of electrolytes on the physical state of cytoplasm is open to conflicting interpretation (see Heilbrunn, 1956), it appears that divalent cations can cause axoplasmic dispersal (Hodgkin & Katz, 1949) and a reduction in viscosity in the order Ca & Sr > Mg > Ba (Chambers & Kao, 1952). Calcium uptake is known to occur on cellular stimulation in a variety of cells, for example, nerve (Hodgkin & Keynes, 1957), muscle (Bianchi & Shanes, 1959), endocrine, and neuroendocrine

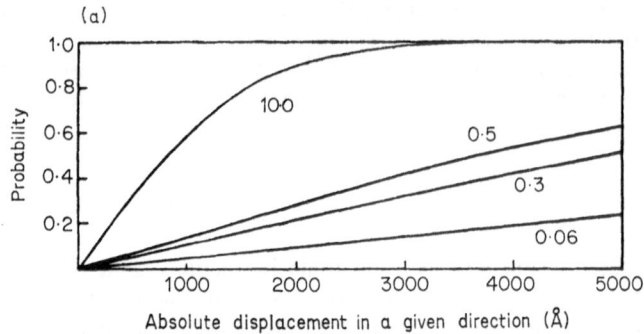

Absolute displacement in a given direction (Å)

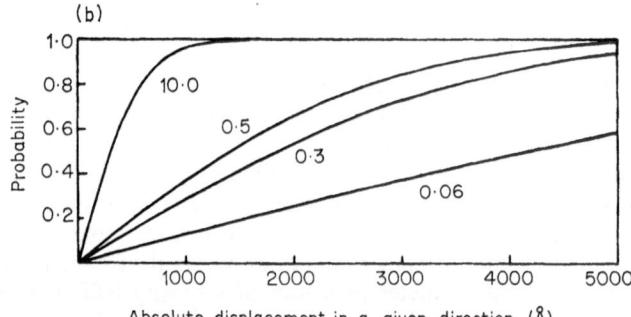

Absolute displacement in a given direction (Å)

FIG. 13. Brownian motion. Cumulative probability distributions of absolute displacements in a given direction of particles of diameter 60 mμ (a) and 400 mμ (b) in 1 sec; for viscosities of 0·06, 0·3, 0·5, and 10·0 poises. (From Shea & Karnovsky, 1966.)

cells (Douglas & Poisner, 1962; 1964). The site of uptake is unknown but it seems possible that stimulation of a secretory cell (which may involve depolarization) could result in a reduction of cytoplasmic viscosity particularly in the important and critical cortical region of the cell. Such a process would extend the mean free path traversed by a given granule and so promote a greater displacement in unit time. Greater mobility would in turn increase the likelihood that a granule from the cell interior 'sees' the cell membrane. The probability of exocytosis occurring may thereby be raised. A more rigorous interpretation along these lines is probably not justified, however, until consideration can be given to the modifying influence of particle 'crowding' on the viscosity of granule suspensions, to the presence of absorbing barriers on the random walk process, and to application of target theory (Matthews, unpublished).

It may be noted here that no account has been taken of convective cytoplasmic streaming which, if it occurs, would partially direct granule movement; introduction of such a positive vectorial component would require some modification of the initial assumptions. Granule traffic flow of an essential random nature could, nevertheless, adequately account for movement of a granule to a strategic position from whence it is able to reach the cell membrane for exocytosis. The time course of translocation is certainly not incompatible with that of secretion.

Whether or not a granule making a close approach to the inner surface of the cell membrane successfully fuses with it for release of the hormone depends on a number of factors. Similar factors may govern the release of neurotransmitter from the nerve terminal (Blioch, Glagloleva, Liberman & Nenashev, 1968).

Consideration must be given primarily to the question of surface charge on granule and cell membrane. The cell membrane constitutes a charged mosaic. All cells under physiological conditions possess a net negative surface charge, with electrophoretic potentials in the range -9 to -25 mV (Pethica, 1961). The predominant negative charge probably arises from ionized carboxyl and mono- and dibasic phosphate groups of membrane protein and phospholipid, respectively. Because of its symmetrical arrangement there seems little reason to doubt that a similar charge exists on the internal surface of the membrane although its magnitude may not precisely correspond to that of the external surface. The inner surface of the squid axon membrane for example possesses a potential of -17 mV (Meves, 1966).

The hormone storage granules are enclosed by a limiting membrane which probably also carries a net negative surface charge. This is certainly so in the case of the granules from chromaffin and neurohypophyseal cells (Banks, 1966; Poisner & Douglas, 1968).

Adhesion between granule and cell membrane surfaces will depend ultimately on the balance of attractive and repulsive forces between them. This is best considered in terms of Verwey & Overbeek's theory (1948) of colloidal stability which suggests that surface contacts are stabilized when attractive London–van der Waal's dispersion forces counterbalance the electrostatic repulsive forces arising from interaction of the diffuse electrical double layer at each surface. The electrostatic double layer originates from the ionic atmosphere surrounding a charged surface and its 'thickness' (designated $1/\kappa$) is dependent on ionic valency and concentration.

At low surface potentials (i.e., $\psi_0 < 25$ mV) the Debye–Hückel constant κ is given by:

$$\kappa = \sqrt{\frac{8\pi Ne^2c}{1000DkT}}$$

where N is Avogadro's number, e the electronic charge, c the ionic strength, D the solvent dielectric constant, k Boltzmann's constant, and T the absolute temperature.

Now, at 37°C and assuming $D = 74$ (water)

$$\kappa = 0.332 \times 10^8 \sqrt{c}$$

where c is the molar concentration of electrolyte. Thus, assuming a univalent cytoplasmic ion concentration of 0·16 M (for example, KCl) and complete ionization, the double layer thickness $1/\kappa = 7.52 \times 10^{-8}$ cm or 7·5 Å. A granule approaching the cell membrane will thus encounter a repulsive force as its electrical double layer begins to interact with that of the cell membrane, at a distance of separation of some 15 Å (2×7.5 Å), making the simplifying assumption that the surfaces are identically charged. In fact, if the surface potential of either cell membrane (ψM) or granule membrane (ψG) is much greater than -25 mV the distance of separation at which repulsive interaction occurs will be greater, that is, the repulsions at long range are increased as either ψM or ψG increases. For two surfaces of identical potential the electrostatic repulsion at all ranges depends on ψ^2.

If the attractive forces exceed the repulsive forces only at very close approach, adhesion will occur only if the potential energy barrier of repulsive forces can be surmounted. The granule can overcome this barrier if its kinetic energy derived from Brownian motion is greater than that of the potential energy barrier. The average kinetic energy (calculated from the Maxwell–Boltzmann distribution) is $\frac{3}{2}kT = 6.4 \times 10^{-14}$ erg at 37°C. In the non-stimulated cell this is probably insufficient to overcome the potential energy barrier generated by the cell membrane. Occasionally granules of 50–400 mμ may acquire translational energies of up to $5kT$ (and very rarely $50kT$), however, which would be sufficient to carry them through quite large repulsive energy barriers, for example, $1000kT$ for the whole cell surface (see Curtis, 1967), because the area of interaction between cell membrane and granule is such a small fraction of the total membrane area. This may account for the 'quantal' or 'basal' release of transmitter or hormone. During stimulation of a secretory cell the probability that exocytosis occurs markedly increases. In order to facilitate exocytosis the high potential energy barrier to granule/cell membrane interaction must be reduced. This requirement could best be met by a local decrease in the surface potential and compression of the electrostatic double layer of the cell, and possibly also of the granule, membrane. To avoid excessive and abortive intracellular aggregation of granules in the vicinity of the cell membrane it is necessary to suppose a lesser effect on the granule membrane; alternatively this membrane may in any case possess a different charge array and a greater charge density.

Now it is known from the work of Overbeek (1952) that divalent (and trivalent) ions are particularly effective in reducing ψ and $1/\kappa$. In terms of promoting adhesion the molar ratio for mono-, di-, and trivalent ions is approximately:

$$100 : 1.6 : 0.13$$

The participation of calcium ions in the secretory event might then be envisaged as follows. For a secretory cell with a transmembrane potential of -20 to -70 mV developed across a membrane thickness of 100 Å, the high field strength of 2 to 7×10^5 V m^{-2} may be expected to maintain dipole orientation and ionization of macromolecular fixed charges, particularly at the exposed membrane surfaces. Ion pair formation would also tend to be prevented by the intense electric field encouraging charge separation. Within the cytoplasm the ionized calcium level is normally low. In these circumstances calcium ions are generally unavailable for promotion of granule/cell membrane adhesion although a few isolated 'target' areas might exist. Depolarization of the cell by an appropriate stimulus will lead to a collapse of the field towards zero. Calcium ions previously structurally associated with integral polymeric complexes of the cell membrane matrix may now be displaced and become available at the inner surface of the membrane together with those entering the cell down the electrochemical gradient. The net negative charge of the inner surface of the membrane, arising mainly from exposed carboxyl and phosphate groups of constituent polydentate macromolecules, will decrease as calcium (more effectively than Na$^+$ or K$^+$ with which it may compete) binds to these ligands. The surface potential ψM therefore now tends to zero, and may even transitorily reverse, the electrostatic double layer disappears, and the potential energy barrier to granule-cell membrane interaction collapses. The probability that an encounter of granule with cell membrane now results successfully in adhesion is thereby greatly increased. As the granule can now penetrate to very close approach, e.g., 5 Å, adhesion may be aided by cross-linking of cell and granule membranes resulting from ion triplet formation, for example,

$$-COO^- \cdots Ca^{2+} \cdots {}^-OOC-$$

Desolvation occurs as the divalent ion reduces the hydration of the acidic carboxyl groups, and other ligands, leading to a more effective interaction of local dispersion forces—greater attraction. This promotes stabilization of the adhesion and resistance to the tendency of kinetic translational energy to dislodge the granule from the cell membrane. Fusion is thus encouraged and the hormone is released into the extracellular space through the cell membrane. It seems more than likely, however, that some enzymatic event is involved in the final stage of exocytosis. This arises from the discovery that (i) chromaffin and neurohypophyseal granules contain an ATPase, and (ii) ATP releases hormone from isolated granules (Poisner & Trifaró, 1967; Poisner & Douglas, 1968). With this evidence in mind Poisner & Trifaró (1967) proposed a model for the release of catecholamines from the adrenal medulla which involves enzyme activation at the chromaffin cell membrane as a final essential step in the hormone release sequence (Fig. 14). Such a model does not entirely explain the specificity of calcium in the secretory process because in physico-chemical properties the calcium

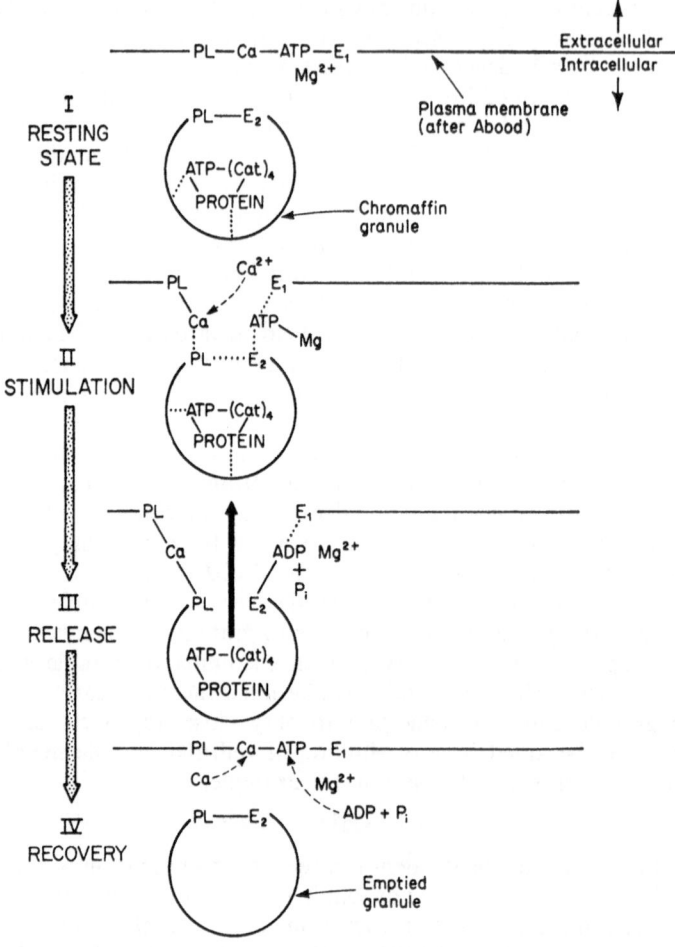

FIG. 14. Model for the molecular events associated with catecholamine secretion from the adrenal medulla. Cell membrane contains phospholipid (PL), ATP, Ca, Mg, and a Na + K activated ATPase (E_1). Granule membrane contains phospholipid (PL) and a Mg-activated ATPase (E_2). Catecholamines (Cat) stored within granule complexed with ATP and protein; complex weakly bound to inner surface of granule membrane. (From Poisner & Trífaró, 1967.)

and magnesium ions are very similar, yet magnesium powerfully blocks the secretory process. Magnesium may compete with calcium for uptake (Rubin, Feinstein, Jaanus & Paimre, 1967), possibly at the initial site of entry at the cell membrane, and it may conceivably interfere with the utilization of calcium as an essential cofactor in some enzymatic reaction at or near the membrane. Full explanation of this subtle ionic interplay must await determination of the precise geometric arrangement of the discriminatory receptive ligands which form so vital a part of the membrane mosaic.

REFERENCES

BANKS, P. (1966). *Biochem. J.*, **101**, 18c

BIANCHI, C. P. & SHANES, A. M. (1959). *J. gen. Physiol.*, **42**, 803

BINDLER, E., LABELLA, F. S. & SANWAL, M. (1967). *J. cell Biol.*, **34**, 185

BLIOCH, Z. L., GLAGOLEVA, I. M., LIBERMAN, E. A. & NENASHEV, V. A. (1968). *J. Physiol., Lond.*, **199**, 11

DEL CASTILLO, J. & STARK, L. (1952). *J. Physiol., Lond.*, **116**, 507

CHAMBERS, R. & KAO, C. Y. (1952). *Exp. cell Res.*, **3**, 564

COUPLAND, R. E. (1965). In *The Natural History of the Chromaffin Cell*. London: Longmans, Green

CURTIS, A. S. G. (1967). In *The Cell Surface: its Molecular Role in Morphogenesis*. London: Logos Press; Academic Press

DANIEL, A. R. & LEDERIS, K. (1963). *Gen. comp. Endocr.*, **3**, 693

DEAN, P. M. & MATTHEWS, E. K. (1968). *Nature, Lond.*, **219**, 389

DEAN, P. M. & MATTHEWS, E. K. (1969). *J. Physiol., Lond.*, in the Press

DICKER, S. E. (1966). *J. Physiol., Lond.*, **185**, 429

DOUGLAS, W. W. (1968). *Br. J. Pharmac.*, **34**, 451

DOUGLAS, W. W. & POISNER, A. M. (1962). *J. Physiol., Lond.*, **162**, 385

DOUGLAS, W. W. & POISNER, A. M. (1964). *J. Physiol., Lond.*, **172**, 19

DOUGLAS, W. W. & POISNER, A. M. (1966). *J. Physiol., Lond.*, **183**, 236

DOUGLAS, W. W. & RUBIN, R. P. (1961). *J. Physiol., Lond.*, **159**, 40

GINSBURG, M. & IRELAND, M. (1966). *J. Endocrinol.*, **35**, 289

GRODSKY, G. M. & BENNETT, L. L. (1966). *Diabetes*, **15**, 910

HAGIWARA, S. & NAKAJIMA, S. (1966). *J. gen. Physiol.*, **49**, 793

HALES, C. N. & MILNER, R. D. G. (1968). *J. Physiol., Lond.*, **199**, 177

HARVEY, A. M. & MACINTOSH, F. C. (1940). *J. Physiol., Lond.*, **97**, 408

HEILBRUNN, L. V. (1956). In *The Dynamics of Living Protoplasm*. New York: Academic Press

HODGKIN, A. L. & KATZ, B. (1949). *J. exp. Biol.*, **26**, 292

HODGKIN, A. L. & KEYNES, R. D. (1957). *J. Physiol., Lond.*, **138**, 253

HÖKFELT, T. (1968). *Z. Zellforsch.*, **91**, 1.

HUBBARD, J. I. & KWANBUNBUMPEN, S. (1968). *J. Physiol., Lond.*, **194**, 407

HUKOVIČ, S. & MUSCHOLL, E. (1962). *Arch. exp. Path. Pharmak.*, **244**, 81

HYMER, W. C. & MCSHAN, W. H. (1963). *J. cell Biol.*, **17**, 67

ISHIDA, A. (1968). *Jap. J. Physiol.*, **18**, 471.

KIRPEKAR, S. M. & MISU, Y. (1967). *J. Physiol., Lond.*, **188**, 219

KIRSCHNER, N., SAGE, H. J. & SMITH, W. J. (1967). *Mol. Pharmac.*, **3**, 254

LACY, P. E., HOWELL, S. L., YOUNG, D. A. & FINK, C. J. (1968). *Nature, Lond.*, **219**, 1177

LAZARUS, S. S., SHAPIRO, S. N. & VOLK, B. W. (1967). *Lab. Invest.*, **16**, 330

MALAMED, S., POISNER, A. M., TRIFARÓ, J. M. & DOUGLAS, W. W. (1968). *Biochem. Pharmac.*, **17**, 241

MATTHEWS, E. K. & DEAN, P. M. (1969). In *The Structure and Function of Pancreatic Islets*, ed. FALKMER, S., HELLMAN, B. & TÄLJEDAL, I.-B. London: Pergamon, in the Press

MEVES, H. (1966). *Ann. N.Y. Acad. Sci.*, **137**, 807

MEYER, J. & BENCOSME, S. A. (1965). *Rev. can. Biol.*, **24**, 179

MILNER, R. D. G. & HALES, C. N. (1967). *Diabetologia*, 3, 47

OVERBEEK, J. TH. G. (1952). In *Colloid Science I*, ed. KRUYT, H. M. Amsterdam: Elsevier

PETHICA, B. A. (1961). *Exp. cell Res. Suppl.*, **8**, 123

PERDUE, J. F. & MCSHAN, W. H. (1966). *Endocrinology*, **78**, 406

POISNER, A. M. & DOUGLAS, W. W. (1968). *Mol. Pharmac.*, **4**, 531

POISNER, A. M. & TRÎFARÓ, J. M. (1967). *Mol. Pharmac.*, **3**, 561

PORTER, K. R. (1966). In *Principles of Biomolecular Organization*, ed. WOLSTEN-HOLME, G. E. W. & O'CONNOR, M., p. 308. London: Churchill

RIESER, P. (1949). *Biol. Bull.*, **97**, 245

RUBIN, R. P., FEINSTEIN, M. B., JAANUS, S. D. & PAIMRE, M. (1967). *J. Pharmac. exp. Ther.*, **155**, 463

SACHS, H. (1967). *Am. J. Med.*, **42**, 687

SACHS, H. & HALLER, E. W. (1968). *Endocrinology*, **83**, 251

SAMLI, M. H. & GESCHWIND, I. I. (1968). *Endocrinology*, **82**, 225

SCHNEIDER, F. H., SMITH, A. D. & WINKLER, H. (1967). *Br. J. Pharmac. Chemother.*, **31**, 94

SHEA, S. M. & KARNOVSKY, M. J. (1966). *Nature, Lond.*, **212**, 253

SIPERSTEIN, E. R. (1963). *J. cell Biol.*, **17**, 521

SIPERSTEIN, E. R. & ALLISON, V. F. (1965). *Endocrinology*, **76**, 70

THORN, N. A. (1965). *Acta endocr., Copenh.*, **50**, 357

VALE, W., BURGUS, R. & GUILLEMIN, R. (1967). *Experientia*, **13**, 853

VALE, W. & GUILLEMIN, R. (1967). *Experientia*, **13**, 855

VERVEY, E. J. W. & OVERBEEK, J. TH. G. (1948). *Theory of the Stability of Lyophobic Colloids.* Amsterdam: Elsevier

WHITTAKER, V. P. (1965). *Progr. Biophys. Mol. Biol.*, **15**, 39

WHITTAKER, V. P. (1968). *Br. med. Bull.*, **24**, 101

SITE OF PROTEIN SECRETION AND CALCIUM ACCUMULATION IN THE POLYMORPHONUCLEAR LEUCOCYTE TREATED WITH LEUCOCIDIN

A. M. WOODIN & ANTONNETTE A. WIENEKE

External Staff of the Medical Research Council
Sir William Dunn School of Pathology, University of Oxford

The two proteins that constitute leucocidin induce changes in the polymorphonuclear leucocytes of rabbit and man that mimic those in excitable or secreting tissues during membrane depolarization or stimulation by hormones (Woodin, 1968). This was known to be true in a descriptive sense several years ago (Woodin & Wieneke, 1966a), and recent work suggests that it is also true of the mechanisms. Thus diisopropyl phosphorofluoridate (DFP) and tetraethylammonium ions modify the effects of leucocidin and the axon in similar conditions (Woodin & Wieneke, 1968a, 1969).

Calcium is not involved in the primary action of leucocidin on the leucocyte although, if it is present in the medium, it enables DFP to enhance the primary action of leucocidin some 4-fold. Calcium also modifies the subsequent response of the cell. These effects stem in the main from the secretion of the granule proteins with the conversion of the granules to vesicles in which calcium is accumulated. The analogy between these phenomena, secretion from exocrine glands and accumulation of calcium in muscle has been drawn elsewhere (Woodin, 1968). This article is concerned with the site at which calcium acts on the cell. A brief outline of the leucocidin-treated cell is given and then the mechanism of the three calcium-mediated reactions is described.

Character of the leucocidin-treated cell

The sequence of events when leucocytes are treated with leucocidin is shown in Fig. 1. Here the cell at the top left-hand corner represents the normal leucocyte, the leucocyte cell membrane is represented at the top right-hand side, and the other two cells are leucocytes in various stages of responding to leucocidin,

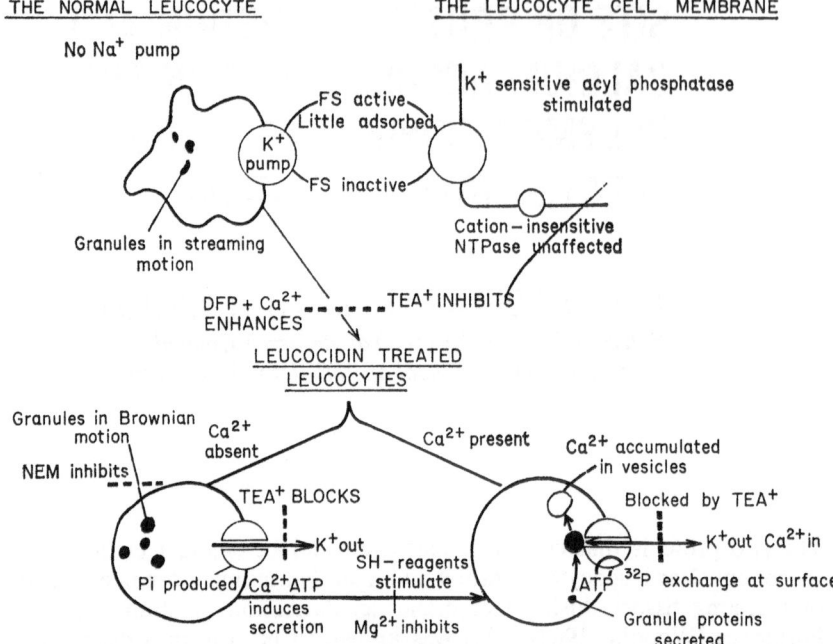

FIG. 1. Properties of the leucocidin-treated leucocyte. For explanation see text.

(a) *The normal leucocyte.* This lacks mitochondria or an endoplasmic reticulum but possesses numerous granules with the enzymic properties of lysosomes. The regulation of the electrolyte balance is of particular relevance to the action of leucocidin. Elsbach & Schwartz (1959) showed that leucocytes at 0°C lose potassium and gain sodium and that subsequent incubation restores the potassium content but does not completely extrude the sodium; the cells remain somewhat swollen. The leucocyte appears to have a defective sodium pump and controls its electrolyte balance by regulating the potassium content. The properties of the cation-sensitive phosphatases of the isolated membrane correlate with these peculiarities (Woodin & Wieneke, 1968b). The ATPase is a non-specific nucleoside triphosphatase or diphosphatase which is insensitive to sodium or potassium or ouabain. The membrane possesses an acyl phosphatase, however, which is stimulated by potassium but loses this sensitivity in the presence of ouabain.

In this article the concept of the 'ion pump' will be used to refer to the physiological mechanism for moving ions. The phosphatases of the cell membrane are considered to be only part of this mechanism. Indeed, the action of leucocidin, DFP and tetraethylammonium ions suggests that there is coupling in the potassium pump between a phosphatase, a phosphorylase and an ion channel. The concept of 'ion channel' is used to refer to a pathway for ion movement without specifying if movement is by thermal diffusion or otherwise.

(*b*) *The primary action of leucocidin.* Leucocidin produces a structural change in the potassium pump of the leucocyte cell membrane. In the isolated membrane this manifests itself as a stimulated potassium-sensitive acyl phosphatase. In the intact cell it is manifested as an increased permeability to cations and an accumulation of orthophosphate in the cytoplasm. In the course of acting on the potassium pump leucocidin is converted to an inactive form but little is adsorbed; the inactivated leucocidin remains in solution and has the same sedimentation behaviour as the active form. There is no evidence for an alteration of the covalent structure of either the membrane or leucocidin, and it is probable that the interaction is a reciprocal conformational change. The only compound found to simulate the inactivation of leucocidin by the leucocyte membrane is triphosphoinositide and the evidence that this is present at the site of action of leucocidin will be discussed later.

The hypothesis that the interaction of leucocidin and the cell membrane is conformational is supported by the shape of the dose response curve (Fig. 2). This shows a plateau of 'no response' for low amounts of leucocidin. It seems that the cell is able to reverse the effect produced by small amounts of leucocidin, and because collision of leucocidin with the cell leads to leucocidin inactivation, multiple collisions cannot mimic the effect of high concentrations.

(*c*) *The response of the cell to leucocidin.* While the leucocidin-treated cell is permeable to cations, in the presence of calcium many small molecules are retained within the cell at 5–10 times the concentration in the medium. In the presence of EDTA the cell becomes slightly leaky and small amounts of cytoplasmic protein can be found in the medium (Woodin, 1961).

Morphologically the earliest change concerns granule movement. Light microscopy shows that the granules of the normal cell move in an orderly streaming fashion and that immediately leucocidin is added this changes to a random Brownian motion. The granules remain visible under the light microscope for more than 10 min if EDTA is present but they then become invisible. If calcium is present the disappearance of the granules is complete in a minute or so. Electron microscopy shows that the granules of the leucocidin-treated cell in the presence of EDTA are intact though swollen and filled with protein, the granules of leucocidin-treated cells in the presence of calcium are absent, and the cytoplasm contains numerous vesicles (Woodin, French & Marchesi, 1963).

The movement of the granules in the leucocidin-treated cell is modified in an interesting fashion by N-ethyl maleimide. This reagent (1 mM)

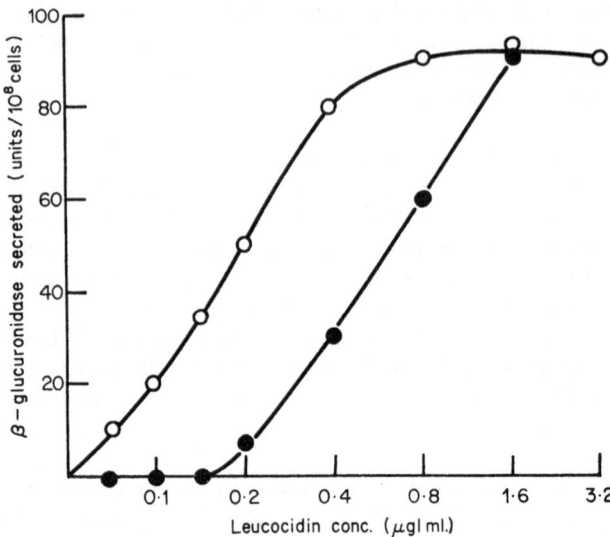

FIG. 2. Enchancement of the secretion of β-glucuronidase by DFP. Leucocytes in Hanks BSS were maintained alone (●), or with 5 mM DFP (○), for 20 min at room temperature and then added to tubes containing serially diluted leucocidin. After 10 min at 37°C the β-glucuronidase in the cell supernatant was determined.

changes the state of the cytoplasm if calcium is absent from the medium; the granules stop moving, they do not swell and are visible, immobilized in the cytoplasm for at least 20 min. This reaction may require the penetration of N-ethyl maleimide into the cytoplasm, for when calcium is present and the normal impermeability is maintained the granules are not immobilized by N-ethyl maleimide; in fact this reagent stimulates the secretion of the granule proteins in these conditions (Woodin & Wieneke, 1966a).

The chief biochemical changes observed when calcium is present are the secretion of the granule proteins, the accumulation of calcium within the cell, and a nucleotide-orthophosphate reaction at the cell surface. None of these reactions occurs if cells are treated with leucocidin in the absence of calcium but they can be induced by calcium addition. This device enables reagents that act on the mechanism of protein secretion, for example, to be distinguished from those acting on the primary action of leucocidin. Magnesium and sulphydryl reagents have been shown to modify protein secretion in this way (Woodin & Wieneke, 1964, 1966a). The same experimental device enables demonstration of the non-participation of leucocidin in the Ca-dependent mechanisms. Thus if leucocytes are treated with leucocidin in the absence of calcium and then the leucocidin is neutralized with antibody, subsequent addition of calcium induces secretion of protein, accumulation of calcium, and establishes the orthophosphate-nucleotide exchange reaction. In the leucocidin-treated cell the molecular events associated with secretion can be completely separated from those associated with excitation.

(*d*) *Inhibition and stimulation of leucocidin.* The device used to identify inhibitors of the calcium-dependent reactions in the cell also enables inhibitors and stimulators of leucocidin to be identified. Tetraethylammonium ions inhibit all the effects of leucocidin if they are present before leucocidin is added but not if they are added after leucocidin but before the calcium-dependent reactions are induced. Similarly, DFP enhances all the actions of leucocidin if it is present at the same time as leucocidin but not if it is added afterwards. The fact that all the effects of leucocidin are changed by these reagents is taken as evidence that DFP and tetraethylammonium ions have actions at the same site as leucocidin, and that they act on the potassium pump. The action of tetraethylammonium ions does not require the presence of calcium and will not be discussed here.

Reactions mediated by calcium in the leucocidin-treated leucocyte

(*a*) *The enhancement of leucocidin by DFP.* DFP acts on the leucocidin-treated leucocyte in conditions similar to those in which it inhibits conduction in the axon and chemotaxis by leucocytes. In all these phenomena the concentration of DFP (1–5 mM) is several orders higher than that required to inhibit esterases and the effect is partly reversible. If DFP is washed away from leucocytes the enhancement of leucocidin is greatly reduced, the inhibition of chemotaxis is reduced and, if it is washed away from the axon, conduction proceeds.

Figure 2 shows that DFP produces a 4-fold enhancement of the release of β-glucuronidase induced by leucocidin. A similar result was found when the action of leucocidin was assessed from the production of orthophosphate or the accumulation of calcium. DFP enhanced leucocidin only when calcium was present in the medium. It acts on the primary action of leucocidin. If DFP is added to the cells after leucocidin is neutralized with antibody no change is observed. The shape of the dose-response curve is altered by DFP (Fig. 2) and as the plateau of 'no response' at low leucocidin concentrations is lost it appears that DFP prevents the cell from reversing the action of low leucocidin concentrations.

The analysis of the effect of DFP on the action of leucocidin indicated that DFP acted on the potassium pump. This was confirmed by showing that DFP reduced the rate of potassium reaccumulation in normal leucocytes which was depleted by cooling. DFP does not alter the activity of the acyl phosphatase of the isolated cell membrane, and it acts on the pump in a different fashion from leucocidin.

These observations led us to suggest that DFP acts by competing with a natural substrate in a phosphorylation reaction. Because calcium is required for the enhancement of leucocidin it is possible that it is necessary for the phosphorylation of the leucocyte potassium pump. This will remain speculative until the proposed phosphorylation site with which DFP reacts can be identified. A detailed account of the effect of DFP is given by Woodin & Wieneke (1969).

(b) *Protein secretion in the leucocyte.* In early experiments it was found that although the leucocidin-treated cell was impermeable to many small molecules, large amounts of protein were present in the cell supernatant. Protein was not found in the supernatant of macrophages after leucocidin treatment and, as these cells do not possess the characteristic granules of the leucocyte, this suggested that the secreted protein came from these granules. This was confirmed by showing that the specific activity of β-glucuronidase, ribonuclease, deoxyribonuclease, peroxidase, and the antibacterial substance phagocytin in the leucocidin-treated cell supernatant was similar to that of lysed granules.

There is little doubt that the route of secretion is by exocytosis (fusion of the granules with the cell surface membrane) and not by intracellular lysis followed by diffusion through the cell membrane. Homogenates of leucocidin-treated cells do not contain the enzymes found in the granules of the normal cell and, judged by the distribution of aldolase, cytoplasmic proteins do not leak out of the cell. The supernatant of the leucocidin-treated cell contains the enzyme peroxidase, yet lysis of the granules does not solubilize it, and if granules from normal cells are added to the supernatant of leucocidin-treated cells all the peroxidase is adsorbed. It is difficult to understand how secretion following intracellular lysis could by-pass this adsorption. Further evidence for the exocytosis route is provided by the inhibition of secretion when N-ethyl maleimide prevents contact of the granules and the cell surface; yet this reagent stimulates secretion when calcium is present and immobilization of the granules does not occur.

Morphological evidence for exocytosis has been provided. Under the light microscope the granules in the leucocidin-treated cell are in random motion but this stops more rapidly when calcium is added. Electron micrographs by Woodin, French & Marchesi (1963) and Woodin & Wieneke (1964) showed some examples of granules fused to the cell surface membrane, and in some cases the membrane at the point of contact had broken. These were not numerous, and Woodin, French & Marchesi (1963) pointed out that the final appearance of exocytosis in a free cell as opposed to a fixed tissue would be difficult to distinguish from an invagination of the surface. The rarity of the sites of fusion would be immediately explained by the suggestion made below that fusion occurs only at the potassium pump of the cell membrane. It has been suggested that in erythrocytes, for example, there may be as few as 1000 sites per cell concerned with the transport of ions.

It may not be necessary for breakage of the fused membranes to precede exocytosis. Lowenstein (1966) has shown that the membranes separating adjacent cells are freely permeable to protein, provided they are kept free from calcium. It is possible that all that is necessary is the elimination of calcium from the site of fusion of the granule and the cell membrane. Woodin & Wieneke (1964) had already concluded that elimination of calcium from the site of fusion is necessary. Secretion from a permeable but unbroken granule in contact with the cell membrane would also permit some selectivity in the size of the proteins that are allowed to escape. Secretion without membrane breakage also explains the observation of Woodin, French & Marchesi (1963) that if ferritin or colloidal gold are included in the medium into which the leucocidin-treated cell secretes protein, then these large particles are not found in the vesicles that re-enter the cytoplasm.

(c) *The participation of calcium in secretion.* The secretion of the granule proteins is dependent on the presence of calcium. If leucocidin-treated cells are prepared in the absence of calcium, secretion can be induced by adding calcium. The efficiency of calcium as an inducer of protein secretion decreases as the time of incubating the leucocidin-treated cells in calcium-free media is increased, but the efficiency can be restored by adding certain nucleoside triphosphates or diphosphates. Nucleoside monophosphates are not active. There is a complex relationship between the concentrations of calcium and ATP which are required to give maximum secretion; although both are required, both inhibit in high concentrations. These observations enabled Woodin & Wieneke (1964) to propose that when calcium is present in the medium, sites are produced for the adherence of the granules and the surface membrane, but that when adherence has occurred calcium must be removed from the site of contact. In this way a localized region of reduced mechanical strength and increased permeability is produced at the site of contact. The ATP was considered to function in the removal of the calcium by providing a high concentration of ortho-phosphate following its hydrolysis by the granule ATPase. Evidence for the production of calcium phosphate is given below. When secretion was stimulated with ATP^{32} no radioactivity was found associated with the lipids or proteins but most was recovered as orthophosphate. Now that the rarity of the sites of fusion of the granules and the membrane is clear, however, a further investigation will be necessary to show that a rare membrane component such as triphosphoinositide is not phosphorylated when labelled ATP is applied in this way.

(d) *Inhibition of protein secretion.* The parallel between secretion in the leucocyte and in exocrine glands is well illustrated by the action of inhibitors. Reagents such as cholesterol or cortisone that stabilize lysosomes against the lytic action of vitamin A or streptolysin O are without effect on the secretion of the leucocyte granule proteins, while magnesium which inhibits acetylcholine release, and sulphydryl reagents that stimulate vaso-pressin secretion, have parallel actions on the leucocyte. These effects are found when the reagents are added after leucocidin has been neutralized by antibody and before secretion is induced by calcium. DFP and TEA are without effect when tested in this way. Their action when added before leucocidin is due to an action on the direct effects of leucocidin and not on the mechanism of protein secretion.

(e) *The fate of the granules.* Electron micrographs of leucocidin-treated cells prepared in the presence of calcium and fixed with osmium show vesicles in the cytoplasm although they are not as numerous as the granules (see Fig. 1 of Woodin, French & Marchesi, 1963). More vesicles can be seen in preparations prepared by fixation with permanganate and lead (see Plate 1 of Woodin & Wieneke, 1964). It is probable that the vesicles do not survive fixation with osmium when they are in the cell. Vesicles can be recovered from leucocidin-treated cells by subcellular fractionation (Plate 1), however, and analysis of their enzyme content is consistent with their being derived from granules which have secreted some of their pro-teins. Thus, compared with the granules, the vesicles have little β-glucu-ronidase or ribonuclease but nearly as much acid and alkaline phosphatase. This correlates with the high β-glucuronidase and ribonuclease activity and the low phosphatase activity of the leucocidin-treated cell supernatant (Woodin, 1962).

It is difficult to conceive of a different origin for the vesicles. The normal leucocyte has little membrane material in the cytoplasm, the cell does not contract during secretion and the nuclear membrane remains intact. The time scale of secretion completely excludes the possibility of *de novo* synthesis.

The vesicles were found to be rich in calcium phosphate and, as calcium is essential for protein secretion, this suggested that the site of intercon-version of the granules and vesicles might be determined by studying the uptake of calcium.

(f) *The accumulation of calcium in the leucocyte.* When leucocytes are treated with leucocidin in the presence of calcium this is accumulated within the cell (1–2 μg Ca/mg dry weight). A similar amount of calcium is bound by disintegrated leucocytes, and this led to the suggestion that the accumulation was due to entry of calcium into the cytoplasm and its being bound by structural components in an indiscriminate way (Woodin & Wieneke, 1963b). It is now clear that this interpretation is wrong, and the correlation between the binding of calcium by disintegrated and leucocidin-treated leucocytes is fortuitous.

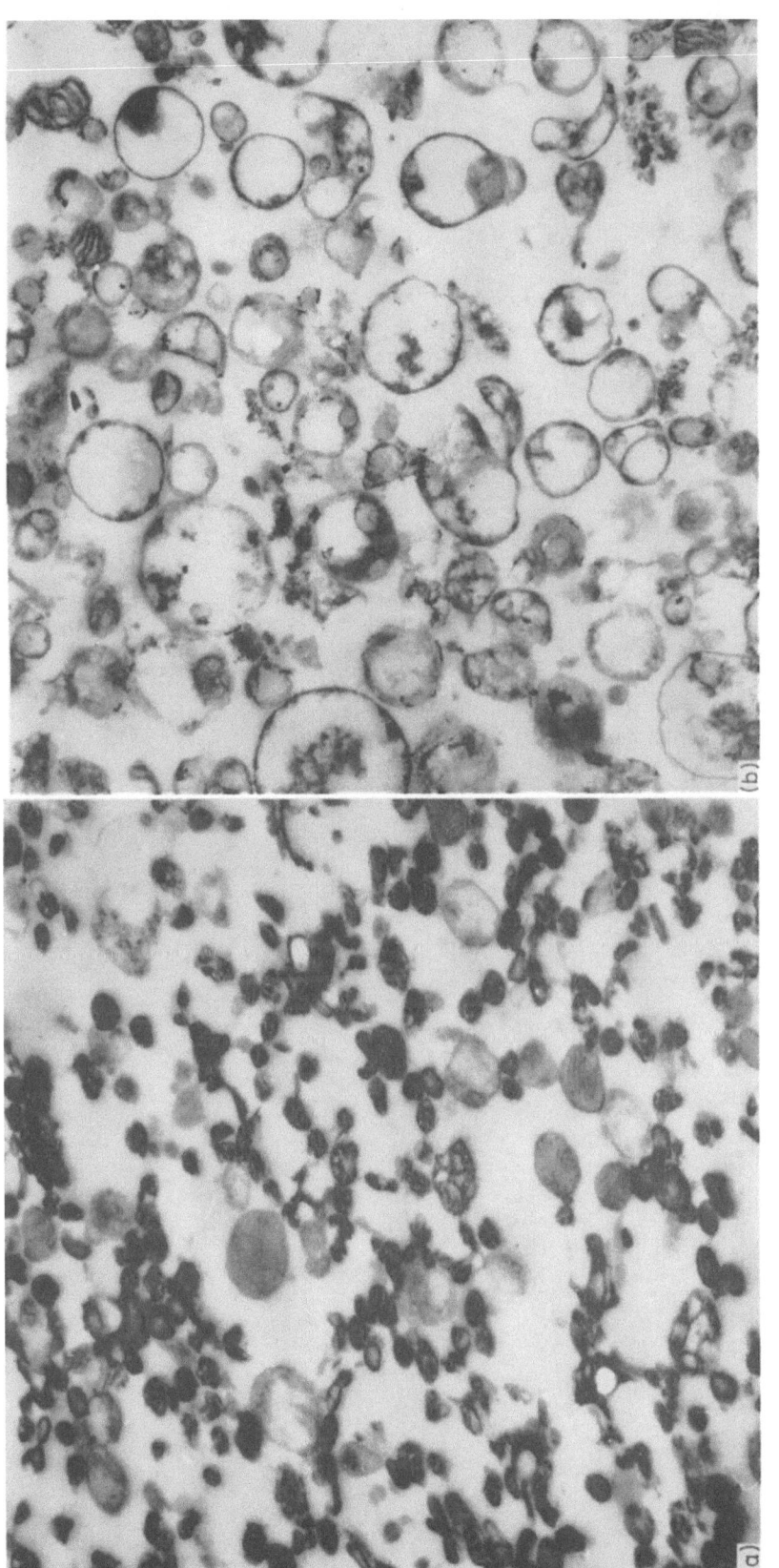

PLATE 1. The granules of the normal leucocyte (a) and the vesicles of the leucocidin-treated leucocyte (b). Fixed in buffered osmium solution.

The calcium that is accumulated by leucocidin-treated cells is confined to the vesicles. Woodin & Wieneke (1964) showed that vesicles derived from leucocidin-treated cells were rich in calcium and orthophosphate. It is not possible to prepare the vesicles or granules quantitatively after the leucocytes have been incubated in media containing calcium; some of the granules and vesicles stick to material that sediments at a lower speed. Table 1 describes an experiment where the recovery of accumulated calcium is correlated with the distribution of the granule enzyme peroxidase and suggests that the accumulation of calcium could be accounted for exclusively in the vesicle fraction. Electron microscopy has shown that this is not an artefact of the isolation procedure. If leucocytes or leucocidin-treated leucocytes prepared in the absence of calcium are fixed with permanganate and stained with lead, the granules appear empty. Either the protein does not react with permanganate and lead or the fixed protein dissolves under the very alkaline conditions of the lead staining. Cells treated with leucocidin in the presence of calcium and prepared in this way, however, show under the electron microscope very electron-dense material in the granules that is absent if the sections are extracted with dilute citric acid before the lead staining (Plate 1, Woodin & Wieneke, 1964).

The accumulation of calcium in the vesicles of the leucocidin-treated cell is not due to a general increase of permeability and penetration of calcium into the cytoplasm. Figure 3 shows that on prolonging the incubation of leucocidin-treated cells in the absence of calcium the amount of calcium that is accumulated on its subsequent addition is decreased. If the calcium uptake were due to increased permeability as a consequence of cell injury the reverse would be expected. After incubation in the absence

TABLE 1. *The distribution of calcium in subcellular fractions of leucocytes*

Subcellular fraction	Normal cells		Leucocidin-treated cells	
	Ca content, $nmol/10^8$ cells	Peroxidase content, % amount recovered	Ca content, $nmol/10^8$ cells	Peroxidase content, % amount recovered
Soluble fraction	Less than 1	Less than 1	3	Less than 1
Granules	7	82	—	—
Vesicles	—	—	85	68
Low speed residue	1·6	18	40	32

Leucocytes (10^8 cells) in 1 ml. of Hanks balanced salt solution containing 1·65 mM ^{45}Ca were incubated alone or with leucocidin (1 μg of each component) for 10 min and then washed with Hanks medium, calcium-free Hanks medium, and sucrose. Subcellular fractions were prepared by the method of Woodin (1961) and the radioactivity and peroxidase activity determined.

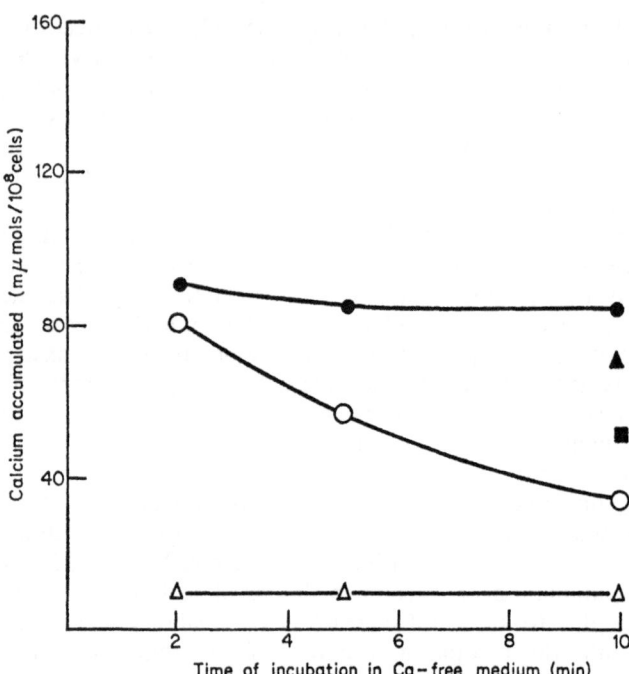

FIG. 3. Uptake of calcium by leucocytes. Leucocytes (5×10^7 cells) in 1 ml. of calcium-free Hanks medium were incubated alone or with 0·2 μg of leucocidin for various times and then 0·1 ml. of 10 mM $^{45}CaCl_2$ or 0·1 ml. of 10 mM ^{45}Ca-ATP added. The mixtures were incubated for a further 10 min and the radio-activity in the washed cell pellet determined. A further sample of cells was homogenized ultrasonically and treated with calcium at 0°C. The radioactivity in the washed particulate fraction was determined. △, Normal cells; ○, leucocidin-treated cells, $^{45}CaCl_2$ added; ●, leucocidin-treated cells, ^{45}Ca-ATP added; ■, leucocidin-treated cells, $^{45}CaCl_2$ added after homogenization; ▲, leucocidin-treated cells, $^{45}CaCl_2$ added at 0°C.

of calcium the level of calcium uptake can be restored by ATP. Figure 3 also shows that after incubation in the absence of calcium the uptake can be increased if the cells are homogenized or if the calcium is added at 0°C. The leucocidin-treated cell is apparently more permeable at 0° than at 37°C. The leucocidin-treated cell has ample binding sites for calcium in the cytoplasm, and as these are not occupied it follows that calcium does not penetrate generally into the cytoplasm. Following leucocidin action calcium occupies sites on the inside of the cell membrane, and this permits fusion with the granules.

The function of ATP in restoring the calcium uptake is not quite clear. Vesicles accumulate calcium at 0°C and so ATP is probably not involved in the entry of calcium into these. More probably ATP is involved in the removal of calcium from the inside of the cell surface membrane. Protein is not secreted at 0°C and, as secretion is stimulated by ATP, it is likely that removal of calcium from the membrane following ATP hydrolysis enables protein to pass out. This was postulated by Woodin & Wieneke (1964) and, because it is now clear that the vesicles are derived from the granules, the removal of calcium from the site of fusion would seem to be essential for the migration of the vesicles into the cytoplasm.

Mammalian cells generally are not permeable to nucleotides and, as the efflux of phosphate esters is not increased in the leucocidin-treated cell, (Woodin, 1961) it is probable that the added ATP is utilized at the surface. There is further evidence that the fusion of the granule and cell membrane enables the utilization of external substrates. If ^{32}P orthophosphate is present in the medium, the nucleotides of the leucocidin-treated cell prepared in the presence of calcium become 20- to 40-fold more radioactive than those of the normal cell. This is not due to increased turnover; if labelled cells are incubated in unlabelled medium the nucleotides of the leucocidin-treated cells lose radioactivity more slowly. The fusion of the granule and the membrane sets up a new orthophosphate-nucleotide exchange reaction utilizing external orthophosphate (Woodin & Wieneke, 1963a).

The pathway of calcium entry is suggested by the effect of tetraethylammonium ions. These (at a concentration of 80 mM) inhibit the accumulation of calcium by the leucocidin-treated cell, and that this is not due to inhibition of leucocidin action is shown by the competition between calcium and tetraethylammonium ions when they are added to leucocidin-treated cells prepared in the absence of calcium. Table 2 shows that the inhibition of calcium accumulation by tetraethylammonium ions is greater when the calcium concentration is smaller. Tetraethylammonium ions do not inhibit the nucleotide-orthophosphate exchange reaction, and it is probable that the reduction in calcium accumulation in the presence of tetraethylammonium ions is due to obstruction of the passage of calcium.

There is good evidence that the vesicles are derived from the granules and that the uptake of calcium in the leucocidin-treated cell can be accounted for by its being confined to the vesicles. There is evidence that the leucocidin-treated cell is not generally permeable to calcium so that accumulation of calcium occurs through contact with the cell surface membrane. As calcium is necessary for protein secretion, and ATP at similar concentrations stimulates secretion and calcium accumulation, it is probable that they occur at the same site.

It is possible that this site is close to the site of action of leucocidin. Tetraethylammonium ions inhibit all the actions of leucocidin, and this is taken as evidence that they have identical sites of action. If tetraethylammonium ions are added after leucocidin they impede the entry of calcium, and it is possible that the entry is by the channel induced by leucocidin. This evidence is admittedly slender. There are, however, other

TABLE 2. *Inhibition of calcium uptake by tetraethylammonium ions*

Calcium conc. (mM)	Calcium accumulated (nmol/10^8 cells)		% inhibition by TEA
	NaCl	TEA Cl	
0·1	6·8	2·6	62
0·2	12	5·8	52
0·5	21	9·0	56
1·0	27	19	29
2·0	37	38	0

Leucocytes (5 × 10^7 cells) in 0·25 ml. of calcium-free Hanks medium were incubated with 0·2 μg of each component of leucocidin for 10 min and then diluted with 0·75 ml. of Hanks medium or Hanks medium in which the NaCl had been replaced by tetraethylammonium chloride, both containing 0·2 mM $^{45}CaCl_2$. The mixtures were incubated for a further 10 min and the cells washed with 3 × 5 ml. cold Hanks medium. The radioactivity in the washed cell pellet was determined.

considerations in its favour. The fusion of the cell membrane and the granule membrane requires that the membrane should be unstable, and it is reasonable that this instability should be at the site of the excitatory agent.

(*g*) *A possible physiological function for triphosphoinositide.* As calcium does not penetrate into the cytoplasm but creates sites on the inside of the cell membrane for granule adherence, there must be a molecule on the inside of the cell membrane which is capable of complexing with calcium and impeding its further passage. Moreover, if as I have suggested this channel is in the potassium pump, the molecule should be rare, ubiquitous, and interact with tetraethylammonium ions and with leucocidin.

The last of these requirements provides a clue to the chemical nature of the receptor. Leucocyte cell suspensions and isolated cell membranes convert leucocidin to an inactive form of which little is adsorbed; the relative amounts of leucocidin and cell membrane causing inactivation of leucocidin and cytotoxic effects in the cell suggest that the inactivation reflects the change induced by leucocidin in the membrane (Woodin & Wieneke, 1966b). Woodin & Wieneke (1967a) found that several phospholipids can undergo complex interactions with leucocidin but that the only material to mimic the inactivation reaction is triphosphoinositide. This converts the S component of leucocidin to an inactive form at physiological salt concentrations but does not adsorb it. The F component of leucocidin is inactivated at low salt concentrations but as there is good reason to believe that the F component of leucocidin interacts with esterified fatty acids, this can be because only at low salt concentrations is access to the interior of the micelle possible. It is possible that triphosphoinositide is responsible for inactivation of leucocidin by the leucocyte membrane. Circumstantial evidence has also been provided that the properties of triphosphoinositide in the membrane are modified in a reciprocal fashion (Woodin & Wieneke, 1968a).

Triphosphoinositide is a rare molecule in the leucocyte membrane (0·6 mmol/mol phospholipid). Its rarity in other tissues has led others (cf. Prottey & Hawthorne, 1966; Sheltawney & Dawson, 1966) to suggest it is absent or restricted to myelin. However, as we have pointed out (Woodin & Wieneke, 1967b, 1968a), analyses of such rare molecules are not reliable unless they are controlled by isotope dilution methods.

Dawson (1970) has provided evidence elsewhere in this volume for the interaction of triphosphoinositide with calcium in preference to magnesium, potassium, and sodium in monolayers. It also occurs in micelles. When a solution containing a mixture of 2 μmol of potassium triphosphoinositide and 0·02 μmol of calcium triphosphoinositide labelled with ^{45}Ca was dialysed against 0·145 M NaCl solution which also contained a dialysis sac filled with potassium triphosphoinositide, equilibration of the ^{45}Ca ion between the triphosphoinositide solutions required 3 days while equilibration of ^{45}CaCl solution was complete in less than 2 hr.

Triphosphoinositide also interacts with tetraethylammonium ions in a novel way. If the calcium salt of triphosphoinositide is made by the Folch procedure the product contains 2 equiv. Ca/mol triphosphoinositide. If the free acid of triphosphoinositide is made by the Folch procedure, neutralized with NaOH or KOH, dissolved in water and dialysed, the products contain 4 equiv. cations/mol triphosphoinositide. However, if the tetraethylammonium salt is made in the same way, the product has only 1 equiv. tetraethylammonium ion/mol triphosphoinositide. Titration shows that the remainder of the phosphate groups are not ionized. It appears that tetraethylammonium ions are capable of altering the dissociation constants of the phosphate groups of the triphosphoinositide micelle in water.

One way in which this could arise would be if tetraethylammonium ions cross-linked the inositol phosphate moiety of triphosphoinositide with the esterified fatty acid side chains. Such cross-linking would change the correlation time of the esterified fatty acid side chains, an effect that could be detected by nuclear magnetic resonance spectroscopy.

It is possible that triphosphoinositide is the molecule responsible for the retention of calcium on the inside of the cell membrane. An essential preliminary to further investigation of this will be the demonstration of the ubiquity of triphosphoinositide by methods controlled by isotope dilution analysis.

(*h*) *The relevance of the leucocidin-treated cell to the properties of the normal leucocyte.* Protein secretion and calcium accumulation do not require the direct participation of leucocidin. so it follows that the normal cell must have the necessary reagents. The only conditions under which the granule proteins are secreted are injurious to the leucocyte; a reagent which can induce secretion of the granule proteins to the exterior under physiological conditions of the leucocyte has not been found. It is probable that the degranulation that follows phagocytosis corresponds physiologically to secretion induced by leucocidin. In phagocytosis the cell membrane surrounds the foreign particle to produce a vesicle which then moves into the cytoplasm. The granules then fuse with and inject their enzymes into the phagocytosis vesicles (Hirsch & Cohn, 1960; Zucker-Franklin & Hirsch, 1964). The accumulation of calcium observed in the leucocidin-treated cell may be of no physiological significance. In the degranulation after phagocytosis contact will not be made with the medium, and the only calcium to be eliminated would be that at the site of fusion of the granule and phagocytosis vesicle.

The function assigned to calcium in inducing secretion in the leucocyte could be served in other secreting tissues provided that the rate of secretion were faster than the rate of removal of calcium from the site of fusion. The intracellular accumulation of large amounts of calcium phosphate may be injurious to a cell in a physiological state and probably does not occur. The alternative to the hypothesis described here, the entry of calcium into the cytoplasm, and the interaction with the granule would require the demonstration that the granules can compete successfully with the mitochondria and nucleus in binding calcium. If this cannot be shown, it would seem to be obligatory to assume that calcium does not penetrate generally into the cytoplasm when secretion is stimulated.

In summary, an outline of the response of the leucocyte to leucocidin is given. There are three reactions mediated by calcium. In the presence of calcium DFP produces an enhancement of the primary action of leucocidin, the proteins of the cytoplasmic granules are secreted, and calcium is accumulated in vesicles. Evidence is described that calcium produces sites on the interior of the cell surface membrane with which the membranes of the granules fuse, and that calcium accumulation occurs into the vesicles produced from the granules after secretion of their proteins. The possible involvement of triphosphoinositide is discussed.

REFERENCES

DAWSON, R. M. C. & HAUSER, H. (1970). In *Calcium and Cellular Function*, ed. CUTHBERT, A., p. 17, London: Macmillan
ELSBACH, P. & SCHWARTZ, I. L. (1959). *J. gen. Physiol.*, **42**, 883
HIRSCH, J. G. & COHN, Z. A. (1960). *J. exp. Med.*, **112**, 1005
PROTTEY, C. & HAWTHORNE, J. N. (1966). *Biochem. J.*, **101**, 191
LOEWENSTEIN, W. R. (1966). *Ann. N.Y. Acad. Sci.*, **137**, 441
SHELTAWNEY, A. & DAWSON, R. M. C. (1966). *Biochem. J.*, **100**, 12
WOODIN, A. M. (1961). *Biochem. J.*, **80**, 562
WOODIN, A. M. (1962). *Biochem. J.*, **82**, 9

WOODIN, A. M. (1968). In *The Biological Basis of Medicine*, ed. BITTAR, E. & BITTAR, N., vol. 2, p. 373. London: Academic Press

WOODIN, A. M., FRENCH, J. E. & MARCHESI, V. T. (1963). *Biochem. J.*, 87, 567

WOODIN, A. M. & WIENEKE, A. A. (1963a). *Biochem. J.*, 87, 480

WOODIN, A. M. & WIENEKE, A. A. (1963b). *Biochem. J.*, 87, 487

WOODIN, A. M. & WIENEKE, A. A. (1964). *Biochem. J.*, 90, 498

WOODIN, A. M. & WIENEKE, A. A. (1966a). *Biochem. J.*, 99, 469

WOODIN, A. M. & WIENEKE, A. A. (1966b). *Biochem. J.*, 99, 479

WOODIN, A. M. & WIENEKE, A. A. (1967a). *Biochem. J.*, 105, 1029

WOODIN, A. M. & WIENEKE, A. A. (1967b). *Biochem. J.*, 105, 1039

WOODIN, A. M. & WIENEKE, A. A. (1968a). *Nature, Lond.*, 220, 283

WOODIN, A. M. & WIENEKE, A. A. (1968b). *Biochem. biophys. res. Comm.*, 33, 558

WOODIN, A. M. & WIENEKE, A. A. (1969). *Br. J. exp. Path.*, 50, 295

ZUCKER-FRANKLIN, D. & HIRSCH, J. G. (1964). *J. exp. Med.*, 120, 569

HORMONES, CELL CALCIUM AND CYCLIC AMP

H. RASMUSSEN & N. NAGATA

Department of Biochemistry, University of Pennsylvania, Philadelphia, Pennsylvania

During the past eight to ten years there has been increasing evidence that hormones, both steroid and polypeptide, influence either the cellular exchange or intracellular distribution of many ions including calcium. In the case of the steroid hormones, cortisol and vitamin D, there is considerable *in vitro* evidence that prior treatment of animals with either of these hormones leads to changes in transcellular transport of calcium (Schacter, 1970; Harrison & Harrison, 1960) as well as to changes in the uptake and release of calcium ions by isolated liver and kidney mitochondria (DeLuca, Engstrom & Rasmussen, 1962; Kimberg & Goldstein, 1966).

Aspects of both these subjects are discussed elsewhere in this symposium and will not be considered further here. The one aspect which has not received attention, however, is the relationship of changes in mitochondrial calcium accumulation and possible changes in cell function.

There is a variety of data which suggests that a redistribution of calcium may be involved in the regulation of gluconeogenesis in the liver, and the proposal has been made that a shift of calcium from mitochondria to cytoplasm could lead simultaneously to an activation of the mitochondrial enzyme, pyruvate carboxylase, and an inhibition of the cytosol enzyme pyruvate kinase, two key changes necessary for switching from glycolysis to gluconeogenesis (Gevers & Krebs, 1966; Bygrave, 1966, 1967; Kimmich & Rasmussen, 1969; Friedmann & Park, 1968).

This hypothesis is based upon the following evidence: (1) both enzymes are inhibited by Ca^{2+}; (2) cortisone alters the ability of isolated mitochondria to accumulate Ca^{2+} and also enhances hepatic gluconeogenesis; (3) glucagon enhances hepatic gluconeogenesis and also brings about the release of ^{45}Ca from cells of the prelabelled isolated perfused liver. All this evidence is circumferential, however, and it has proved exceedingly difficult to obtain direct evidence of shifts of calcium from one cellular compartment to another. It should be added that shifts of other ions such as Mg^{2+} and K^+ (Papavasiliou, Miller & Cotzias, 1968; Friedmann & Park, 1968; Finder, Boyme & Shoemaker, 1964) also occur after changes in hormone concentration and may thus be either a direct consequence of the hormal effect or an accompaniment of changes in calcium fluxes. Nonetheless, they may also be of physiological significance. None of the data is sufficiently rigorous or extensive to establish the hypothesis that steroid hormones alter metabolism by altering either cellular calcium exchange or the redistribution of calcium within the cell; but on the other hand, the available evidence does not rule out such a possibility.

The situation in regard to polypeptide and amine hormones is a little clearer. A number of these do influence cellular calcium exchange (Rasmussen, 1968) and in several instances these changes are clearly important in controlling intracellular events (Table 1). This list is not inclusive because there are several other examples—ACTH and the adrenal cortex (Birmingham, Elliot & Valere, 1953), and TSH and thyroxine release (Zor, Lowe, Bloom & Field, 1968)—where there is suggestive but inconclusive evidence of a similar relationship. The important point is that in many of these cases calcium seems to be at least one of the important coupling factors between a membrane event, peptide hormone-receptor interaction, and an intracellular event. Similar evidence for a coupling role of calcium in neurosecretion (Eccles, 1964), endocrine secretion

TABLE 1. *Polypeptide and amine hormones which have been shown to influence cellular calcium exchange and the intracellular event presumed to be controlled by changes in intracellular calcium*

Hormone	Cell	Physiological event
Oxytocin	Myometrium	Contraction[a]
Angiotensin	Adrenal glomerulosa	Steroidogenesis
LHRF	Adenohypophysis	LH release
TRF	Adenohypophysis	TSH release
Adrenaline	Heart	Inotropic response
Acetylcholine	Salivary gland	Enzyme release
Acetylcholine	Adrenal medulla	Epinephrine release
Vasopressin	Toad bladder	H_2O permeability
MSH	Melanocytes	Melanin dispersal
Leucocidin	Leucocyte	Granule release
PTH	Renal tubule	Gluconeogenesis
Acetylcholine	Exocrine pancreas	Enzyme release
Glucagon	Liver	Gluconeogenesis

[a] References in order of table: Csapo (1959); Daniels, Gevers & Buckley (1967); Samli & Geschwind (1968); Vale, Burgus & Gullemin (1967); Namm, Mayer & Maltbie (1968); Douglas & Poisner (1962); Douglas & Rubin (1961); Schwartz & Walter (1968); Dikstein, Weller & Sulman (1963); Woodin & Wieneke (1964); Nagata & Rasmussen (1968); Hokin (1966); Friedmann & Park (1968).

TABLE 2. *Relationship between Ca²⁺ and 3'5' AMP in physiological systems*

System	Ca²⁺	3'5'-AMP	Stimulus
Neurotransmission	+	+(?)	Electrical, KCl[a]
Amylase (SG)	+(?)	+	Adrenaline, KCl[b]
Growth hormone release	+	+	GHRF[c]
Insulin release	+	+	Glucose; Tolbutamide[d]
LH release	+	+	LHRF; KCl[e]
TSH release	+	+	TRF; KCl[f]
Steroid release	+	+	ACTH[g]
Glucose release	+(?)	+	Glucagon[h]
Thyroxine release	+(?)	+	TSH[i]
Heart			
glycogenolysis	+	+	Adrenaline[j]
intropy	+	+(?)	
H₂O permeability	+	+	AVP[k]
Renal tubule			
gluconeogenesis	+	+	PTH[l]
Ca²⁺ transport	+	+	
Melanin dispersal	+	+	MSH[m]
Slime mould aggregation	+(?)	+	?[n]
Sea urchin egg	+	+	Fertilization[o]

[a] Eccles (1964); DeRobertis *et al.* (1967).
[b] Rasmussen & Tenenhouse (1968); Bdolah & Schramm (1965).
[c] Schofield (1967).
[d] Grodsky & Bennett (1966); Hales & Milner (1968); Turtle *et al.* (1967); Malaisse *et al.* (1967).
[e] Samli & Geschwind (1968).
[f] Vale & Guillemin (1967); Vale *et al.* (1967).
[g] Birmingham *et al.* (1953); Sutherland *et al.* (1965).
[h] Sutherland *et al.* (1965); Friedmann & Park (1968).
[i] Zor *et al.* (1968); Gilman & Rall (1966).
[j] Sutherland *et al.* (1965); Cheung & Williamson (1965); Namm *et al.* (1968); Ozawa & Ebashi (1967); Meyer *et al.* (1964).
[k] Orloff & Handler (1962); Schwartz & Walter (1968); Thorn (1961).
[l] Chase & Aurbach (1967); Nagata & Rasmussen (1968); Wells & Lloyd (1967).
[m] Abe *et al.* (1969); Dikstein, Weller & Sulman (1963).
[n] Konijn, de Meene, Bonner & Barkley (1967); Gingell & Garrod (1969).
[o] Castañeda & Tyler (1968).

(Milner & Hales, 1967), and muscular contraction (Bianchi, 1961) has been available for some years. But what has struck us as a possible important relationship is that of cyclic AMP and Ca²⁺. In many of these systems (Table 2) there seems to be a requirement for Ca²⁺ and an activation of adenyl cyclase, although in some cases the evidence is not yet conclusive.

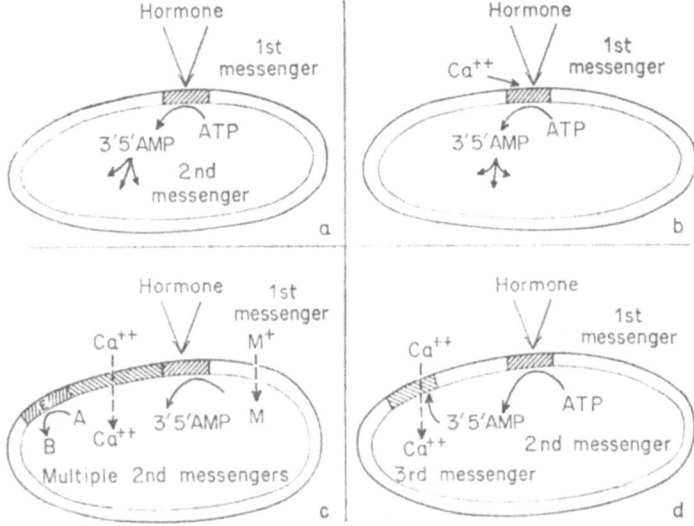

FIG. 1. Relationship of hormone receptor interaction to cell calcium and 3'5'-AMP. (a) The second messenger concept of Sutherland in which specific hormones interact with adenyl cyclases on specific cells to cause an increase level of 3'5'-AMP, a second messenger, within the cell which in turn brings about the physiological response; (b) the possibility that Ca^{2+} functions in this system by being required for hormone-receptor interaction; (c) the possibility that several simultaneous membrane events result from a single hormone-receptor interaction resulting in several simultaneous second messengers including an increase in both 3'5'-AMP and Ca^{2+} within the cell; (d) the possibility that 3'5'-AMP is the sole second messenger which in turn induces additional membrane events, for example, a change in Ca^{2+} permeability with a resultant third messenger, Ca^{2+}, being altered.

The questions raised by this relationship can be discussed in the context of the 3'5'-AMP being the second messenger in a number of hormonal control systems (Fig. 1a) (Sutherland, Oye & Butcher, 1965). In this concept a hormone interacts with its receptor site, a specific adenyl cyclase on the cell surface. This leads to the increase of 3'5'-AMP within the cell, and thus to a specific physiological consequence depending on the nature of the cell. In such a system 3'5'-AMP and Ca^{2+} could be related in one of several ways. The first would be that the specific adenyl cyclases require Ca^{2+} for activity (Fig. 1b). Present evidence indicates, however, that adenyl cyclase is a Mg^{2+}-dependent enzyme (Sutherland *et al.*, 1965) and is actually inhibited by Ca^{2+}.

A second possibility would be that most peptide hormones, when they interact with the cell surface, bring about a number of simultaneous changes in the permeability of the plasma membrane and in the activity of membrane bound enzymes; that is, there are several simultaneous second messages (Fig. 1c). In a system of this kind it would then be possible for the separate second messages to act sequentially in the cell and bring about considerable amplification of the response.

It is worthwhile noting that, in such a system, an increase of intracellular 3'5'-AMP, induced by a change in extracellular concentration, could cause an increase in the measured physiological response and this need not imply or prove that 3'5'-AMP is the sole normal second messenger. It is also noteworthy that the best characterized extracellular enzyme sequence, that involved in blood clotting, involves sequential activations and several key positive feedback loops to achieve the desired amplification. This extracellular system probably represents a reasonable model of what happens intracellularly when a hormone acts on a cell. If this is a valid analogy, then the model of hormone-cell interaction depicted in Fig. 1c is much more likely to correspond to reality than that depicted in Fig. 1a. At present, however, experimental data in most systems are insufficient to decide between them.

A third possible relationship between calcium and 3'5'-AMP is that shown in Fig. 1d. In this model 3'5'-AMP is the initial second message which elicits in turn a change in the Ca^{2+} permeability of the membrane, and Ca^{2+} then is a third messenger. In the systems listed in Table 2, it has not yet been possible to exclude this model, but, as indicated by the difference in the lists shown in Tables 1 and 2, there are systems in which the physiological signal, the first messenger, leads to a change in Ca^{2+} permeability without any evidence of an activation of adenyl cyclase. This would be difficult to account for by model 1d but would be easily explained by model 1c. On the other hand, it has been rather clearly established that caffeine and theophylline are both inhibitors of the phosphodiesterase that catalyses the hydrolysis of 3'5'-AMP (Sutherland & Rall, 1960). This implies a structural similarity between theophylline and 3'5'-AMP. In the present context theophylline and caffeine have been found directly to alter the permeability of cellular and subcellular membranes to Ca^{2+} (Bianchi, 1961). Hence the possibility exists that 3'5'-AMP also influences these membranes in a similar fashion, but at present there is no direct proof that this is the case.

Two additional possible relationships between Ca^{2+} and 3'5'-AMP are worthy of comment. The first is that ATP but not 3'5'-AMP is a good chelator of calcium; hence the hydrolysis of ATP leads to the following:

$$Ca\text{-}ATP \rightarrow 3'5'\text{-}AMP + Ca^{2+}$$

This type of reaction might well represent a means of suddenly changing the calcium ion concentration beneath the cell surface; it might be of great importance in the activation of secretory processes. The other relationship depends on the fact that adenyl cyclase is an ATPase, so its activation could lead to a decrease in ATP concentration at the plasma membrane and a decline in the activity of other membrane-bound ATPases, particularly the Na/K$^+$-activated ATPase, and the Ca^{2+} activated ATPase (Schatzmann, 1970). The resulting changes in monovalent cations brought about by a decreased activity of the Na$^+$ pump, as discussed by Baker & Blaustein (1968), lead to an increase in intracellular calcium. All these possible relationships are summarized in Fig. 2. The figure emphasizes the complexity of this relationship, but also the multiple sites of possible control.

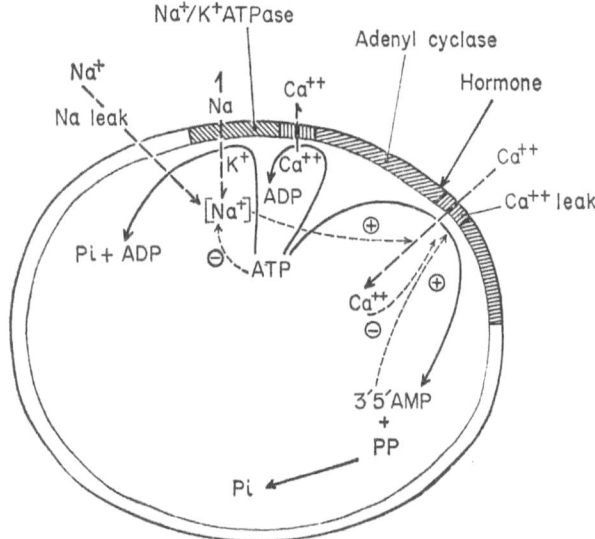

FIG. 2. Complex interrelationships between 3'5'-AMP, and calcium at the cell surface. Showing the Na$^+$ pump, the calcium pump, the adenyl cyclase and the leaks of both Na$^+$ and Ca^{2+}. A rise in intracellular Ca^{2+} may result from: (1) an increase in Ca^{2+} leak; (2) a fall in the activity of the Ca^{2+}-ATPase; and (3) an increase in intracellular Na$^+$.

With these considerations as a background, we now examine present evidence as to this relationship. Selection between these alternatives is not yet possible in the case of many of the systems, but some of the alternatives appear less likely on the basis of more extensive evidence in two of them: adrenaline and the heart, and parathyroid hormone and the renal tubule.

Adrenaline and the heart
It has long been known that perfusion of the heart with adrenaline leads to both an inotropic response and an increase in glycogenolysis (Cheung & Williamson, 1966). The former effect had been attributed to some ill-defined change in calcium distribution or uptake, whereas the

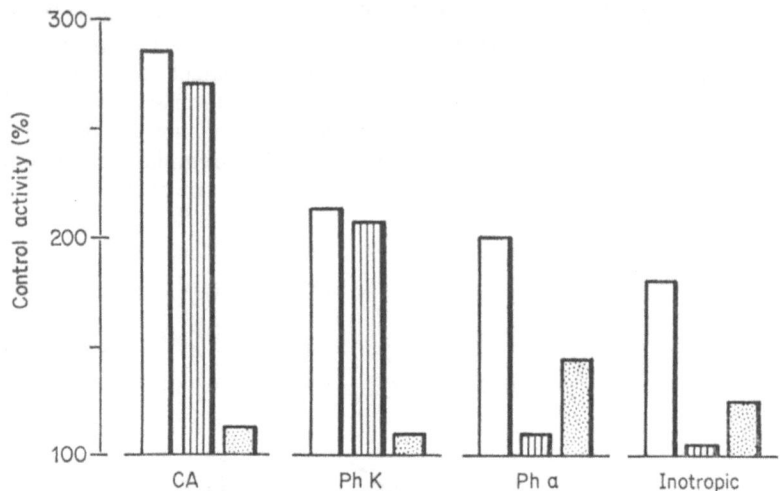

FIG. 3. Effect of adrenaline in the presence (open bars) and absence (vertically striped bars) of external Ca^{2+}, and of high calcium alone (stippled bars) upon the concentration of 3'5'-AMP (CA), the activities of phosphorylase *b* kinase (PhK) and phosphorylase *a* (Ph a) and force of contraction (inotropic) of the isolated perfused heart. (Modified from Namm *et al.*, 1968).

latter had been attributed to an activation of the phosphorylase system by 3'5'-AMP. However, more recent *in vitro* studies of several of the isolated enzymes involved in the activation of phosphorylase (Meyer, Fischer & Krebs, 1964; Mayer and Krebs, 1968; Ozawa and Ebashi, 1967), as well as the work of Namm, Mayer & Maltbie (1968) on the isolated perfused rat heart (Fig. 3) have shown that: (1) 3'5'-AMP increases after administration of adrenaline in both the presence and absence of Ca^{2+}; (2) in the absence of external Ca^{2+} neither the glycogenolytic nor inotropic response to adrenaline seem to occur; (3) Ca^{2+} is required for and stimulates the activity of phosphorylase *b* kinase; and (4) 3'5'-AMP causes an activation of phosphorylase *b* kinase in the presence or absence of calcium. These results can be fitted into the scheme outlined in Fig. 4 which shows that Ca^{2+} is involved in a critical fashion both in the glycogenolytic and inotropic effects of adrenaline on the heart; also two second messengers, 3'5'-AMP and Ca^{2+}, are required for the sequential activation of two steps in phosphorylase activation.

These data show that Ca^{2+} does not exert its effect by regulating adenyl cyclase activity, thus eliminating the scheme in 1b. They do not, however, allow a distinction between the alternate possibilities of a simultaneous change in calcium permeability and an activation of adenyl cyclase (Fig. 1c); or a sequential change, adenyl cyclase activation followed by a 3'5'-AMP-induced change in calcium permeability (Fig. 1d). In this context

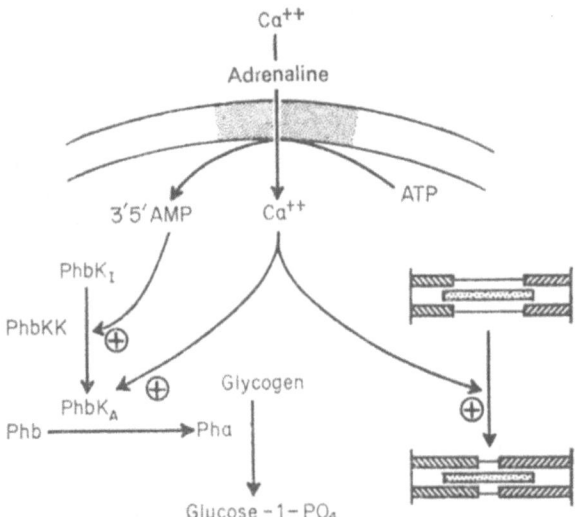

FIG. 4. Schematic summary of the relationship between adrenaline, Ca^{2+}, 3'5'-AMP and changes in the mechanical and metabolic activity of the heart. $PhbK_I$, Phosphorylase *b* kinase; PhbKK, phosphorylase *b* kinase kinase; Phb, phosphorylase *b*; Pha, phosphorylase *a*.

it should be pointed out that mechanisms 1c and 1d are not mutually exclusive, and in fact control in this system might invoke both mechanisms.

Parathyroid hormone and the renal tubule

It has long been known that parathyroid hormone (PTH) has important effects on the metabolism of extracellular calcium (Rasmussen, 1961), but only recently has evidence been obtained concerning two important aspects of its action which are germane to this discussion. The first is that 3'5'-AMP is an intermediate in the action of this hormone on kidney and bone (Chase & Aurbach, 1967; Wells & Lloyd, 1967; Rasmussen, Pechet & Fast, 1968), and the second, that an increase in intracellular calcium may be an important consequence of its action on these cells (Nagata & Rasmussen, 1968; Borle, 1968a,b; Talmage, 1967). In view of these findings and our initial *in vivo* studies concerning parathyroid hormone and renal cell metabolism (Nagata & Rasmussen, 1968), we have gone on to develop an *in vitro* system in which to study the relationship between PTH, Ca^{2+}, cyclic AMP and renal gluconeogenesis (Nagata & Rasmussen, unpublished.)

As shown in Fig. 5, when isolated renal tubules, primarily segments of nephrons, are incubated with malate as a substrate there is a nearly linear increase in glucose formation over a period of 30–45 min. Furthermore, this rate is altered if extracellular calcium concentration is changed in a reversible manner. Thus, if incubated in a media containing no added calcium for 15 min, the rate of gluconeogenesis is slow but measureable, and is increased if Ca^{2+} is then added and the incubation continued for another 15 min. This means that in the absence of added calcium the tubules remain functionally intact and can respond in a normal manner to the later addition of Ca^{2+}.

The importance of this point is shown in Fig. 6. PTH, as well as calcium, will increase the rate of gluconeogenesis from malate, but only if external

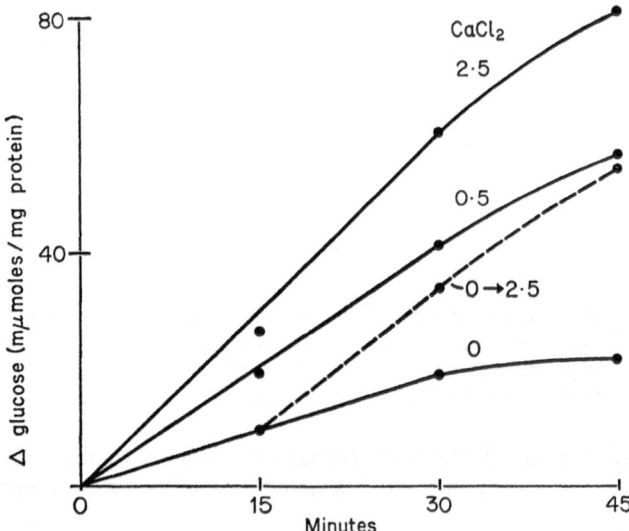

Fig. 5. Rate of glucose production (●—●) from malate as a function of calcium concentration (mM) in the incubation medium. The effect of calcium can be seen (●-- ●) even after a preincubation in the absence of this cation.

calcium is present. Also shown is the fact that PTH is without effect on gluconeogenesis from glycerol, and that when glycerol is the substrate calcium has an effect only in going from a no calcium (0·50 mM EGTA $\simeq$ 10^{-8} M Ca) to a low calcium, as little as 0·05 mM Ca^{2+}, but then has no further effect. These data would place the major effect of both PTH and increased extracellular calcium at some step between malate and triose phosphates in the gluconeogenic sequence (Fig. 7). They indicate that extra-cellular calcium is required for the stimulation of renal gluconeogenesis by PTH. As shown in Fig. 8, however, PTH induces a significant and rapid rise in 3′5′-AMP within this tubule and this rise is not influenced by the presence or absence of extracellular calcium. Nor for that matter does an increase in extracellular calcium cause any change in 3′5′-AMP concentra-tion even though it has a significant effect on renal gluconeogenesis. Thus, just as with adrenaline and the heart, the hormone causes an activation

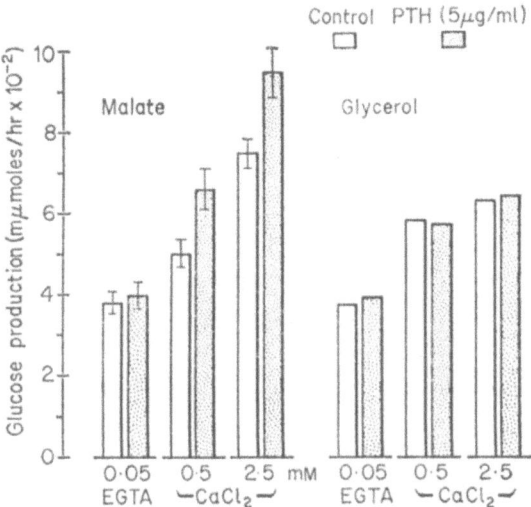

FIG. 6. Effect of Ca^{2+} and parathyroid hormone (PTH) on gluconeogenesis from malate and glycerol.

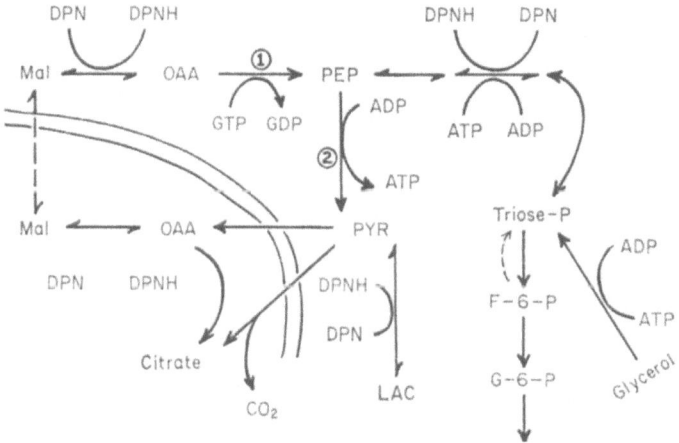

FIG. 7. Sequence of reaction steps from malate to glucose, and possible control sites. (1) phosphoenol pyruvate carboxykinase, and (2) pyruvate kinase.

of adenyl cyclase in the absence of extracellular calcium; but in spite of the rise in 3'5'-AMP the normal change in metabolism is not seen.

These data, taken together with the recent work of Borle (1968a,b) and our previous data (Nagata & Rasmussen, 1968) showing that PTH increases the uptake of calcium by both HeLa and kidney cells grown in

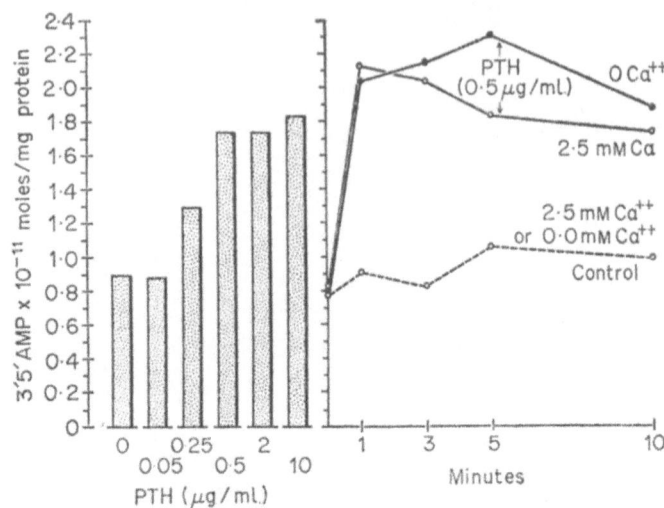

FIG. 8. Effect of PTH concentration upon 3'5'-AMP levels in renal tubules (left); and the time course of change in 3'5'-AMP concentrations after PTH addition in the presence and absence of Ca^{2+}.

tissue culture, suggest that both an increase in extracellular calcium and the addition of PTH exert their effects on isolated renal tubules by a common mechanism, that of increasing the calcium ion concentration in one or more intracellular compartments.

In an effort to define more precisely the compartment within the cell as well as the enzymatic steps controlled by this change, we have investigated the changes in the concentrations of key metabolites under two conditions: (1) 0·25 mM external calcium plus PTH; and (2) 0·25 mM to 2·5 mM calcium. In the latter experiments ouabain was included because a change in extracellular calcium is known to alter Na^+ permeability which in turn alters the energy metabolism of the cell, and this was largely eliminated by ouabain addition.

As seen in Fig. 9a and b, both a shift from 0·25 to 2·5 mM Ca^{2+}, and the addition of PTH with 0·25 mM $CaCl_2$ led to similar changes in the intracellular concentration of oxaloacetate, phosphenolpyruvate, pyruvate, and to increased glucose production. In both instances glucose production increased, OAA fell, PEP rose, and pyruvate fell. Also as shown in Fig. 9c, in the absence of extracellular calcium PTH did not increase glucose production or change the levels of any of these key intermediates in spite of the fact that 3'5'-AMP rose to the same extent (Fig. 8) as in the case when 0·25 mM calcium was present.

From these data it is clear that both PTH and an increase in extracellular calcium regulate gluconeogenesis by regulating one or possibly two enzy-

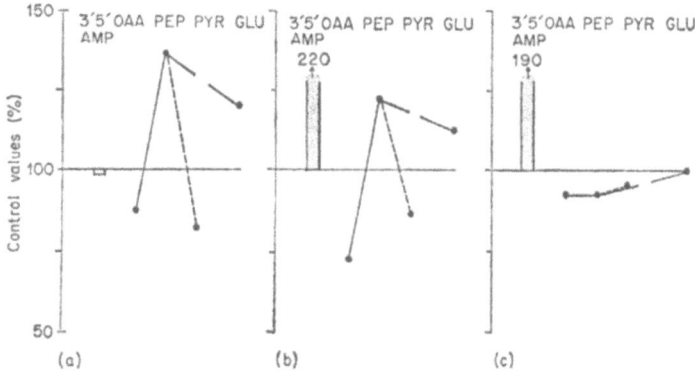

FIG. 9. Relative changes in the concentrations of oxaloacetate (OAA), phosphoenol pyruvate (PEP), pyruvate (Pyr), glucose (Glu) and 3'5'-AMP when: (a) external calcium concentration is changed from 0·25 mM to 2·5 mM in the presence of ouabain; (b) PTH (0·5 μg/ml.) is added to a medium containing 0·25 mM Ca^{2+}; and (c) PTH is added to a medium containing no external Ca^{2+}.

matic steps—pyruvate kinase (PK) and phosphoenol pyruvatecarboxy-kinase (PEPCK)—inhibiting the first and activating the latter (Fig. 7). It is our working hypothesis that these effects are brought about by changing the concentration of ionized calcium within the cytosol of the renal tubular cells (Fig. 10), and that similar changes in the other target cells reponsive to parathyroid hormone are responsible for many if not all the major effects of this hormone upon calcium and phosphate transport, and cell metabolism.

As yet unanswered, as is the case with adrenaline in the heart, is the relationship between the change in intracellular 3'5'-AMP and the increase in intracellular calcium. One of two possibilities exist. Either the activation of adenyl cyclase and the change in calcium permeability are simultaneous events following hormone receptor interaction (Fig. 1c), or the rise in 3'5'-AMP as a consequence of adenyl cyclase activation leads to the change in calcium accumulation (Fig. 1d). In spite of several different approaches to this problem, including the effects of adding dibutyryl 3'5'-AMP extra-cellularly and a study of the effects of 3'5'-AMP on the calcium transport and binding properties of subcellular membranes, no clear choice can yet be made between these alternatives.

Finally, brief mention should be made of the possible relationship of calcium, 3'5'-AMP, and microtubular function in two systems: insulin release from the beta cells of the pancreas and the effect of melanocyte stimulating hormone (MSH) upon melanin dispersal in frog melanocytes. In both cases evidence has been produced that Ca^{2+} ions are required (Dikstein, Weller & Sulman, 1963; Milner & Hales, 1967), that colchicine

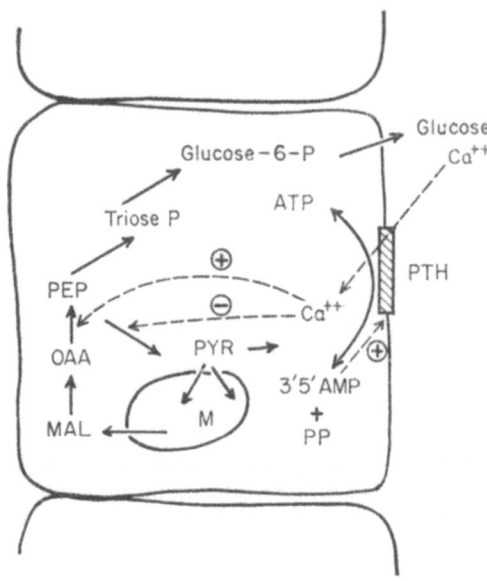

FIG. 10. Schematic representation of the effect of parathyroid hormone (PTH) on intracellular calcium and gluconeogenesis in the isolated renal tubule.

blocks (Malawista, 1965; Lacy, Howell, Young & Fink, 1968), and that 3'5'-AMP is involved (Abe, Butcher, Nicholson, Baird, Liddle & Liddle, 1969; Turtle, Littleton & Kipnis, 1967; Malaisse, Malaisse-Lagae & Mayhew, 1967). Also in both cases there is morphological evidence that microtubules within the cell may be important in the movement of the melanin-containing granules (Green, 1968) and the insulin-containing secretory granules (Lacy et al., 1968). It is known from other work that the major protein of microtubules is similar to actin (Adelman, Borisy, Shelanski, Weisenberg & Taylor, 1968). It is thus tempting to speculate that the movement of these granules involves a mechanism similar in its requirements to that of muscle contraction and that Ca^{2+} play a key regulatory role. A similar relationship is also seen in the effects of colchicine and the role of calcium and microtubules in mitosis (Borisy & Taylor, 1967; Perris & Whitfield, 1967). Much additional information is needed, however, before such speculation assumes the nature of a hypothesis.

Summary and conclusions

Many hormones alter the movement of calcium across cellular and subcellular membranes. Of particular interest are the large number of hormonal agents which appear to alter both the Ca^{2+} permeability and adenyl cyclase activity of the cell membrane. In many of these systems a change in Ca^{2+} concentration within one or more cellular compartments plays an important regulatory role. In two systems, adrenaline and the heart, and parathyroid hormone and the renal tubule, it is now clear the Ca^{2+} is not required for the activation of adenyl cyclase but is required for the ensuing changes in metabolism. However, it is not yet possible to decide whether: (a) the change in adenyl cyclase activity and in calcium permeability are simultaneous consequences of hormone-receptor interaction, (b) the change in adenyl cyclase activity leads to an increase in 3'5'-AMP which in turn brings about the change in Ca^{2+} permeability, or (c) both mechanisms are operative. A possible relationship between Ca^{2+}, 3'5'-AMP, and actin-like proteins in microtubules is briefly discussed.

This work was supported by grants from the U.S. Atomic Energy Commission AT(30–1)3489 and the U.S. Public Health Service (AM-09650).

REFERENCES

ABE, K., BUTCHER, R. W., NICHOLSON, W. E., BAIRD, C. E., LIDDLE, R. A. & LIDDLE, G. W. (1969). *Endocrinology*, **84**, 362

ADELMAN, M. R., BORISY, G. G., SHELANSKI, M. L., WEISENBERG, R. C. & TAYLOR, E. W. (1968). *Fedn Proc.*, **27**, 1186

BAKER, P. F. & BLAUSTEIN, M. P. (1968). *Biochim. biophys. Acta*, **150**, 167

BDOLAH, A. & SCHRAMM, M. (1965). *Biochem. Biophys. res. Comm.*, **18**, 452

BIANCHI, C. P. (1961). *Circulation*, **24**, 518

BIRMINGHAM, M. K., ELLIOTT, F. H. & VALERE, P. (1953). *Endocrinology*, **53**, 687

BORISY, G. G. & TAYLOR, E. W. (1967). *J. cell Biol.*, **34**, 525

BORLE, A. B. (1968a). *J. cell Biol.*, **36**, 567

BORLE, A. (1968b). *Endocrinology*, **83**, 1316

BYGRAVE, F. L. (1966). *Biochem. J.*, **101**, 480

BYGRAVE, F. L. (1967). *Nature, Lond.*, **214**, 667

CASTAÑEDA, M. & TYLER, A. (1968). *Biochem. Biophys. res. Comm.*, **33**, 782

CHASE, L. R. & AURBACH, G. D. (1967). *Proc. natl. Acad. Sci. U.S.*, **58**, 518

CHEUNG, W. Y. & WILLIAMSON, J. R. (1966). *Nature, Lond.*, **207**, 979

CSAPO, A. (1959). *Ann. N.Y. Acad. Sci.*, **75**, 790

DANIELS, A. E., GEVERS, W. B. & BUCKLEY, J. P. (1967). *Life Sci., Oxford*, **6**, 545

DeLUCA, H. F., ENGSTROM, G. W. & RASMUSSEN, H. (1962). *Proc. natl. Acad. Sci. U.S.*, **48**, 1604

DE ROBERTIS, E., ARNAIZ, G. R. D., ABRECHT, M., BUTCHER, R. W. & SUTHERLAND, E. W. (1967). *J. biol. Chem.*, **242**, 3487

DIKSTEIN, S., WELLER, C. P. & SULMAN, F. G. (1963). *Nature, Lond.*, **200**, 1106

DOUGLAS, W. & POISNER, A. M. (1962). *Nature, Lond.*, **196**, 379

DOUGLAS, W. W. & RUBIN, R. P. (1961). *J. Physiol.*, **159**, 40

ECCLES, J. C. (1964). *The Physiology of Synapses*. New York: Academic Press

FINDER, A. G., BOYME, T. & SHOEMAKER, W. C. (1964). *Am. J. Physiol.*, **206**, 738

FRIEDMANN, N. & PARK, C. R. (1968). *Proc. natl. Acad. Sci. U.S.* **61**, 504

8—C.C.F.

GEVERS, W. & KREBS, H. A. (1966). *Biochem. J.*, **98**, 720
GILMAN, G. A. & RALL, T. W. (1966). *Fedn Proc.*, **25**, 617
GINGELL, D. & GARROD, D. R. (1969). *Nature, Lond.*, **221**, 192
GREEN, L. (1968). *Proc. natl. Acad. Sci. U.S.*, **59**, 1179
GRODSKY, G. M. & BENNETT, L. L. (1966). *Diabetes*, **15**, 910
HALES, C. N. & MILNER, R. D. G. (1968). *J. Physiol., Lond.*, **194**, 725
HARRISON, H. E. & HARRISON, H. C. (1960). *Am. J. Physiol.*, **199**, 265
HOKIN, L. (1966). *Biochim. biophys. Acta*, **115**, 219
KIMBERG, D. V. & GOLDSTEIN, S. A. (1966). *J. biol. Chem.*, **241**, 95
KIMMICH, G. A. & RASMUSSEN, H. (1969). *J. biol. Chem.*, **244**, 190
KONIJN, T. M., VAN DE MEENE, J. G. C., BONNER, J. T. & BARKLEY, D. S. (1967). *Proc. natl. Acad. Sci. U.S.*, **58**, 1152
LACY, P. E., HOWELL, S. L., YOUNG, D. A. & FINK, C. J. (1968). *Nature, Lond.*, **219**, 1177
MALAISSE, W. J., MALAISSE-LAGAE, F. & MAYHEW, D. (1967). *J. clin. Invest.*, **46**, 1724
MALAWISTA, S. E. (1965). *J. exp. Med.*, **122**, 361
MAYER, S. E. & KREBS, E. G. (1968). *Fedn Proc.*, **27**, 352
MEYER, W. L., FISCHER, E. H. & KREBS, E. G. (1964). *Biochemistry*, **3**, 1033
MILNER, R. D. G. & HALES, C. M. (1967). *Diabetologia*, **3**, 478
NAGATA, N. & RASMUSSEN, H. (1968). *Biochemistry*, **7**, 3728
NAMM, D. H., MAYER, S. E. & MALTBIE, M. (1968). *Mol. Pharmac.*, **4**, 522
ORLOFF, J. & HANDLER, J. S. (1962). *J. clin. Invest.*, **41**, 702
OZAWA, E. & EBASHI, S. (1967). *J. Biochem.*, **62**, 285
PAPAVASILIOU, P. S., MILLER, S. T. & COTZIAS, G. C. (1968). *Nature, Lond.*, **220**, 74
PERRIS, A. D. & WHITFIELD, J. F. (1967). *Nature, Lond.*, **216**, 1350
RASMUSSEN, H. (1961). *Am. J. Med.*, **30**, 112
RASMUSSEN, H. (1968). *Protein and Polypeptide Hormones*, Excerpta Medica Intern. Cong. Series No. 161. vol. 1, p. 247
RASMUSSEN, H. (1968). *Protein and Polypeptide Hormones*. Excerpta Medica Intern. Cong. Series No. 161, vol. 3, in the Press
RASMUSSEN, H., PECHET, M. & FAST, D. (1968). *J. clin. Invest.*, **47**, 1843
RASMUSSEN, H. & TENENHOUSE, A. (1968). *Proc. natl. Acad. Sci. U.S.*, **59**, 1364
SAMLI, M. H. & GESCHWIND, I. I. (1968). *Endocrinology*, **82**, 225
SCHACHTER, D. (1970). In *Calcium and Cellular Function*, ed. CUTHBERT, A. W. London: Macmillan
SCHATZMANN, H. J. (1970). In *Calcium and Cellular Function*, ed. CUTHBERT, A. W. London: Macmillan
SCHOFIELD, J. G. (1967). *Nature, Lond.*, **215**, 1382
SCHWARTZ, I. L. & WALTER, R. (1968). In *Protein and Polypeptide Hormones*, ed. MARGOULIES, M., part I, p. 264. Amsterdam: Excerpta Medica Foundation
SUTHERLAND, E. W., OYE, I. & BUTCHER, R. W. (1965). *Rec. Prog. Hormone Res.*, **21**, 623
SUTHERLAND, E. W. & RALL, T. W. (1960). *Pharmac. Rev.*, **12**, 265
TALMAGE, R. V. (1967). *Am. Zoologist*, **7**, 825
THORN, N. A. (1961). *Acta Endocr.*, **38**, 563

TURTLE, J. R., LITTLETON, G. K. & KIPNIS, D. M. (1967). *Nature, Lond.*, **213**, 727

VALE, W., BURGUS, R. & GUILLEMIN, R. (1967). *Experientia*, **23**, 853

VALE, W. & GUILLEMIN, R. (1967). *Experientia*, **23**, 855

WELLS, H. & LLOYD, W. (1967). *Endocrinology*, **81**, 139

WOODIN, A. M. & WIENEKE, A. A. (1964). *Biochem. J.*, **90**, 498

ZOR, U., LOWE, I. P., BLOOM, G. & FIELD, J. B. (1968). *Biochem. biophys. res. Commun.*, **33**, 649

THE S–S POLYPEPTIDE RECEPTOR AS A METAL RECEPTOR

H. O. SCHILD

Department of Pharmacology, University College London, Gower Street, London, W.C.1

It has long been known that the magnesium content of Ringer solution affects the activity of S–S polypeptides (Fraser, 1939; Stewart, 1949). Thus, when the activity ratio of oxytocin/vasopressin is measured in the isolated uterus in the absence and presence of magnesium, it is found that the activity ratio changes in favour of vasopressin in solutions which contain magnesium although vasopressin remains less active than oxytocin on a molar basis. It seems probable that this effect is due to a selective potentiation of vasopressin by magnesium. Magnesium, although it improves the activity of several synthetic S–S polypeptides relative to oxytocin, has no effect on the activity of desaminooxytocin, one synthetic polypeptide which is definitely more active than oxytocin (Munsick, 1968.)

The object of the present investigation, which is reported in more detail elsewhere (Schild, 1969), was to make a comparative study of the effects of various metals on the S–S polypeptide receptor and to compare these effects with those exerted by metals on certain enzymes (Vallee & Coleman, 1964). That magnesium can be substituted by certain other metals—for example, manganese—to potentiate vasopressin effects on the uterus has previously been reported (Bentley, 1965). The test preparation used in the present work is the potassium-depolarized rat uterus: an isolated smooth muscle immersed in potassium sulphate Ringer instead of the usual sodium chloride Ringer (Evans, Schild & Thesleff, 1958). This type of preparation has no membrane potential but it responds to drugs such as the S–S polypeptides by a slow graded contraction which can be readily quantitated. The responses of depolarized muscle are probably mediated through the same receptors as are responses in normal, polarized smooth muscle. Calcium (0·1 or 1·0 mM) was always added to the solution because it is an essential requirement for all drug induced contractile effects in depolarized smooth muscle (Schild, 1963).

Effect of magnesium

Magnesium potentiates the responses of depolarized uterus to vasopressin (Fig. 1), producing a parallel shift of the log dose-response curve with dose ratio of about 10. The magnesium effect is rapidly established and it decays rapidly. The extent of potentiation is a function of magnesium concentration in the bath fluid.

In comparison with metal enzymes the S–S polypeptide-magnesium system shows similarity with reversible 'metal-enzyme complexes' (Vallee & Coleman, 1964) rather than with stable 'metallo-enzymes'. Some metal enzymes lose all activity when deprived of metal while others (arginase)

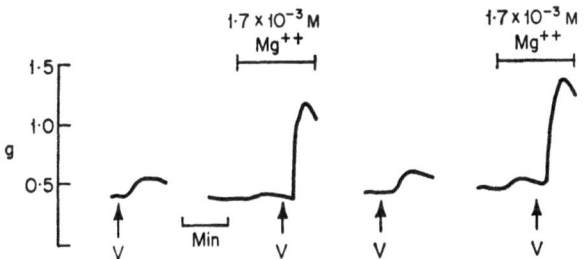

FIG. 1.   Response to a constant dose of lysine-vasopressin in the absence and presence of Mg^{2+}. Successive doses. Note the reversibility of potentiation by Mg^{2+}. Isometric contraction of depolarized rat uterus.

cannot be reduced to zero activity even after drastic procedures for the removal of metal. It was found that the vasopressin receptor retained reduced activity even after thorough washing in magnesium-free solution and this suggested that the receptor can function without metal. Nevertheless the difficulties of applying drastic measures for metal removal to the living tissue mean that the question of whether the receptor can function in the absence of metal cannot be regarded as settled.

Specificity of potentiation
Different S–S polypeptides are potentiated to varying degrees by magnesium. Lysine-, arginine-, and ornithine-vasopressin were strongly potentiated but oxytocin was only weakly potentiated and in some experiments actually antagonized by magnesium. Magnesium failed to potentiate the contractile effects of acetylcholine or of the polypeptides hypertensin and bradykinin in the depolarized uterus. The specificity of the potentiating effect suggests an effect of the metals at receptor level.

Other metals
A variety of metals can substitute for magnesium in potentiating S–S polypeptides (Table 1). Among these the transition metals manganese, cobalt, and nickel are considerably more active than magnesium. Trivalent compounds also seemed active, although less so than divalent; it cannot be excluded, however, that they are reduced to the divalent form before producing their effect. Strongly complexed ions such as cobaltic hexammine were inactive.

TABLE 1. *Activity of metals (means) relative to magnesium*

Co^{2+}	18	Fe^{2+}	0·5
$(Co^{3+}$	15)	$(Fe^{3+}$	0·1)
Mn^{2+}	10	Ca^{2+}	0·1
Ni^{2+}	8	Be^{2+}	0
$(Mn^{3+}$	6)	Sr^{2+}	0
Zn^{2+}	1	Ba^{2+}	0
Mg^{2+}	1	Cu^{2+}	0

Zinc and ferrous were about equiactive with magnesium; cupric sulphate was inactive. Alkaline earth metals were inactive with the exception of magnesium. Calcium may have some slight specific potentiating effect on the S–S polypeptide receptor but this was difficult to disentangle from its essential role in the contractile mechanism.

Mechanism of potentiation

Assuming that the potentiating effect is exerted at receptor level, a wide variety of possible mechanisms exist, none of which can be completely excluded. For example, the metal might produce a change in the conformation of the receptor or an 'allosteric' effect without any direct interaction with the active receptor site. Alternatively the metal may react primarily with the polypeptide 'substrate' or with the active site of the receptor; it might even interact with both and form a tertiary complex (Klotz & Ming, 1954).

Assuming that the various S–S polypeptides act on a common receptor —and there is some experimental evidence for this from experiments with antagonists (Martin & Schild, 1965)—any mechanistic scheme should be able to account for the finding that the degree of potentiation of different S–S polypeptides varies, the weakly active ones being potentiated by metals to a greater extent than the strongly active. Another relevant finding is the complete reversibility of the potentiating effect; and this is coupled with a time factor which might suggest some interaction between metal and receptor (Schild, 1969).

The following scheme assumes that metals produce their effect by coordination with electron donor groups present in the receptor surface and by simultaneous coordination with one or more electron donor groups in the S–S polypeptide molecule. It is assumed that the substituents that determine the specificity of the S–S polypeptide molecule are bound directly to the receptor surface whether or not this contains metal. The function of the metal would be to improve alignment between drug and receptor; in the case of highly active S–S polypeptides which are already well aligned the metal would contribute little to activity, but with relatively inactive ones it will aid apposition to the receptor and thus increase activity.

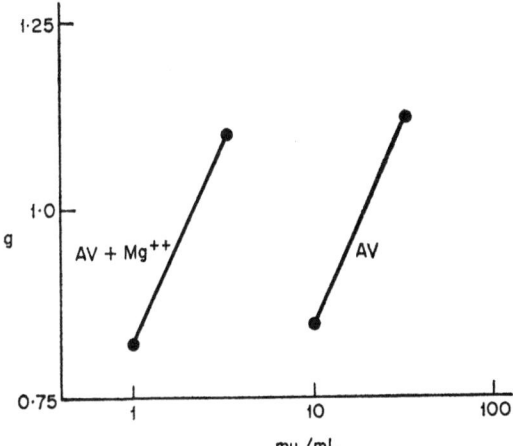

FIG. 2. Parallel shift of log dose-response curve of arginine vasopressin (AV) by 2mM Mg^{++}

The corresponding mass action equilibria for this scheme are given below:

$$P + R \rightleftharpoons PR$$

$$M + R \rightleftharpoons MR$$

$$P + MR \rightleftharpoons PMR$$

where P stands for S–S polypeptide, R for receptor, M for metal, and MR for metal-occupied receptor. The equation giving fractional receptor occupation, y, in terms of concentration of polypeptide (P), and metal (M), is given by

$$y = \frac{K_1(P) + K_3(P)K_2(M)}{1 + K_2(M) + K_1(P) + K_3(P)K_2(M)}$$

It can be shown that this equation gives parallel lines when log dose–response curves are plotted in the absence and presence of metals. It has generally been found that such lines are indeed parallel (Bentley, 1965; Krejci & Polacek, 1968) as shown in Fig. 2.

REFERENCES

BENTLEY, P. J. (1965). *J. Endocr.*, **32**, 215
EVANS, D. H. L., SCHILD, H. O. & THESLEFF, S. (1958). *J. Physiol., Lond.*, **143**, 474
FRASER, A. M. (1939). *J. Pharmac. exp. Ther.*, **66**, 85
KLOTZ, I. M. & MING, U. C. L. (1954). *J. Am. chem. Soc.*, **76**, 805
KREJCI, J. & POLACEK, J. (1968). *Eur. J. Pharmac.*, **2**, 393

MARTIN, P. J. & SCHILD, H. O. (1965). *Br. J. Pharmac. Chemother.*, **25**, 418
MUNSICK, R. A. (1968). *Handbook of Experimental Pharmacology*, **23**, 443
SCHILD, H. O. (1963). *Proc. 2nd Int. Pharmacological Meeting, Prague*, 95
SCHILD, H. O. (1969). *Br. J. Pharmac.*, **36**, 329
STEWART, G. A. (1949). *J. Pharm. Pharmac.*, **1**, 436
VALLEE, B. L. & COLEMAN, J. E. (1964). *Comprehensive Biochemistry*, **12**, 165

INTERACTION OF LOCAL ANAESTHETICS AND CALCIUM WITH ERYTHROCYTE MEMBRANES

J. C. METCALFE

M.R.C. Molecular Pharmacology Research Unit,
Department of Pharmacology, University of Cambridge

At present there is little definitive information about the way in which the components of cell membranes are organized, and new experimental approaches to the problem are required. Recent work has been aimed at detecting perturbations in the structure of cell membranes by physical techniques; for example, Cohen, Keynes & Hille (1968) have followed changes in the birefringence and light scattering of nerve membranes during the action potential. Alternatively, probe molecules with suitable spectroscopic properties have been used to report indirectly on the perturbations which they sense within the membrane. These probes may also be expected to provide information about the structure in which they are inserted. Here we describe two probe techniques which make use of magnetic resonance spectroscopy to detect changes in molecular motions within the membrane. Although they depend on quite distinct magnetic resonance phenomena, they provide directly comparable data.

Molecules with electron spin resonance (ESR) spectra were first applied as membrane probes by Hubbell & McConnell (1968, 1969). The discovery of sterically protected nitroxides ($>$N—O) which are stable in aqueous solutions has resulted in the synthesis of a wide range of nitroxide analogues of biological compounds, termed 'spin labels'. The unpaired electron of the nitroxide group provides a simple ESR spectrum which is sensitive to the molecular motion of the spin label. Thus changes in the molecular motion of spin labels bound to membranes in response to perturbations can readily be detected and compared for a range of spin labels. The labels may be lipid soluble molecules which are analogues, for example, of anaesthetics, of membrane components such as phospholipids or cholesterol, or, on the other hand, of covalent protein reagents.

When the partition coefficient of the spin label is sufficiently low to give a significant concentration of unbound nitroxide, this unbound fraction can generally be distinguished from the fraction which is bound to the membrane. This property of the spectra was used by Hubbell & McConnell (1968) to study the excitable membrane of isolated nerves using the spin label 2,2,6,6-tetramethylpiperidine-1-oxyl (TEMPO).

The spectrum consists of two components which are sufficiently resolved to distinguish TEMPO in the aqueous bathing medium and TEMPO bound in a fluid hydrophobic region of the nerve membrane. Addition of local anaesthetics markedly increased the partition of TEMPO into the nerve membrane, and Hubbell & McConnell interpreted this to mean that the anaesthetic 'melted out' or disordered the membrane to increase the fluid hydrophobic region available to TEMPO. Conversely, cholesterol and calcium ions decreased the binding of TEMPO in the nerve, and it was suggested that this resulted from an increased packing or ordering of the membrane by these agents. Indirect evidence was put forward to show that TEMPO is a specific probe for lipid regions of the membrane, and that it is the fluidity of the lipid moiety which directly determines the partition coefficient of TEMPO. Recent studies by Hubbell & McConnell (1969) have made use of spin labels which are more restricted than TEMPO in their motion in the membrane and which show spectral changes which can be directly interpreted as increased molecular motion in the presence of anaesthetics.

A problem in the use of probe molecules in cells or tissues containing several kinds of membranes is the difficulty of determining in which membranes the probe is localized and, therefore, where the perturbations are occurring. For example, in the present state of the art it is not possible to design probes which will be specifically localized in the excitable membrane of a nerve. Experiments on a well defined membrane where all the probe molecules are located in a single type of structure are therefore useful both in characterizing the response of probes to perturbations and in relating this to the structure of the membrane. For this reason, we have examined the interaction of anaesthetics with erythrocyte membranes using spin labels, and we have also been able to study the same interactions by nuclear magnetic resonance (NMR).

In these latter experiments the anaesthetic molecule itself acts as the probe, and perturbations in the membrane structure are inferred from changes in the NMR spectrum of the anaesthetic. When the molecular motion of the anaesthetic is restricted by insertion into the ordered structure of the membrane, there is a broadening of the resonance lines of the anaesthetic, which is a direct measure of the restriction to motion.

With increasing concentrations of the anaesthetic, its motion within the membrane becomes progressively freer and, by inference, this reflects the increasing fluidity of the membrane components. Thus the general description of the effect of anaesthetics on nerve and erythrocyte membranes by both NMR and ESR studies is essentially the same.

The data reviewed here are limited to a comparison by both magnetic resonance techniques of the effects of two anaesthetics and calcium on the erythrocyte membrane. The local anaesthetics selected are benzyl alcohol and xylocaine, to allow comparison of uncharged and cationic agents. We also consider the problem of relating the perturbations to the organization of the structure with which the probes interact. Full accounts of some of this work are published elsewhere (Metcalfe, Seeman & Burgen, 1968; Hubbell, Metcalfe & McConnell, in preparation).

Human red cell ghosts were prepared as described previously (Metcalfe, Seeman & Burgen, 1968), and finally suspended in the appropriate H_2O buffer for the ESR experiments. For NMR experiments, the corresponding 99·8% D_2O buffers were used to minimize the water proton signal. The spin labels were prepared by Dr. Wayne Hubbell (Hubbell & McConnell, 1969).

Nitroxide ESR spectra

The unpaired electron of the nitroxide group is aligned in the direction of an applied magnetic field, and when it absorbs a precise quantum of microwave energy it is flipped into the anti-parallel direction. This absorption of energy from the exciting field is the experimental basis for the observation of resonance.

The structure of the nitroxides is well established and the N, O and adjacent C atoms lie in a plane with the unpaired electron mainly localized in a $2p\pi$ orbital perpendicular to the plane of the atoms (Fig. 1). This corresponds to a polar structure $> \overset{+}{N}—\overset{-}{O}$.

The spectrum of TEMPO tumbling freely in aqueous solutions (at $\sim 10^{11}$ sec^{-1}) is shown in Fig. 2. It consists of three resonance absorption signals, conventionally represented as the first derivative of the simple absorption spectrum. There are three components of the spectrum because the unpaired electron experiences a magnetic field which is perturbed by the nuclear magnetic moment of the nitrogen atom on which it is localized. The nitrogen isotope ^{14}N has a spin quantum number $I = 1$ which allows only three orientations of the nuclear magnetic moment with respect to an applied field (parallel, antiparallel and perpendicular). It is these three perturbing fields which split the ESR spectrum into three equally spaced resonances (Fig. 3).

When the molecular motion of the spin label is restricted, for example, by increasing the viscosity of the medium, the spectra pass through well characterized changes. Three stages in this continuous transition are

J. C. Metcalfe

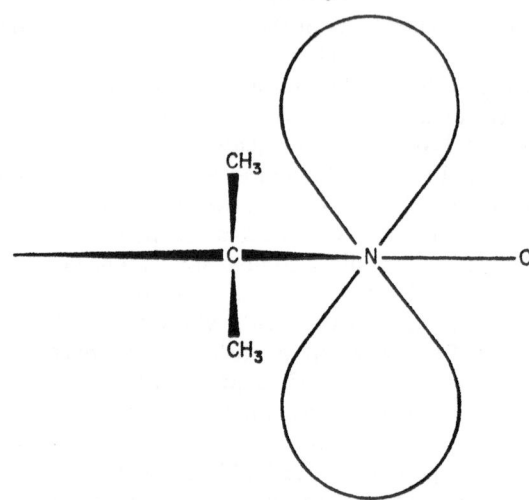

FIG. 1. 2pπ orbital of sterically protected nitroxide radicals.

illustrated in Fig. 4. The first effect is a broadening of the high field resonance, and spectra of this kind are described as weakly immobilized ($\sim 10^9$ sec^{-1}). At lower tumbling rates towards 10^8 sec^{-1}, all three resonances become broadened and asymmetric in the intermediate range of immobilization. Finally, when the nitroxides are highly immobilized, for example, by freezing in a glass-like structure, the spectrum consists of a low field maximum, a central line, and a minimum at high field. These strongly immobilized nitroxide spectra correspond to tumbling rates up to $\sim 10^7$ sec^{-1}. It will be seen from Fig. 4 that the spacing between the signals as well as their shape is determined by the molecular motion of the

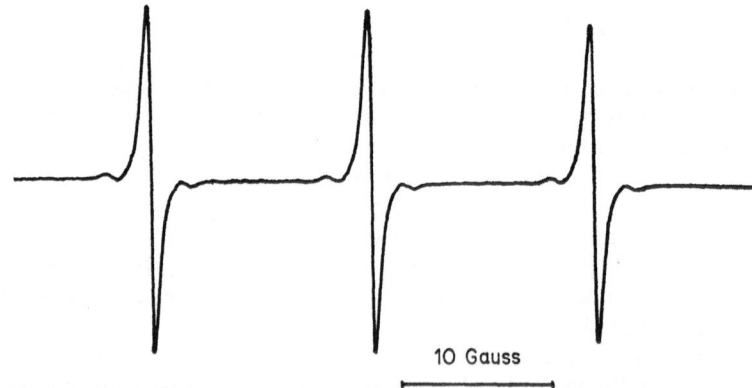

10 Gauss

FIG. 2. Spectrum of 1×10^{-4} M TEMPO in aqueous solution.

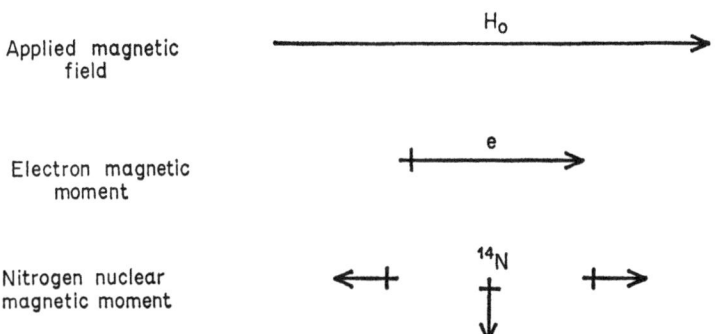

FIG. 3. The nitroxide spectrum is split into a triplet by the hyperfine interaction of the unpaired electron with the nitrogen nuclear magnetic moment.

spin label. This increases from ~ 32 gauss between the high and low field resonances for the freely tumbling nitroxides to ~ 64 gauss between the maxima and minima for highly immobilized nitroxides. Although the spectra look somewhat complex, they are in fact well understood both qualitatively and quantitatively. An excellent introduction to nitroxide spectra and spin labelling is given by Hamilton & McConnell (1968). The spectra in the intermediate immobilization range are particularly sensitive to changes in molecular motion, a change by a factor of 2 being readily detected. In addition, the splitting and the shape of the curves is markedly affected by anisotropy in the motion of the spin labels. Hubbell & McConnell (1969) have detected a preferred axis of rotation of a steroid spin label perpendicular to the membrane interface in nerve membranes and in erythrocytes oriented by hydrodynamic shear. The range of intermediate immobilization is therefore especially useful for probing changes in both orientation and motion within membranes.

A useful property of a free or weakly immobilized nitroxide is that the splitting between its resonances depends on the solvent which affects the polarity of the nitroxide group. It was this property which enabled Hubbell & McConnell (1968) to distinguish between TEMPO in an aqueous medium and TEMPO weakly immobilized within a fluid hydrophobic region of the nerve membrane.

The insertion of a spin label into the membrane may itself cause a perturbation. A practical test of whether the perturbation is significant in a surviving tissue is to determine if the physiological functions of interest are affected by the presence of the spin label. For erythrocyte membranes we regard a perturbation as unimportant if it does not affect the stability of the membrane to hypotonic haemolysis. In practice, this implies a ratio of probe molecules to membrane lipids of less than about 1:100. A further experimental check is to repeat experiments at a range of spin label concentrations (from 5×10^{-4} M to 1×10^{-6} M) and to determine

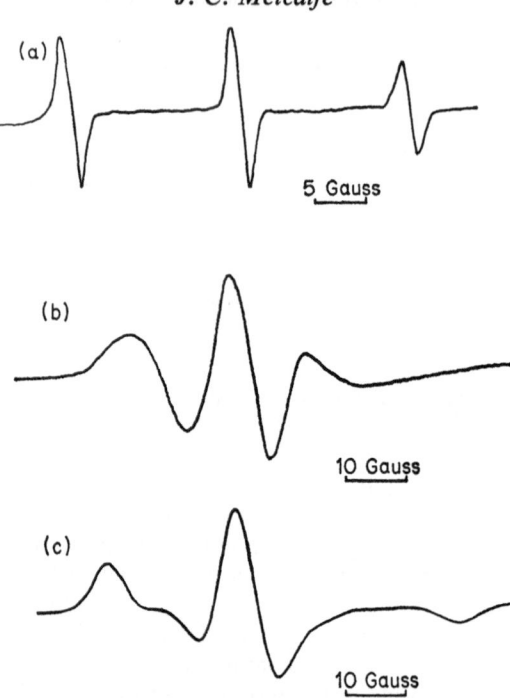

FIG. 4. Effect of progressive immobilization on the spectra of nitroxide radicals. The ranges of immobilization are commonly described as (a) weak; (b) intermediate; and (c) strong immobilization. They correspond to isotropic rotational rates of the order of 10^9 sec^{-1}; 10^8 sec^{-1}; and 10^7 sec^{-1} respectively.

whether the response to perturbations is affected by the concentration of probe within the membrane. We have found that very similar responses are observed in erythrocyte membranes at least up to 1.0×10^{-4} M with the spin labels used here.

A further problem relates to the polar nature of the nitroxide group. When this is inserted into a membrane, and particularly into a hydrophobic region, the nitroxide may induce an abnormal environment in its immediate vicinity and therefore report on perturbations of an already distorted environment. This cannot be completely excluded with a probe technique, but the consistent responses of a range of spin labels, in which the nitroxide is known to be located at different regions of the membrane, suggest that this is not an important limitation with nitroxide spin labels. It is, however, important to note that spin labels which are structural analogues of membrane components may differ significantly in their interactions with the membrane in comparison with the native components. Nitroxide groups biosynthetically incorporated into a membrane (Keith, Waggoner & Griffith, 1968) may reasonably be regarded as reflecting perturbations of native membrane components. It may also be justifiable to extrapolate data from a nitroxide to its unlabelled analogue where a particular biological activity is common to the two structures.

Nuclear magnetic resonance measurements

Information about the motion of an extraneous molecule within the membrane is obtained from the relaxation rate of proton nuclei of the molecule. The line width of the resonance absorption signals from the protons are directly proportional to their relaxation rate $(1/T_2)$, where $(1/T_2) = \pi \Delta \nu_{1/2}$ and $\Delta \nu_{1/2}$ is the line width of the resonance at half height. T_2 is the spin-spin relaxation time.

In the conditions of the present experiments only a small fraction ($<10\%$) of the benzyl alcohol or xylocaine molecules are bound to the membrane at any instant and the observed relaxation rate $(1/T_2)_{\text{obs}}$ for the protons of the anaesthetics is the weighted mean of the free and bound forms. That is,

$$(1/T_2)_{\text{obs}} = \alpha(1/T_2)_{\text{bound}} + (1 - \alpha)(1/T_2)_{\text{free}} \tag{1}$$

where $(1/T_2)_{\text{bound}}$ and $(1/T_2)_{\text{free}}$ are the relaxation rates of protons bound to the membrane and free in solution, respectively, and α is the fraction of molecules bound. If α is known, we can calculate $(1/T_2)_{\text{bound}}$ and compare the relaxation rates of anaesthetic molecules within the membrane at each concentration. T_2 is directly proportional to the rate of rotation of the protons, so we can follow changes in the molecular motion of the anaesthetic molecules in the membrane from changes in $T_{2\text{ bound}}$. It should be noted that equation (1) applies only to molecules exchanging sufficiently

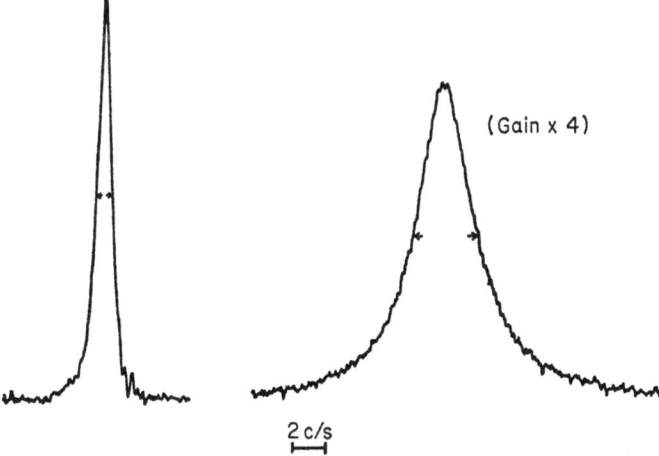

(Gain x 4)

2 c/s

FIG. 5. Aromatic resonance of 150 mM benzyl alcohol in solution (left) and in the presence of 1·0% by weight erythrocyte membranes (right). Temperature 25°C.

rapidly to give an averaged line width for the free and bound forms, and that both benzyl alcohol and xylocaine fulfil this condition.

Partition coefficients were measured by equilibrium dialysis in the same D_2O buffer as the NMR experiments to provide α values directly applicable to the NMR data. Unless otherwise stated, buffer solutions contained 45 mM NaCl, 30 mM Na acetate and 5 mM Na phosphates, at pH 7·4.

Haemolysis experiments were performed as described by Seeman & Weinstock (1966).

Nuclear magnetic relaxation: results with benzyl alcohol

$$\langle\!\!\langle\rangle\!\!\rangle\!\!-CH_2-OH$$

The broadening of the resonance line from the aromatic protons of benzyl alcohol in the presence of an erythrocyte membrane suspension is shown in Fig. 5. From the line width at each alcohol concentration and the corresponding α value, the relaxation rate of the bound molecules $(1/T_2)_{bound}$ is calculated. The relaxation rate decreases with increasing concentration, but there is a pronounced upswing at higher concentrations (Fig. 6).

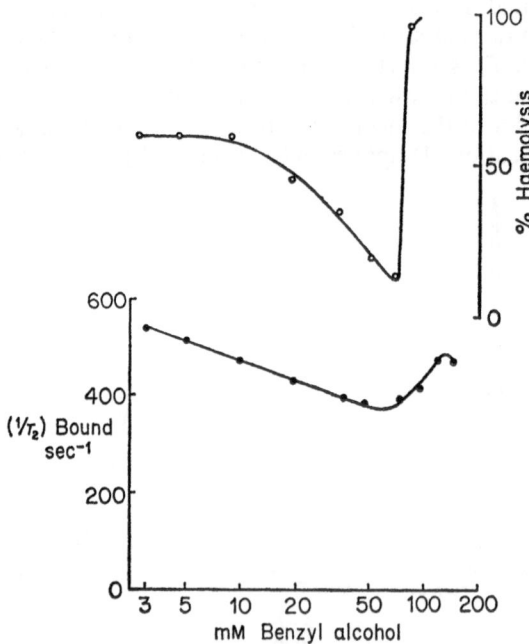

FIG. 6. Lower curve, relaxation rate of the aromatic protons of benzyl alcohol bound to erythrocyte membranes at 25°C. Upper curve, effect of benzyl alcohol on the hypotonic haemolysis of human erythrocytes.

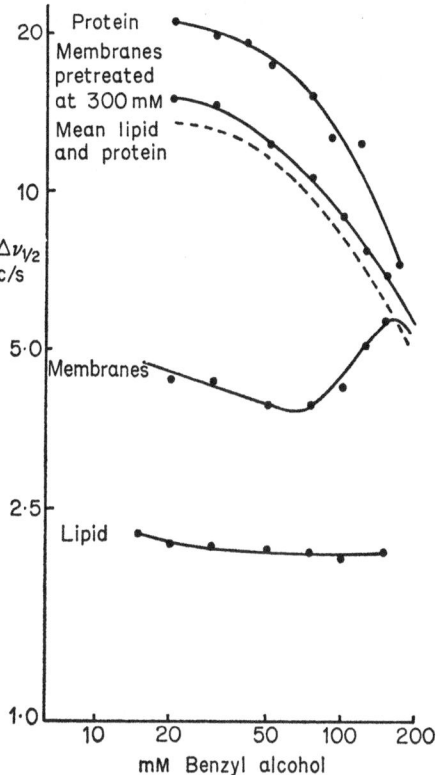

FIG. 7. Line widths of aromatic protons of benzyl alcohol in various membrane preparations (1·0% by weight) at 25°C. The dashed curve is the mean line width calculated from the separated protein and lipid curves, corresponding to the composition of the intact membrane.

The upswing coincides with the lytic concentration range for benzyl alcohol; at a concentration below 80 mM, benzyl alcohol stabilizes the erythrocyte membrane against hypotonic haemolysis, but at higher concentrations it causes lysis (Fig. 6). The correlation between the upswing in the NMR relaxation curve and the onset of lysis has been observed with all the anaesthetics examined and is discussed later.

To begin to separate the contributions of lipid and protein to the overall interaction in the intact membrane, the line width measurements were repeated using the separated membrane lipids in a vesicle suspension, and a soluble preparation of membrane protein (>95%) obtained by the butanol separation method of Maddy (1966). The curves for both membrane lipid and protein are monotonic in contrast to the biphasic form of the intact membrane (Fig. 7). It is clear that the mean line width of the separated components corresponding to the composition of the membrane does not coincide with the curve for the intact structure. In particular, any contribution to the line width from the protein in the intact membrane must be greatly restricted compared with the separated form.

The origin of the upswing became clear on examining the reversibility of the line width changes. In the prelytic concentration range below 80 mM, the curve is entirely reversible. On the other hand, membranes pretreated with lytic concentrations of benzyl alcohol show irreversible line width

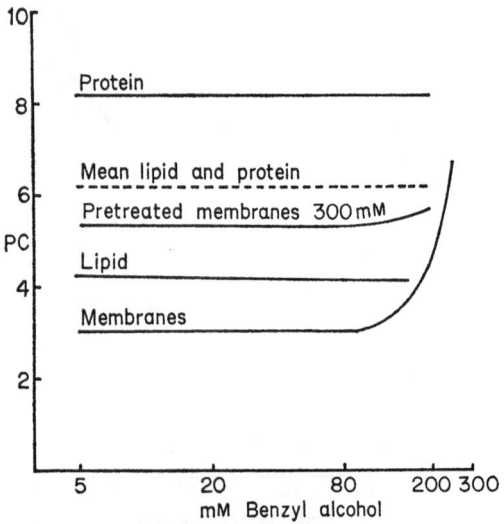

Fig. 8. Partition coefficients (PC) of benzyl alcohol in various membrane preparations at 25°C.

changes. In Fig. 7, the curve for membranes pretreated at 300 mM benzyl alcohol is close to the mean curve for the separated components. The major contribution to the calculated curve is from the protein.

We interpret the decreasing relaxation rate in the prelytic concentration range as indicating that the alcohol molecules are inserted into an increasingly fluid environment as the alcohol concentration increases. At the lytic concentration, the membrane is so perturbed that new protein binding sites are exposed which were protected and inaccessible at lower concentrations. The exposure of these new binding sites marks the onset of irreversible perturbation of the membrane by benzyl alcohol.

The location of the new binding sites on the membrane protein is confirmed by measurements of the partition coefficient for the membrane and the separated components (Fig. 8). The intact membrane has a constant partition coefficient up to 80 mM, which then increases at higher concentrations. Both lipid and protein have higher partition coefficients than the intact membrane. The partition coefficient is unaltered by pretreatment with benzyl alcohol up to 80 mM, but the value for membranes pretreated with 300 mM benzyl alcohol lies close to the mean value for the separated components corresponding to the membrane composition. It is therefore the partition coefficient of the membrane protein which makes the major contribution to the partition coefficient of the pretreated membranes, and which accounts for the increasing partition coefficient of the untreated membranes in the lytic concentration range.

It is concluded that both the NMR and partition coefficient data indicate that in the lytic concentration range the membrane progressively interacts with benzyl alcohol as the sum of its separated components and indeed some gross physical disruption of the structure is observed. In the prelytic concentration range, interaction between the membrane components modifies and restricts interaction of the membrane with benzyl alcohol. This results directly in the low partition coefficient and relaxation rate of benzyl alcohol in the prelytic range for intact membranes.

Results with xylocaine

Xylocaine is a tertiary amine with a pK of 7·4. Experiments were performed at pH 6·4 where $\sim 90\%$ of the xylocaine is in the cationic form. Xylocaine progressively stabilizes the erythrocyte membrane against hypotonic haemolysis with increasing concentration, but eventually causes lysis. The onset of lysis depends both on the pH and the temperature and occurs at 400 mM at 25°C, pH 6·4 (Fig. 9).

Line width measurements were made on the six methyl protons on the aromatic ring. The form of the curve is very similar to that of benzyl alcohol and the upswing again coincides with the lytic concentration range (Fig. 9). When the lytic range is altered by changing the pH or the temperature, there are corresponding shifts in the position of the line width upswing. Irreversible changes in the NMR curve similar to those observed after pretreatment with benzyl alcohol are produced by pretreatment with lytic concentrations of xylocaine.

Effects of calcium ions

So far benzyl alcohol and xylocaine seem to interact in a very similar way with the erythrocyte membrane, as judged by the NMR data. But they show quite distinct responses to the presence of calcium. Concentrations of Ca^{2+} up to 10 mM have very little effect on the line width curve for benzyl alcohol. Calcium also causes only minor changes in the effect of benzyl alcohol on the haemolysis of the erythrocyte membrane. Both observations suggest that calcium has little effect on the interaction of benzyl alcohol with the erythrocyte membrane. On the other hand, the effect of Ca^{2+} on the xylocaine NMR curve is readily detected; 50 mM Ca^{2+} results

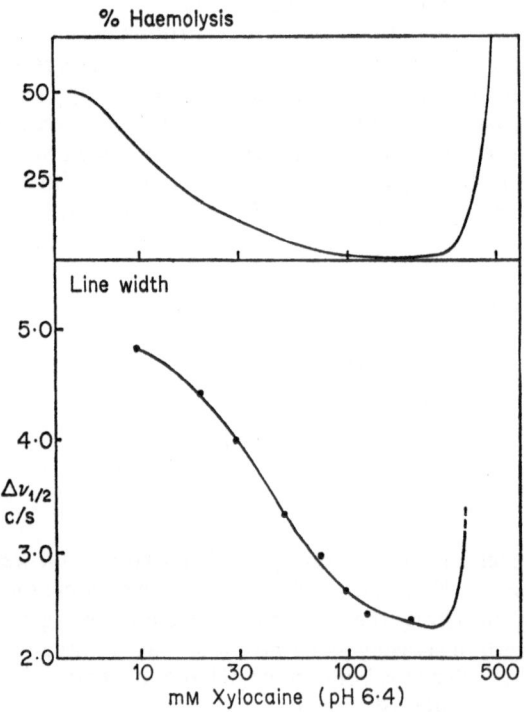

FIG. 9. Lower curve, line widths of the 2,5 methyl protons of xylocaine in 1·0% by weight erythrocyte membranes at 25°C. Upper curve, effect of xylocaine on the hypotonic haemolysis of human erythrocytes.

in a reduction of line width of the methyl protons by up to 40% at low concentrations of xylocaine although the effect decreases with increasing xylocaine concentrations (Fig. 10). The simplest explanation of this observation is that the calcium cations are able to displace xylocaine in its cationic form from the membrane, and that the reduction in the partition coefficient reduces the observed line width. Further evidence that both Ca^{2+} and xylocaine compete for cation binding sites in the membrane is presented later.

A quantitative evaluation of this competition is complicated in this experiment by the presence of both charged and uncharged forms of

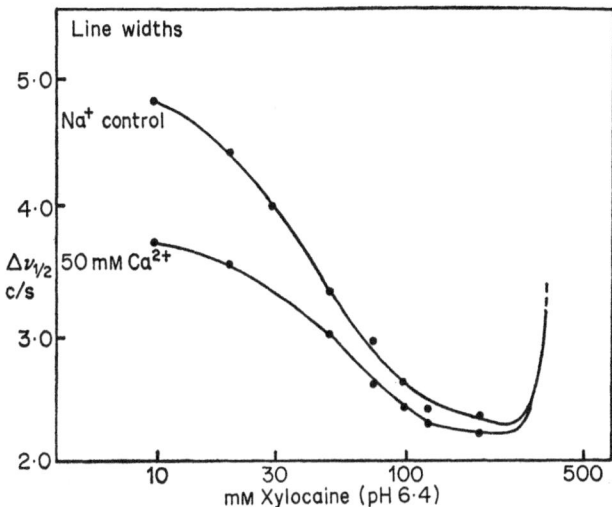

FIG. 10. Line widths of the 2,5 methyl protons of xylocaine in 1·0% by weight erythrocyte membranes at 25°C. The Na^+ control measurements were made in a buffer containing 200 mM NaCl, 30 mM sodium acetate, 10 mM sodium phosphates. The lower curve was obtained by replacing 100 mM NaCl by 50 mM $CaCl_2$.

xylocaine. It is probable that the uncharged form will have a higher partition coefficient into the membrane, and that it will not be displaced by Ca^{2+}.

Relation of the NMR curves to anaesthetic action

It has been shown for a wide range of chemical compounds that the concentration which produces maximum stabilization of the intact erythrocyte against hypotonic haemolysis is directly proportional to the anaesthetic potency (see Seeman, 1966a). For all the anaesthetics which have been examined directly by NMR, the upswing in the line width occurs at the same concentration as the onset of lysis induced by the anaesthetic. The range of anaesthetics with suitable spectral and solubility properties for direct measurement is limited. It has been found, however, that the interaction of other molecules with the membrane can be followed indirectly by a reporter technique. This depends on the observation that the upswing in the NMR curves occurs simultaneously for two anaesthetics present together in the membrane, so that they are detecting essentially the same membrane perturbations. Using a low reporter concentration of benzyl alcohol, it has been possible to show that the correlation between the NMR upswing and lysis holds for a wide range of compounds which could not be examined directly. This is illustrated for the *n*-alkyl alcohols, using benzyl alcohol as the reporter in Fig. 11.

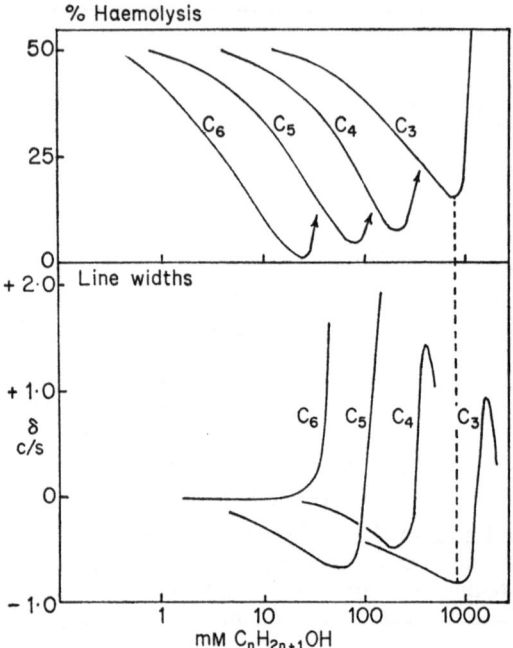

FIG. 11. Lower curves, changes induced in the line width of the aromatic protons of 15 mM benzyl alcohol by the *n*-alkyl alcohols in the presence of 1·0% erythrocyte membranes. Upper curves, corresponding effects of the *n*-alkyl alcohols on the haemolysis of human erythrocytes. (Adapted from Seeman, 1966b.)

ESR spin labels

Dodecyl dimethyl tempoyl ammonium (I)

$$CH_3-(CH_2)_{11}-\overset{\displaystyle CH_3}{\overset{\displaystyle |+}{N}}-CH_3 \qquad (I)$$

This spin label illustrates both the kind of perturbations induced in erythrocyte membranes by local anaesthetics, and also the competition for cation binding sites in the membrane. It is probable that the spin label I will be preferentially oriented in the membrane with the quaternary charge at the polar interface region with the solvent. This is supported by the observation of Hubbell & McConnell (1969) that the nitroxide of this label inserted into phosphatide vesicles is rapidly reduced by aqueous reducing agents.

The spectrum of (I) in an erythrocyte membrane suspension is shown in Fig. 12a. The spectrum is broadened and asymmetrical compared with a freely tumbling nitroxide and the label is in the intermediate immobilization range. There is a small component of free spin label detectable in the spectrum at the high field region. It has been confirmed by centrifugation that this is unbound spin label rather than a weakly bound fraction, and the partition coefficient for spin label I is approximately 2,000.

In the presence of increasing concentrations of benzyl alcohol, the spectrum is progressively sharpened, corresponding to a fluidizing effect of the anaesthetic on the environment of the spin label in the membrane (Fig. 12a). The spectrum shows no evidence of new binding sites in the lytic concentration range. There is a slight decrease in the amount of free spin label with increasing alcohol concentration, and this implies that the partition coefficient is slightly increased in the presence of benzyl alcohol. A related observation by Seeman is that the binding of Ca^{2+} is increased by uncharged anaesthetics, and it seems probable that they cause a general increase in cation binding, of which the Ca^{2+} and spin label effects are examples.

It is of interest to compare the effect of benzyl alcohol on the spectra of the spin label bound to the separated membrane components (Fig. 12b,c). Both lipid and protein have a high binding capacity for the label, and benzyl alcohol fluidizes the environment of the spin label at both types of binding site. The similar response of both components with this particular spin label probably accounts for the continuous nature of the spectral changes observed in the intact membrane over the whole concentration range.

An entirely different response of the spin label to perturbation by xylocaine is observed. Instead of a smooth change in the whole spectral envelope, there is a progressive increase in the concentration of free spin label, superimposed on the spectrum of the bound component, as the concentration of xylocaine is increased (Fig. 13). There is little change in the spectrum of the spin label remaining bound in the membrane, except for the reduction in amplitude corresponding to the amount of displaced spin label.

All the cations which have been examined cause the displacement of the spin label, but it has not been observed with neutral or anionic molecules. The effectiveness of simple cations depends on the valency and follows the order $Ce^{3+} > Ca^{2+} > Na^+$. For monovalent cations in molecules with hydrophobic groups the displacing activity seems to depend on the partition coefficient and hence on the local concentration of the agent in the membrane. For example, both chlorpromazine and the dodecyl trimethyl ammonium ion are much more effective competing

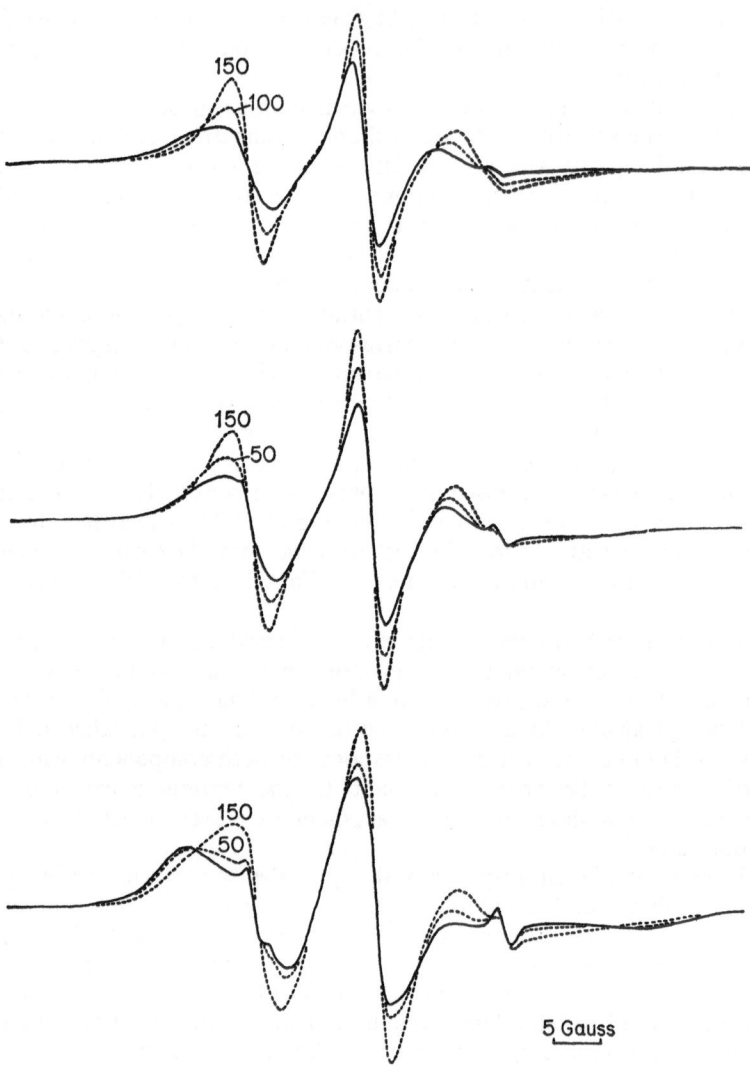

FIG. 12. Effect of benzyl alcohol on membrane preparations containing
1×10^{-4} M spin label I. (————), no benzyl alcohol; (– – – –), with benzyl
alcohol. Temperature 25°C. The first tracings (a) 1·0% erythrocyte membranes, with 0, 100, and 150 mM benzyl alcohol; the second tracings (b) 1·0%
membrane lipids, with 0, 50, and 150 mM benzyl alcohol; the third tracings,
(c) 0·3% membrane protein, with 0, 50, and 150 mM benzyl alcohol.

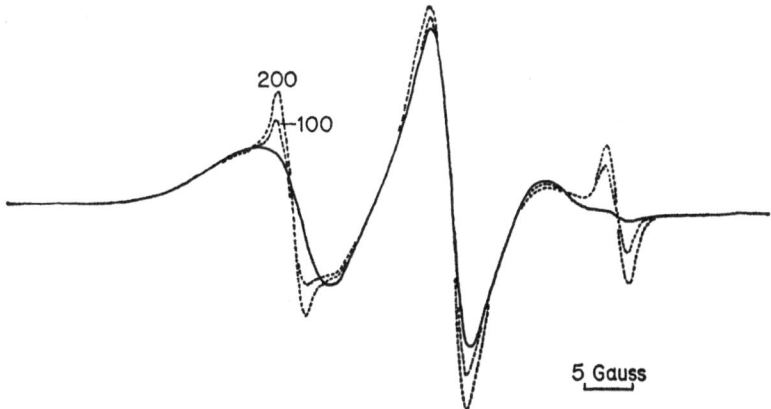

FIG. 13. Same as Fig. 12 (a), but with 0, 100, and 200 mM xylocaine.

agents than xylocaine. The alkyl ammonium compound is an analogue of the spin label in which the tempoyl group carrying the nitroxide function is replaced by a methyl group.

Together these results suggest that there are a limited number of binding sites for the cationic spin label in the membrane and that the label can be displaced competitively by other cations. Competition depends on the valency of the charge and the concentration of the agent in the membrane.

Preliminary experiments with the separated membrane components indicate that Ca^{2+} is more effective at displacing the spin label from the membrane protein than from lipid vesicles. The ability of hydrated cations such as Ca^{2+} to displace the spin label from the membrane suggests that the cationic group on the label must be accessible to the aqueous medium and is consistent with the expected orientation of the label.

Steroid nitroxide (II)

(II)

This label shows a significant anisotropy in its motion in both nerve and erythrocyte membranes (Hubbell & McConnell, 1969). It rotates preferentially about its long axis which is oriented perpendicular to the membrane surface, and is restricted in its rotation about the other two axes.

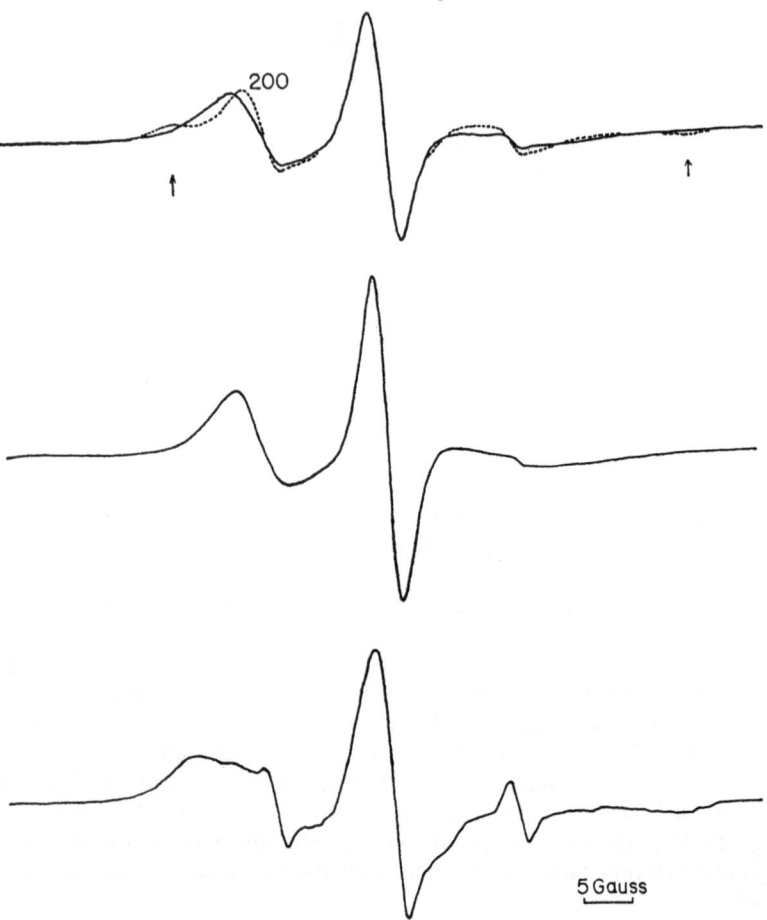

Fig. 14. Membrane preparations containing 1×10^{-4} M spin label II. Temperature 25°C. First (a) 1·0% erythrocyte membranes, with a lytic concentration of benzyl alcohol (200 mM). The arrows indicate the strongly immobilized component of the spectrum which is only observed in the lytic concentration range; second (b) 1·0% membrane lipids; third (c) 0·3% membrane protein. Note especially the strongly immobilized component present in this spectrum.

The spectrum of the label in erythrocyte membranes is shown in Fig. 14a. Prelytic concentrations of benzyl alcohol again cause a progressive sharpening of the spectrum similar to that observed with the previous label I. The environment of the spin label is therefore more fluid and also less anisotropic. At lytic concentrations of benzyl alcohol a new spectral component is observed which corresponds to a fraction of the spin label molecules which are now in a highly immobilized environment (Fig. 14a). The spectra of spin label II bound to the separated membrane components are shown in Fig. 14b and c. It is immediately apparent that there is a highly immobilized component in the membrane protein spectrum which is similar to that observed in the whole membrane at lytic concentrations of alcohol. There is little change in the spectrum of spin label II bound to membrane protein in the presence of benzyl alcohol, in marked contrast to the fluidizing effect on the spectrum of label I bound to membrane protein. The highly immobilized component of the spectrum of label II in the membranes therefore remains clearly resolved in the lytic concentration range, and can be attributed to new protein binding sites for the spin label which were inaccessible at prelytic concentrations. This is precisely comparable with the conclusions from the NMR experiments. The response of the lipid bound spin label II to benzyl alcohol is very similar to that observed with label I and there is simply a progressive fluidizing effect over the whole concentration range.

Xylocaine produces only modest spectral changes with spin label II, qualitatively similar to those of benzyl alcohol. Calcium had no significant effect on the spectrum of this label either in the membrane or the separated components. In general, uncharged spin labels do not distinguish the perturbations produced by neutral and cationic agents.

Conclusions

The most useful feature of the results obtained from the two magnetic resonance techniques is the consistent description of interactions with the membrane which they provide. For example Ca^{2+} has no significant effect on the interaction of neutral molecules such as benzyl alcohol with the membrane, as judged either by NMR or by the use of neutral spin labels. On the other hand, the competitive nature of the binding of Ca^{2+}, xylocaine, and the cationic spin labels is readily detected by both techniques, and it is clear that the membrane has a restricted capacity for the uptake of cations. Quantitatively the effectiveness of cations in competing for the binding sites depends on the valency and the partition coefficient, and it should be possible to determine the distribution of cation binding sites between lipid and protein in the intact membrane. A detailed analysis of cation binding will be presented elsewhere.

The magnetic resonance data also provide information about changes in molecular motion within the membrane when perturbed by anaesthetic agents. In the erythrocyte membrane the environment of the anaesthetic molecules becomes more fluid with increasing anaesthetic concentration in the prelytic concentration range, as evidenced by the NMR relaxation rate of the anaesthetic. The same effect of anaesthetics on a range of spin labels, including those described here, suggests that the increasing mobility of the anaesthetic reflects an increased fluidity of the membrane components themselves. The consistency of these effects obtained by two distinct techniques is powerful support for the interpretation of molecular changes in the membrane.

While the membrane clearly tends to interact with benzyl alcohol as the sum of its separated components in the lytic concentration range, in the prelytic range the structure of the membrane itself modifies and restricts the interaction of each component with benzyl alcohol. The partition coefficient, the NMR relaxation rate, and the ESR spectra in the prelytic range are all characteristic of the intact membrane and provide a stringent set of criteria for the intact structure. All the modifications of the structure which we have examined result in the alteration of one or more of these parameters, and the reversibility of the membrane modifications can be readily determined. Thus it is possible to determine whether a membrane is structurally intact from its interactions with extraneous molecules.

Lytic concentrations of anaesthetics cause irreversible changes in the structure of the membrane. With benzyl alcohol the changes are marked by the increase in partition coefficient, the upswing in the NMR relaxation rate, and the new component in the ESR spectrum of the steroid label. All the evidence indicates that the new binding sites responsible for these changes in the lytic range are predominantly on the membrane protein and are inaccessible at prelytic concentrations. It is concluded that in the lytic concentration range the interactions between membrane components which are responsible for maintaining the structure in a form in which it can be reversibly perturbed are disrupted. The identification of these critical interactions between the membrane components, which are marked by the abnormal protein binding sites, seems to be feasible and should indicate an important structural element in the membrane.

This is one approach to the use of membrane perturbations to provide direct information about the organization of membrane components. More generally the interpretation of perturbations in specific structural terms is likely to be limited unless the probe molecules and perturbations can be precisely localized within the membrane. For example, although it is clear that in the prelytic concentration range the environment of benzyl alcohol molecules becomes increasingly fluid, the components of the membrane in which the perturbations are occurring are not immediately distinguishable. The difficulty in distinguishing the binding of extraneous molecules to lipids and protein in the intact membrane is apparent from the similarity of the spectra of most spin labels when bound to the separated membrane components. In addition, similar spectral changes are observed in response to perturbing agents. These similarities derive from the common nature of the chemical forces involved in the interactions with either component. For this reason, attempts to determine the localization of small molecules in the membrane must generally rest on quantitative rather than qualitative differences in the binding to membrane lipids and protein.

A detailed analysis of the distribution of benzyl alcohol between the lipid and protein in the intact erythrocyte membrane is presented elsewhere (Burgen, Colley, Turner, Metcalfe & Metcalfe, in preparation).

Two independent estimates indicate that the major contribution to the partition coefficient of benzyl alcohol in the intact membrane is from binding to membrane protein.

Methods of localizing probe molecules are particularly important if structural perturbations are to be well defined. Unfortunately, noncovalent probes which are highly specific for one or other component are likely to be rare, because of the similarity of the chemical interactions of both components which has already been mentioned. In this respect the spin label TEMPO is of interest because it seems to be a rather specific probe for the lipid regions of at least some membrane systems (Hubbell & McConnell, 1968). All the other non-covalent spin labels which have been examined bind to both separated lipid and protein from erythrocyte membranes.

Another problem in the interpretation of perturbation effects is the possibility that perturbations may be propagated through the coherent structure of the membrane from their initial location to more distant components. Such propagated perturbations may be regarded as analogous to the well known co-operative interactions between protein conformations, and have been formalized for model membrane structures in some detail by Changeux (see, for example, Changeux, 1969). It seems entirely feasible for example, that the activity of functional membrane proteins might be affected by a hypothetical perturbing agent which is specifically localized in the lipid region of the membrane.

The possibility that perturbations are propagated through this membrane structure greatly increases the difficulty in identifying which of a set of perturbations are directly responsible for a particular effect on membrane function, such as conduction in neural membranes. It is possible that anaesthetics may simultaneously affect the organization and interactions of a number of membrane components, and the structure of associated water. One or more such perturbations may alter the critical geometry of functional membrane binding sites and be sufficient to produce the same physiological end effect.

This prompts the question of whether the particular kind of perturbations of molecular motion examined here in erythrocyte membranes have any relevance to anaesthetic action. It is worth noting that similar perturbations by anaesthetics were detected in nerve membranes in the spin label studies of Hubbell & McConnell (1968, 1969), and it is clear that the response is not an idiosyncracy of the erythrocyte membrane. This may account for the finding of Seeman (1966a) that anaesthetic potency in nerves is directly correlated with the stabilization of the erythrocyte membrane against haemolysis. This suggested that a similar pattern of structural changes may be induced in a range of quite distinct membranes. In support of this, a number of diverse membranes have been found to cause NMR relaxation changes in benzyl alcohol similar to those described here for the erythrocyte membrane (Metcalfe & Burgen, 1968). It seems unlikely, however, that anaesthetic action results uniquely from a fluidizing effect of anaesthetics on the membrane components because some preliminary data suggest that anaesthetic action is also associated with molecules which increase the order or packing of the membrane. At present we consider that a unique mechanism for anaesthetic action is improbable, but suggest that perturbations which can be well defined should provide a new insight into membrane structure.

The author wishes to acknowledge the collaboration of Professor H. M. McConnell, Dr. W. L. Hubbell, and Professor A. S. V. Burgen in parts of the work presented here.

REFERENCES

CHANGEUX, J. P. (1969). In Nobel Symposium II. *Symmetry and Function of Biological Systems at the Macromolecular Level.*, ed. ENGSTRÖM, ARNE & STRANDBERG, BOR, p. 235

COHEN, L. B., KEYNES, R. D. & HILLE, B. (1968). *Nature, Lond.*, **218**, 438

HAMILTON, C. L. & MCCONNELL, H. M. (1968). In *Structural Chemistry and Molecular Biology*, ed. RICH, A. & DAVIDSON, N., p. 115. New York: W. H. Freeman

HUBBELL, W. L. & MCCONNELL, H. M. (1968). *Proc. natl. Acad. Sci. U.S.*, **61**, 12

HUBBELL, W. L. & MCCONNELL, H. M. (1969). *Proc. natl. Acad. Sci. U.S.*, **63**, 16

KEITH, A. D., WAGGONER, H. S. & GRIFFITH, O. H. (1968). *Proc. natl. Acad. Sci. U.S.*, **61**, 819

MADDY, A. H. (1966). *Biochim. biophys. Acta*, **117**, 193

METCALFE, J. C. & BURGEN, A. S. V. (1968). *Nature, Lond.*, **220**, 587

METCALFE, J. C., SEEMAN, P. M. & BURGEN, A. S. V. (1968). *Mol. Pharmac.*, **4**, 87

SEEMAN, P. M. & WEINSTOCK (1966). *Biochem. Pharmac.*, **15**, 1737

SEEMAN, P. M. (1966a). *Int. Rev. Neurobiol.*, **9**, 145

SEEMAN, P. M. (1966b). *Biochem. Pharmac.*, **15**, 1632

CAFFEINE, CALCIUM AND
THE ACTIVATION OF CONTRACTION

H. C. LÜTTGAU
Department of Cell Physiology, Ruhr-University, Bochum, Germany

It has been known for some time that caffeine promotes activation and impedes inactivation of phasic muscle fibres. This occurs despite the fact that low concentrations of the drug have no effect on the electrical characteristics of the excitable membrane and on the actomyosin system of striated muscle fibres (Axelsson & Thesleff, 1958; Korey, 1950; Hasselbach, 1953). This leaves some part of the extensive tubular system as the most probable site of drug action (Sandow, 1965). The experiments with isolated muscle fibres described here provide some further information about the effect of caffeine on the coupling mechanism between excitation and contraction (Lüttgau & Oetliker, 1968). They also show how the drug—because of its specific action—can be useful in elucidating the causes of muscular failure (Lobsiger, 1968). References to further publications can be found in a review article by Sandow (1965).

Figure 1 gives a general description of the action of caffeine on isolated muscle fibres. Concentrations up to 1–2 mM produced merely a potentiation of contraction. With higher concentrations a contracture developed which reached the height of maximal potassium contractures with a half-time of 2–4 sec. Full tension could be maintained for minutes. This shows that the spontaneous relaxation of potassium contractures after only several seconds is not due to an exhaustion of energy reserves of the muscle. It suggests that normal relaxation is caused by an exhaustion of some kind of 'activator' or by an inhibitory mechanism.

The final step in the action of caffeine on excitation-contraction coupling is certainly an increase in the myoplasmic concentration of calcium ions. This has been directly shown by Weber & Herz (1968) and Weber (1968) with isolated fragments of the reticulum from frog and rabbit muscles. Weber tentatively suggests that the observed reduction in the calcium content of the reticulum under caffeine is due to an inhibition of calcium uptake. The experiments, however, probably do not completely exclude the alternative possibility, namely a release of calcium from the reticulum. But apart from these limitations in experimental evidence, there are theoretical considerations which make it difficult to believe that caffeine merely blocks the uptake of calcium. Caffeine can cause a maximal contraction within a few seconds of external application, so this assumption implies a persisting release of large amounts of calcium ions from the reticulum even in the resting state. In the absence of caffeine these quantities

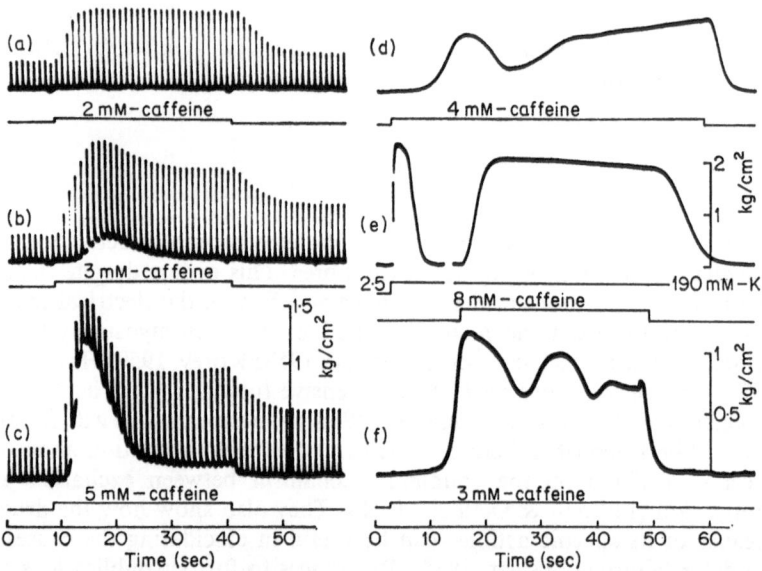

FIG. 1. Left: effect of different caffeine concentrations on isometric twitch tension. Right: contractures induced by the addition of caffeine to Ringer solution (d,f) and to a solution with 190 mM K (e). Three different fibres (diameter: 105, 145, and 164 μ), t = 21°–23°C. Isolated fibres from semitendinosus and iliofibularis muscles of *Rana temporaria* were used in all experiments described in the present article. (From Lüttgau & Oetliker, 1968.)

have to be reabsorbed at the same rate. Rough calculations show that if one ATP is needed for the active uptake of two calcium ions, the estimated energy consumption in the conditions described would become five to ten times larger than that actually measured in the resting state. It therefore seems that, at least in living muscles, the increase in the ionic concentration of calcium caused by caffeine in the myoplasm must to a great extent be due to a 'passive' release of calcium from the reticulum.

It is not unreasonable to assume a different action of caffeine on fragments of the reticulum. This is indicated by the observation that in mammals caffeine causes a release of calcium from isolated fragments of the reticulum although it does not induce a contracture in the whole muscle (Weber & Herz, 1968; Sandow, 1965). A reasonable agreement would be obtained if it is assumed that in the isolated tubular system caffeine acts predominantly on the active uptake and in living fibres it acts on the passive release.

Site of drug action
Experiments with radioactive caffeine indicate that the drug easily crosses cell membranes (Bianchi, 1962). It is therefore feasible that it reaches any part of the tubular system within seconds of application to the external solution. A consideration of how caffeine might cause the release of calcium from the terminal cysternae suggests two sites of action: either caffeine exerts its action in an indirect way in the tubular wall of the T-system or it acts directly on the calcium transport in the membranes of the terminal cysternae.

The tubular system of skeletal muscle fibres is well described elsewhere (Peachey, 1965). A few remarks about the main tubular elements, however, are necessary for the comprehension of the observations and arguments which follow. (1) The transverse tubular system (T-system) is part of the plasma membrane of the cell. There is experimental evidence that even large molecules like ferritin can easily diffuse from the external solution into the transverse tubes. (2) The terminal cysternae are part of the sarcoplasmic reticulum. They are in close contact—though separated by membranes—with the T-system and form most probably the calcium-accumulating part of the triads.

Lüttgau & Oetliker (1968) have put forward some arguments which in their opinion favour the T-system as the site of drug action. Two of them will be considered:

(1) Contractures caused by the addition of caffeine to Ringer solution can rapidly be suppressed solely by an increase in the ionic concentration of external calcium or magnesium, without any change in membrane potential. External calcium 'stabilizes' the contractile mechanism, for example, by shifting the threshold for contraction to less negative potentials, and a time of several seconds seems to be sufficient to adjust the calcium concentration of the T-system to that of the external solution. The calcium content of the myoplasm and of the longitudinal system, however, remains practically unchanged during this time. We therefore believe that external calcium interferes with an activator for contraction in the T-system. Because caffeine contractures can be influenced by calcium in exactly the same way as potassium contractures, it is only reasonable to assume that caffeine acts at the same site.

(2) The amplitude of caffeine contractures depends on membrane potential. We were able to show that fibres with a potential between -50 and -20 mV are most sensitive to caffeine. A caffeine contracture initiated when the membrane potential was in this range could be reduced by shifting the potential to more positive or negative potentials. This observation also favours the T-system as the site of drug action because it is in direct

connection with the surface membrane. The experiments mentioned do not fully exclude the alternative possibility that caffeine acts at the longitudinal system. Such an assumption, however, would imply that the opposition of the two tubular systems allows control of the action of caffeine by external calcium and the membrane potential.

Mode of action

Experiments show that the drug affects both activation and relaxation of the contractile mechanism. This can best be demonstrated in a quantitative manner with low concentrations of caffeine which do not cause a contracture when the resting potential is normal (for example, 1–2 mM, see Fig. 1). We first consider the process of activation (Fig. 2). In normal conditions the development of tension became evident when the membrane potential was suddenly changed to values less negative than -50 mV and maximal tension was reached at about -30 mV. When 1·5 mM caffeine was added the threshold curve shifted by about 15 mV to more negative potentials. 50% of maximal tension was reached at -50 mV; this is the threshold in normal conditions (Fig. 3). The observed contractures showed a time course similar to that of potassium contractures. Their duration, however,

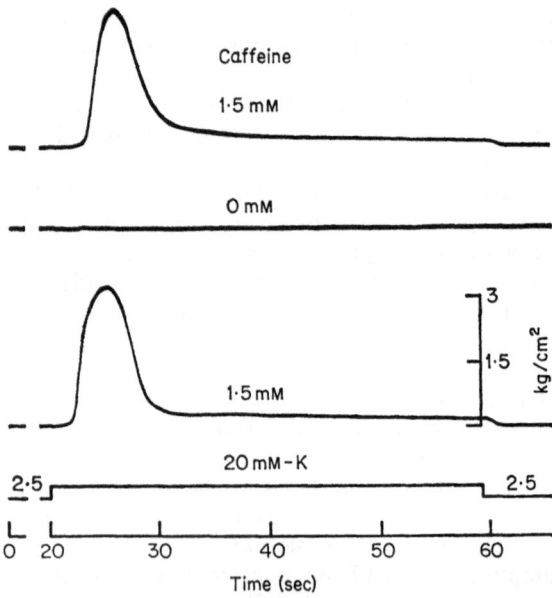

Fig. 2. Effect of caffeine on the mechanical threshold. 20 sec before the immersion of the fibre in the contracture solution with constant [K] [Cl] product the sodium in Ringer solution was replaced by choline. In the first and third run 1·5 mM caffeine was included in these solutions. Fibre 16, diameter 108 μ, maximal tension 5·7 kg/cm², $t = 22.5$°C. (From Lüttgau & Oetliker, 1968.)

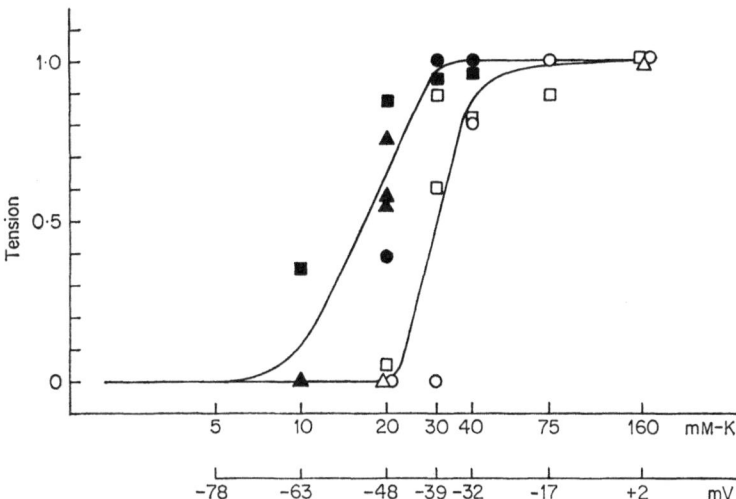

F<small>IG</small>. 3. Relation between peak tension and potassium concentration or membrane potential with 1·5 m<small>M</small> caffeine (filled symbols) and in the absence of caffeine (open symbols). 1·0 on the ordinate is equivalent to the average height of the first and last contracture in 190 m<small>M</small> K. The abscissa gives the potassium concentrations of the test solutions on a logarithmic scale. (A correction for activity coefficient has been made in plotting the right-hand point (190 m<small>M</small> K, 2·5 m<small>M</small> Cl) on the potassium scale. Except for this concentration a constant [K] [Cl] product of 300 (m<small>M</small>)^2 was used throughout.) Corresponding values of membrane potential are plotted on the lower scale. Three different fibres, $t = 22°$–$23°$C. (From Lüttgau & Oetliker, 1968.)

was longer and with higher concentrations of caffeine very often no relaxation took place. This shows that caffeine also impedes processes which cause relaxation. We have tested this in a more direct manner with a procedure described first by Hodgkin & Horowicz (1960). The fibre was depolarized to zero membrane potential by an isotonic solution of potassium sulphate (Fig. 4). When complete relaxation had occurred this solution was replaced by a test solution for restoration, and 20 sec later the restoration was again checked by a second complete depolarization. The test solutions contained 20–75 m<small>M</small> potassium (−50 to −15 mV). In normal conditions the ability to contract was completely restored when the membrane was repolarized to −50 mV. At membrane potentials between −50 and −20 mV only a partial recovery was possible. The degree of restoration was related to membrane potential by an S-shaped curve with a half value at about −35 mV. The addition of caffeine improved restoration. The restoration curve shifted by about 15 mV to less negative potentials when 1·5 m<small>M</small> caffeine was applied (Fig. 5).

The restoration curve presents the steady relation between the membrane potential and the state of the contractile system. A similar curve can

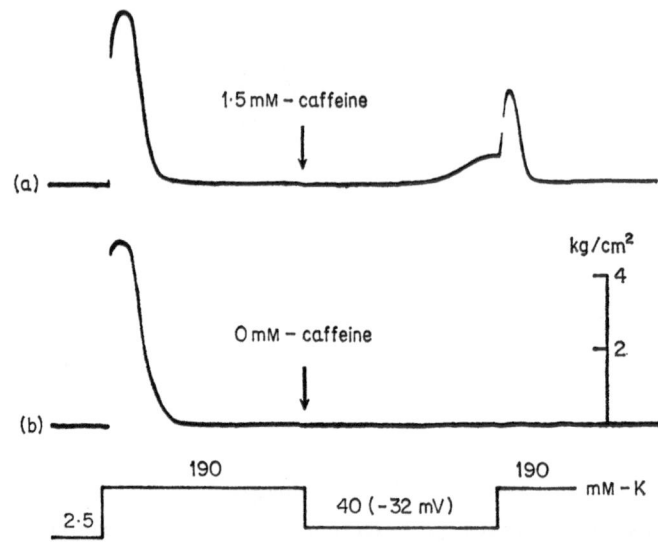

FIG. 4. Effect of caffeine on restoration of contractility. After a phasic contracture caused by a solution with isotonic potassium sulphate, the fibre was allowed to recover for 20 sec in a test solution containing 40 mM K at a constant [K] [Cl] product. Contractility was subsequently checked with a second application of the potassium sulphate solution. In (a) 1·5 mM caffeine was present during the recovery phase and the second exposure to potassium sulphate. Fibre 23, diameter 124 μ, maximal tension 5·0 kg/cm², $t = 22°C$. (From Lüttgau & Oetliker, 1968.)

therefore be obtained when the first complete depolarization is omitted (inactivation curve). An improvement in restoration is thus equivalent to a lesser degree of inactivation.

These observations on threshold and restoration of contraction show that caffeine facilitates activation as well as restoration and impedes relaxation (or 'inactivation') probably by 'easing off' a restraining mechanism which in the living fibre prevents a permanent contractile activity.

Because of its specific properties caffeine can be used as a tool for investigating muscular disorders. This can, for example, be demonstrated with a fatigue phenomenon observed in skeletal muscle fibres (Mashima, Matsumura & Nakayama, 1962; Eberstein & Sandow, 1963; Lobsiger, 1968). When muscle fibres of the frog are stimulated for prolonged periods of time at 1–4 Hz the twitch height decreases towards zero, although the shape of the action potentials changes only slightly and energy reserves are still available. The latter can easily be shown by the initiation of contractile activity with caffeine. Twitch tension often becomes larger than normal even after a previous decline to less than 5–10% of the original value. In

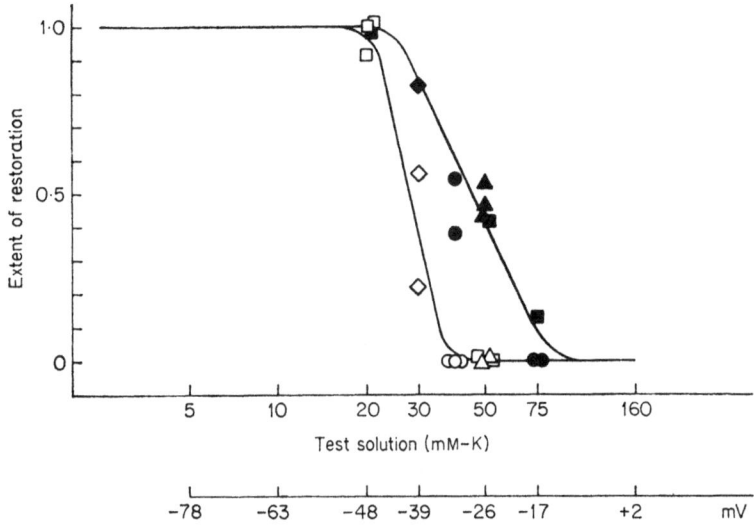

FIG. 5. Relation between potassium concentration and the state of the contractile system. Open symbols: without caffeine; filled symbols: 1·5 mM caffeine. Experimental procedure as described in Fig. 4. On the ordinate is plotted the height of the second contracture, expressed as a percentage of the height of the first one. The abscissa gives the potassium concentration of the test solutions on a logarithmic scale (constant [K] [Cl] product) and the corresponding membrane potentials. Four different fibres, $t = 21°–23°C$. (From Lüttgau & Oetliker, 1968.)

fibres poisoned with metabolic inhibitors fatigue appears earlier, but the addition of caffeine reveals that energy reserves are still available. A shift in threshold for contraction to less negative potentials is probably the main cause of fatigue (Lobsiger, 1968). This result suggests that the preservation of the mechanism which activates contraction is more dependent on metabolic process than the contractile apparatus itself.

The experiments described show that in the coupling of excitation and contraction a control and security mechanism is incorporated between the action potential and the onset of contraction. It is probably needed to protect fibres from losing all their energy reserves. Caffeine, which acts somewhere in the tubular system, is helpful in investigating this mechanism, even though the exact mode of action of the drug has not yet been clarified.

REFERENCES

AXELSSON, J. & THESLEFF, S. (1958). *Acta physiol. scand.*, **44**, 55
BIANCHI, C. P. (1962). *J. Pharmac. exp. Ther.*, **138**, 41
EBERSTEIN, A. & SANDOW, A. (1963). In *The Effect of Use and Disuse on Neuromuscular Functions*, p. 515. Prague.

HASSELBACH, W. (1953). *Z. Naturforsch.*, **8b**, 212
HODGKIN, A. L. & HOROWICZ, P. (1960). *J. Physiol., Lond.*, **153**, 386
KOREY, S. (1950). *Biochim. biophys. Acta*, **4**, 58
LOBSIGER, E. A. (1968). Thesis, Berne University
LÜTTGAU, H. C. & OETLIKER, H. (1968). *J. Physiol., Lond.*, **194**, 51
MASHIMA, H., MATSUMURA, M. & NAKAYAMA, Y. (1962). *Jap. J. Physiol.*, **12**, 324
SANDOW, A. (1965). *Pharmac. Rev.*, **17**, 265
PEACHEY, L. D. (1965). *J. cell Biol.*, **25**, 209
WEBER, A. (1968). *J. gen. Physiol.*, **52**, 760
WEBER, A. & HERZ, R. (1968). *J. gen. Physiol.*, **52**, 750

CALCIUM AND THE ACTION POTENTIAL IN SMOOTH MUSCLE

EDITH BÜLBRING
Department of Pharmacology, University of Oxford

T. TOMITA
Department of Physiology, Kyushu University

In most excitable membranes the action potential is produced by an inward movement of Na ions across the membrane. The maximum rate of rise in nerve is about 500 V/sec and is roughly proportional to the external Na concentration (Hodgkin & Katz, 1949). The action potential is abolished in Na-free solution. It is also known, however, that in some crustacean muscle fibres, Ca ions, not Na ions, carry the inward current for the production of the action potential (Hagiwara & Naka, 1964). The removal of Na ions from the external solution does not, therefore, change the size or the shape of the action potential.

In the smooth muscle of the guinea-pig taenia coli, the maximum rate of rise of the action potential is about 10 V/sec; this is the same order of magnitude as in crustacean muscle (Hagiwara & Nakajima, 1966) but less than one-tenth of that in most other excitable membranes.

When the spontaneous spike activity was studied, it was found that the amplitude of the action potential was not much affected by Na-deficiency, although the rate of rise was reduced (Holman, 1957, 1958; Bülbring & Kuriyama, 1963). In a solution containing less than 5 mM Na the spontaneous electrical activity was abolished, and it could be restored by excess Ca. From these observations, made on the spontaneously generated action potentials, it was concluded (*a*) that the action potential was due to Na entry, but that the inward movement was rather limited, or (*b*) that Ca could replace Na when the external Na concentration was reduced.

When the spikes evoked by electrical stimulation are investigated, the results suggest that the Ca contribution to the action potential is more important than the Na contribution (Brading, Bülbring & Tomita, 1969b).

Briefly, the observations are as follows: When the NaCl in the solution is replaced with sucrose, leaving 10 mM Na contained in the buffers, the membrane is hyperpolarized by about 10 mV and spontaneous spikes stop after

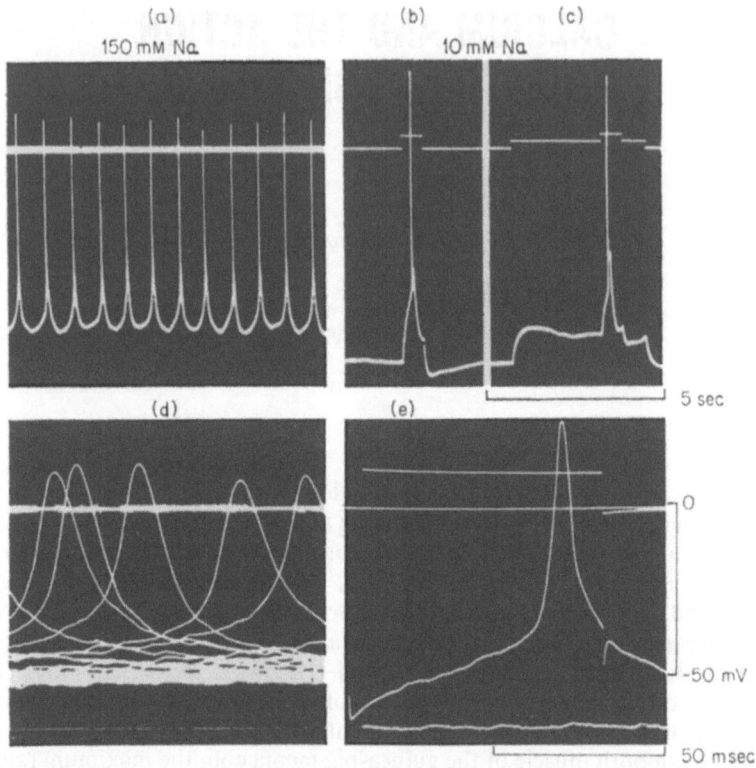

FIG. 1. Effects of a low external Na concentration on the action potential in taenia coli. 35°C. Spontaneous spikes recorded intracellularly in normal Krebs solution with (a) slow sweep and (d) fast sweep. After 1 hour in 10 mM Na solution (sucrose substitution) the spike was evoked by externally applied current, in (b) by a 600 msec pulse taken with slow sweep speed, in (e) evoked by a 75 msec pulse taken with fast sweep. In (c) the spike was evoked during conditioning depolarization. Upper trace show zero potential level and relative current intensity. Note hyperpolarization of the membrane, suppression of spontaneous activity, increase in amplitude and rate of rise of the spike in low Na solution. (From Brading, Bülbring, & Tomita, 1969b.)

about 10 min. Spikes, however, can easily be evoked by a depolarizing current pulse. The spike amplitude is larger and the rate of rise is faster than in normal solution (Fig. 1). Ionic analysis indicates that in the low Na solution the cells do not lose much Na, and it is probable that the Na-equilibrium potential is reversed, while the overshoot of the spike is increased from $+10$ to $+20$ mV.

In this low Na solution, Ca deficiency causes only a small depolarization of the membrane, but the amplitude and the rate of rise of the spike

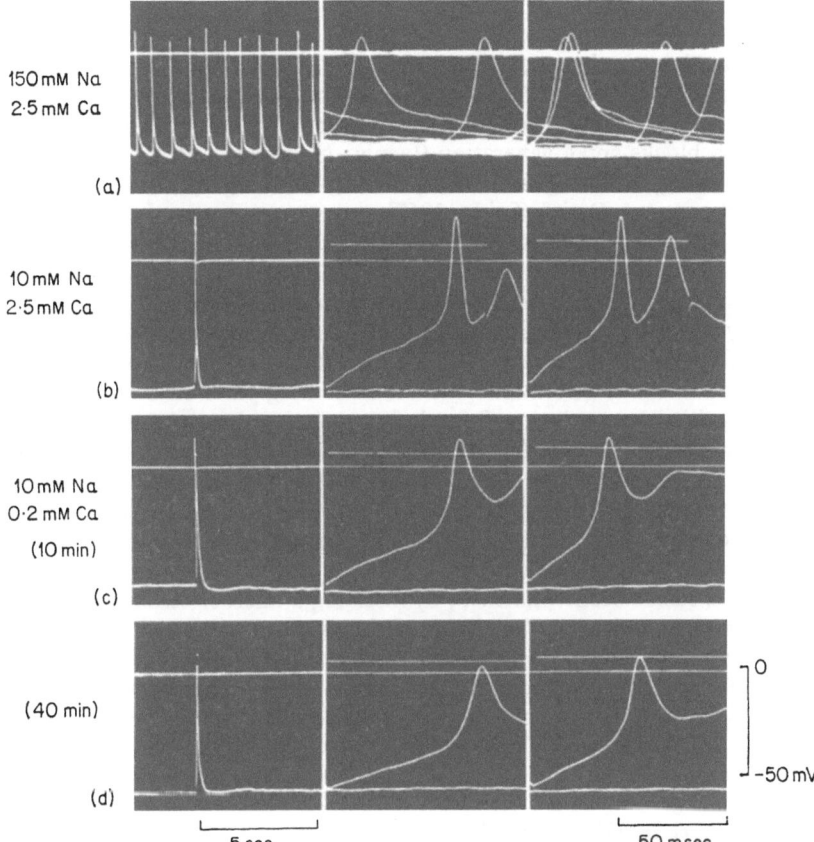

150 mM Na
2·5 mM Ca

(a)

10 mM Na
2·5 mM Ca

(b)

10 mM Na
0·2 mM Ca
(10 min)

(c)

(40 min)

(d)

0

−50 mV

5 sec   50 msec

Fig. 2. Taenia coli. 35°C. Effects of low Ca in the presence of a low external Na concentration (10 mM). (a) Spontaneous spikes recorded intracellularly in normal Krebs solution, at two different sweep speeds; (b) evoked after 30 min in 10 mM Na solution; (c) evoked spikes after 10 min and (d) after 40 min in 0·2 mM Ca and 10 mM Na. Note increase in rate of rise and amplitude of the spike in low Na, and gradual decline after reducing Ca. (From Brading Bülbring, & Tomita, 1969b.)

are considerably decreased (Fig. 2), suggesting that the spike is due to an increase in Ca conductance of the membrane. These changes, induced by reducing the external Ca concentration, are not instantaneous, and it may take about 30 min before the changes of spike configuration are complete.

The possibility that Ca is the main ion involved in the spike of smooth muscle is also suggested by the following observations which are similar to those made on crustacean muscle (Fatt & Ginsborg, 1958; Hagiwara &

Naka, 1964; Hagiwara & Nakajima, 1966) where the spike is due to Ca entry:

(i) Tetrodotoxin has no effect; it does not block the spike (taenia: Kuriyama, Osa & Toida, 1966; Nonomura, Hotta & Ohashi, 1966; Bülbring & Tomita, 1967; ureter: Washizu, 1966; Kuriyama, Osa & Toida, 1967).

(ii) Mn blocks spike generation (taenia: Nonomura, Hotta & Ohashi, 1966; Brading, Bülbring & Tomita, 1969b; Bülbring & Tomita, 1969b; ureter: Kuriyama, Osa & Toida, 1967).

(iii) Ba can substitute for Ca in the spike generation (taenia: Hotta & Tsukui, 1968; Bülbring & Tomita, 1968, 1969b).

The spike activity evoked by electrical stimulation can easily be investigated by external recording using the double sucrose-gap method (Bülbring & Tomita, 1969a). A small part (about 2 mm) in the middle of the tissue is exposed to the test solutions and the remaining parts on both sides are in isosmotic sucrose solution. Stimulating current pulses are applied to one side immersed in sucrose solution through a resistor of 50–300 MΩ. The changes in membrane potential in the centre part are recorded across the sucrose-gap on the other side of the tissue. Because the intensity of the applied current pulse is kept constant, the change in the amplitude of the electrotonic potential is taken as a measure of the change in membrane resistance. It has been confirmed, by inserting a microelectrode into a cell in the centre pool, that the shape of the electrotonic potential and the spike observed with an intracellular electrode is similar to that observed externally across the sucrose-gap (Kuriyama & Tomita, unpublished).

The spike mechanism has recently been further investigated with this method, in the taenia coli and also in the guinea-pig ureter. Some of the experiments to be described were done in collaboration with H. Kuriyama.

Spike activity in Na-free solution

Taenia. Even in a completely Na-free solution, in which NaCl is substituted either by sucrose or by Tris (hydroxymethyl) aminomethane chloride, a spike can still be evoked. The threshold becomes higher, particularly in Tris–Cl solution, but it is always possible to evoke a spike by a strong current pulse or during conditioning depolarization (Fig. 3). In sucrose solution the spikes are, in fact, larger and sharper than in normal solution. In Tris–Cl solution, the spike may be smaller than in normal solution, and usually only one single spike is produced by a long current pulse.

In a Na-free solution containing only 0·25 mM Ca, the spike can still be larger than in normal solution if the Na-substitute is sucrose, but if the Na-substitute is Tris–Cl it is difficult to produce a spike.

Ureter. In contrast to the taenia, the action potential in the ureter consists of two components, the spike and the plateau (Washizu, 1966; Kuriyama,

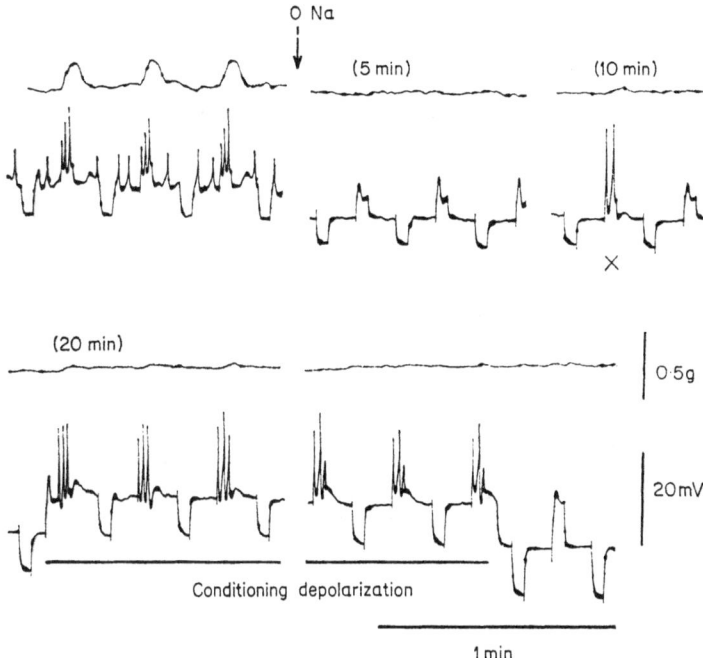

FIG. 3. Taenia coli. Locke solution, 35°C. External recording with double sucrose gap method. Constant current pulses with alternating polarities of 3 sec duration applied every 10 sec. The effect of removing Na (at arrow) is to hyperpolarize the membrane and to abolish spontaneous activity. The threshold is raised, but spikes can always be evoked with strong current pulses (at ×) or during conditioning depolarization.

Osa & Toida, 1967). Repetitive spikes usually occur on the plateau. In Na-free solution the spike component becomes larger and the plateau component becomes smaller. The tendency for repetitive activity is less in Tris–Cl than in sucrose solution. The observations suggest that, in the ureter, the spike is due to Ca entry (as in the taenia), while the plateau may be due to an increase in the Na conductance of the membrane.

The next step was therefore to study the effect of removing Ca.

Effects of exposure to Ca-free solution

Taenia. In the taenia, when Ca is removed from Locke solution (which contains no Mg), the membrane is depolarized and a high frequency spike discharge appears, which becomes finally an oscillatory activity. This change is temperature dependent (Brading, Bülbring & Tomita, 1969a). At 37°C all the electrical activity is lost rapidly, while at about 30°C the activity may continue for a very long time (more than 30 min).

The depolarization is larger and the spike activity disappears more completely and more rapidly in Ca-free Locke than in Ca-free Krebs solution. This difference is probably due to the absence of Mg in Locke solution, as will be described later. Locke solution also differs from Krebs solution in containing no phosphate and less bicarbonate.

Even after the spontaneous activity ceases, it is usually possible to produce some electrical oscillatory activity by hyperpolarizing the membrane with inward current pulses. When a chelating agent, EDTA or EGTA, is added to Ca-free solution, the oscillatory activity disappears quickly and there is a progressive decline in the effect of inward current. Finally activity disappears.

The observation that removal of Ca abolishes spike activity only slowly and that the effect is temperature dependent, suggests that bound Ca in or near the membrane, may be more important for the maintenance of spike activity than free Ca ions in the external solution.

When the membrane has been depolarized by Ca-free solution and when, in this condition, Na is also removed by replacing NaCl with sucrose or with Tris–Cl, the membrane is quickly repolarized and the membrane resistance is increased (Fig. 4). From this observation it is concluded that the removal of Ca from the membrane increases the Na conductance and thereby causes the depolarization of the membrane. This depolarization is not affected by tetrodotoxin.

Addition of only about one-tenth of the normal Ca concentration is enough to produce recovery of spontaneous and evoked spike activity when there is no Na in the external solution. It takes more than 10 min to reach a steady state after addition of Ca, suggesting that some accumulation of Ca at or near the membrane is necessary for the process of recovery. If the Ca concentration in the external solution is further increased to more than 1 mM, the spontaneous activity is quickly suppressed and, because of an increase in the threshold, a higher intensity of stimulating current is required to produce the spike. In a solution with no Na but containing enough Ca, it is always possible to evoke a spike if a stimulus of sufficiently high intensity is applied (see Fig. 3).

The spike from the crayfish muscle is thought to be due to Ca entry across the membrane and is insensitive to tetrodotoxin (Ozeki, Freeman & Grundfest, 1966). When the external Ca concentration is reduced, the crayfish muscle fibre becomes selectively permeable to Na and produces a prolonged action potential which is blocked by tetrodotoxin, 5×10^{-7} g/ml. (Reuben, Brandt, Girardier & Grundfest, 1967).

In the taenia, a prolonged action potential can be evoked during conditioning hyperpolarization in Ca-free solution. It is sensitive to Na ions, as in the crayfish muscle, but it is not affected by tetrodotoxin (2×10^{-6} g/ml.).

Ureter. Compared with the taenia, the depolarization of the membrane in Ca-free solution is small and gradual in the ureter. The repetitive spike

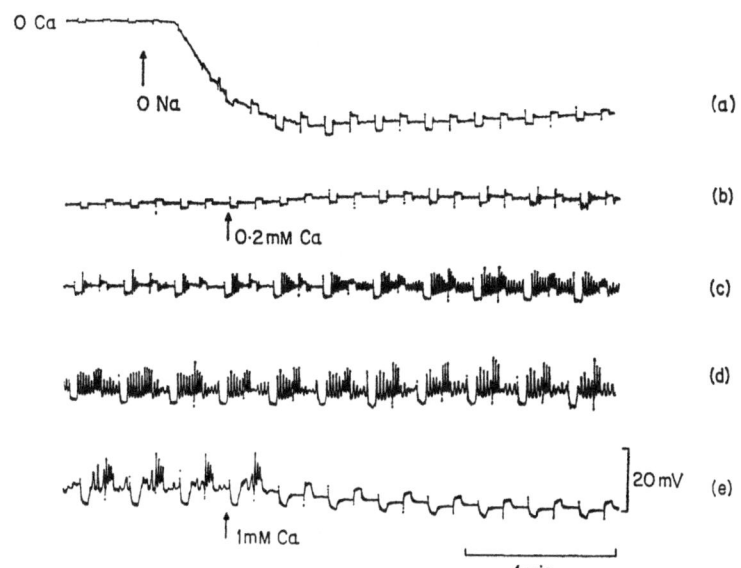

FIG. 4. Taenia coli. External recording with double sucrose gap method. 35°C. Constant current pulses with alternating polarities (3 sec duration, every 10 sec) applied throughout. (a) State of depolarization and low membrane resistance after 15 min exposure to Locke solution containing zero Ca. At arrow, removal of Na (sucrose substitution), causing repolarization and recovery of membrane resistance. (b) After 7 min in zero Na and zero Ca. At arrow: 0·2 mM Ca added. (c) and (d) Continuous records following (b). (e) After 20 min in 0·2 mM Ca and zero Na, at arrow, 1 mM Ca added. Note slow recovery of spike activity by 0·2 mM Ca but suppression of spikes by 1 mM Ca in zero Na. (From Brading, Bülbring, & Tomita, 1969b.)

activity is suppressed (Bennett, Burnstock, Holman & Walker, 1962), however, and the slow potential (plateau) is prolonged up to 10 sec duration. Tetrodotoxin (10^{-6} g/ml.) has no effect on this prolonged plateau in Ca-free solution, nor has it any effect on the normal action potential—either on the spikes or on the plateau.

Effects of Mg added to Ca-free solution
In the presence of a normal Ca concentration, a change in the external Mg concentration has only a small effect on the electrical activity (Bülbring & Kuriyama, 1963), Mg has, however, a dramatic stabilizing effect in Ca-free solution.

In the taenia, when a small amount of Mg (less than 1 mM) is added to Ca-free solution, the membrane is repolarized and the membrane resistance recovers. After several minutes in this solution, it usually becomes possible to evoke a spike whose amplitude increases gradually, with time, towards

Taenia

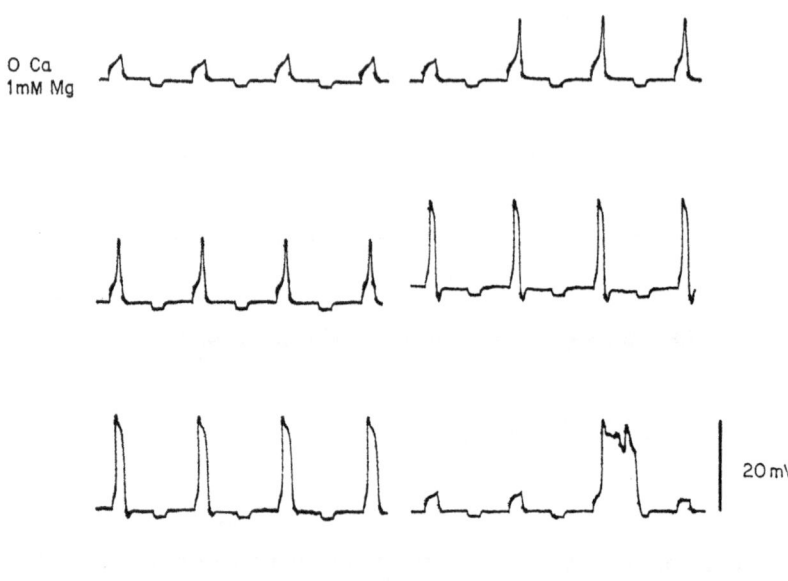

O Ca
1mM Mg

20 mV

1 min

FIG. 5. Taenia coli. External recording with the double sucrose gap method. Locke solution, 30°C. Constant current pulses with alternating polarity applied throughout. At the beginning of the record, the muscle had been exposed for 15 min to Ca-free solution and for 5 min to 1 mM Mg. The records were taken at 5 min intervals. Note the gradual appearance of slow activity evoked by the depolarizing pulse.

normal (see Fig. 5). This process is slow, as is the effect of removing Ca or of adding small amounts of Ca to Ca-depleted tissue. These spikes, which can be evoked in Ca-free solution containing a small amount of Mg, have a slower time course than the normal spikes. They are blocked by removing Na, but they are not affected by tetrodotoxin.

When the external Mg concentration is increased (1–5 mM) in Ca-free solution, no spike can be evoked although the solution contains the normal Na concentration. If Ca is now reintroduced, a spike can be elicited. This observation indicates that Mg, like Ca, decreases the Na conductance of the membrane. But, unlike Ca, Mg is probably unable to cross the membrane to produce the spike. It may be, if the external Mg concentration is low, that the Mg in the membrane which controls Na conductance can be displaced by depolarization of the membrane, and that the slow spike in this condition is due to Na-entry.

When Ca is removed, and at the same time Mg in a high concentration (2–5 mM) is added, there is no depolarization of the membrane and the membrane resistance remains more or less normal; yet the spike activity completely disappears. The same results are obtained in the ureter. They indicate a competition between Ca and Mg for a membrane site which controls the Na conductance.

Effects of Sr and Ba

Taenia. Sr can substitute for Ca without affecting the spike (Hotta & Tsukui, 1968). Ba, on the other hand, prolongs the spike duration forming a plateau (Bülbring & Kuriyama, 1963; Bülbring & Tomita, 1968, 1969b) (Fig. 6). The Ba plateau formation is highly temperature-dependent. There seems to be a competition between Ca and Ba. As the external Ca concentration is increased the Ba effect becomes less, and when the Ca

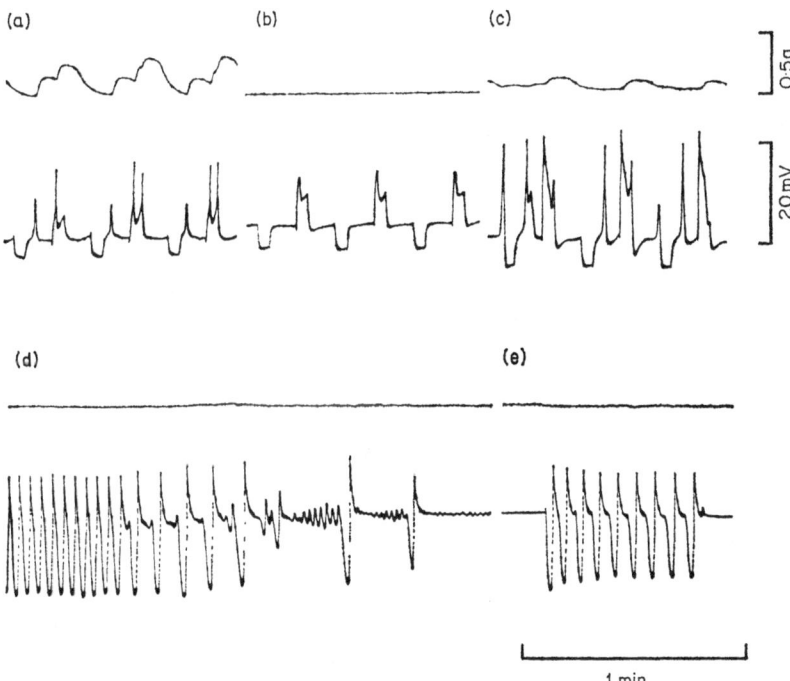

FIG. 6. Taenia coli. Records of tension (top) and electrical activity (bottom) using double sucrose-gap method. 31°C. Upper record: responses to alternately applied inward and outward current pulses (a) in Krebs solution, 2·5 mM Ca; (b) 7 min in 0 Ca; (c) 10 min in 0 Ca + 0·25 mM Ba. Lower record: (d) spontaneous activity in 0 Ca + 1 mM Ba. At dot, Ba concentration was increased to 2·5 mM Ba. Note gradual plateau formation until depolarization was maintained. (e) Spikes were produced by sustained inward current application. (From Bülbring & Tomita, 1968.)

concentration is reduced the Ba effect becomes greater. When the solution contains Na, the plateau is very prolonged and finally a sustained depolarization is developed; this depolarization does not develop in a solution which does not contain Na.

Ureter. Substitution of Ca with Sr or Ba prolongs the plateau. The plateau is longer in Ba solution than in Sr solution, and the duration may be as long as several minutes. During the plateau the membrane resistance is very much reduced. Preliminary experiments suggest that Ba can prolong the plateau only in the presence of Na. When Na is replaced with sucrose or Tris–Cl in Sr solution the plateau disappears and only the spike component is produced by electrical stimulation. Thus Sr can replace Ca for the spike but it is less efficient than Ca in reducing Na conductance. Ba seems, however, to be unable to replace Ca for the spike component.

Discussion

The spike can be evoked in Na-free solution, and Na is therefore not essential for spike generation in the taenia or in the ureter. Furthermore, the spike is actually larger and sharper without Na ions. Therefore it may be that, if enough Ca ions are present, the inward movement of Na ions does not contribute to the spike, and that Ca is the main carrier of the inward current during the spike even when Na ions are available.

It is difficult to abolish the electrical activity completely by removing Ca from the external solution. The free intracellular Ca ion concentration in relaxed muscle is probably very low (Portzehl, Caldwell & Rüegg, 1964), so it is possible that any trace of Ca left in the external solution may still give a high enough concentration gradient to produce the activity. Because the total tissue Ca remains fairly high in Ca-free solution (Brading, personal communication), it is also possible that Ca is bound near the membrane, or at a site in the membrane where its presence is essential for spike generation, and its removal may be rather limited even in Ca-free solution. In addition, the slow disappearance of spike activity in Ca-free solution may also be due to the continued supply of Ca by a transport mechanism from a storage site inside the cell.

It may be, as postulated for the crustacean muscle fibre by Hagiwara & Takahashi (1967), that the amount of bound Ca at some site, rather than the free-external Ca concentration, is the most important factor controlling the spike configuration. This idea is supported in smooth muscle by the following facts. (*a*) Blockade of the spike by removing the external Ca is very much slower than can be explained by loss of Ca through diffusion from the extracellular space (Brading & Jones, 1969). (*b*) The effects of removal of Ca are highly temperature-dependent, while ion diffusion from the tissue is little affected by temperature change (Jones, unpublished). (*c*) The effects of adding small amounts of Ca or Mg appear gradually, suggesting that some accumulation of divalent cations is necessary for the spike activity. (*d*) It is difficult to explain competition between Ca and other ions on membrane activity without postulating some binding site (Goodford, 1967).

There are three possible sources of Ca available to carry the charge across the membrane: (1) free Ca ions in the external solution, (2) loosely bound Ca near the membrane, and (3) bound Ca at the membrane. The results described favour bound Ca as the source. If it is Ca bound to the membrane that provides inward current for the spike, then this Ca must be released when the membrane is depolarized by stimulation. The increase in amplitude and in the rate of rise of the spike in low Na solution could then be explained by an increase in the amount of bound Ca when the concentration of competing Na ions is reduced, as suggested for cardiac muscle (Lüttgau & Niedergerke, 1958; Niedergerke & Orkand, 1966).

If the bound Ca is only transiently released from the membrane, for example, by electrical stimulation, the Ca permeability of the membrane at this site would be increased, and Ca ions cross the membrane to produce the action potential with a short duration. The most likely explanation of the depolarization in Ca-free solution is that it is due to an increase in Na permeability of the membrane, caused by the removal of Ca from the membrane, as in crustacean muscle fibres (Reuben, Brandt, Girardier & Grundfest, 1967). If the site remains unoccupied by Ca for a while, then the Na permeability of the membrane is increased, and Na ions carry the inward current. This could be an explanation of the plateau in the ureter. Repetitive spikes can be produced during the plateau in normal solution, so there are probably two mechanisms, one for the spike generation (involving Ca entry) and another for the plateau formation (involving Na entry).

Mg seems to have a stabilizing effect on the membrane, by which the Na permeability is reduced. The spike, produced by a depolarizing pulse in Ca-free solution containing a small amount of Mg and the normal Na concentration, may be explained by assuming that depolarization displaces Mg from the gate in the membrane for Na ions, opening it, which results in an inward movement of Na ions across the membrane. Excess Mg, by making the displacement of Mg from the membrane difficult, and zero Na, by causing a lack of current carriers across the membrane, both suppress the electrical activity.

Our tentative hypothesis may be summarized as follows. Some divalent cations (Ca, Sr, Ba, in the taenia; Ca, Sr, in the ureter) can carry an inward current to produce a spike without the plateau. The amplitude and the rate of rise of the spike are augmented by increasing the concentration of the divalent cation, but at the same time the threshold also becomes higher. If the concentration of any divalent cation which has a stabilizing action is properly adjusted, a slow potential change (plateau-like spike) can be evoked which is due to an increase of the Na conductance of the membrane. But it becomes difficult to elicit this type of spike when there is a high concentration of divalent cation.

It should be emphasized that this hypothesis is based on experimental evidence obtained on only two types of smooth muscle, the taenia and the ureter. The electrical activity in these two muscle types is different, and different mechanisms seem to be involved under physiological conditions. It could be dangerous, therefore, to generalize the conclusions to include all smooth muscles, though the fundamental mechanism by which Ca controls the ion permeability of membrane may be similar.

REFERENCES

BENNETT, M. R., BURNSTOCK, G., HOLMAN, M. E. & WALKER, J. W. (1962). *J. Physiol., Lond.*, **161**, 47

BRADING, A. F., BÜLBRING, E. & TOMITA, T. (1969a). *J. Physiol., Lond.*, **200**, 621

BRADING, A. F., BÜLBRING, E. & TOMITA, T. (1969b). *J. Physiol., Lond.*, **200**, 637

BRADING, A. F. & JONES, A. W. (1969). *J. Physiol., Lond.*, **200**, 387

BÜLBRING, E. & KURIYAMA, H. (1963). *J. Physiol., Lond.*, **166**, 29

BÜLBRING, E. & TOMITA, T. (1967). *J. Physiol., Lond.*, **189**, 299

BÜLBRING, E. & TOMITA, T. (1968). *J. Physiol., Lond.*, **196**, 137

BÜLBRING, E. & TOMITA, T. (1969a). *Proc. R. Soc., B*, **172**, 89

BÜLBRING, E. & TOMITA, T. (1969b). *Proc. R. Soc., B*, **172**, 121

FATT, P. & GINSBORG, B. L. (1958). *J. Physiol., Lond.*, **142**, 516

GOODFORD, P. J. (1967). *J. Physiol., Lond.*, **192**, 145

HAGIWARA, S. & NAKA, K.-I. (1964). *J. gen. Physiol.*, **48**, 141

HAGIWARA, S. & NAKAJIMA, S. (1966). *J. gen. Physiol.*, **49**, 793

HAGIWARA, S. & TAKAHASHI, K. (1967). *J. gen. Physiol.*, **50**, 583

HODGKIN, A. L. & KATZ, B. (1949). *J. Physiol., Lond.*, **108**, 37

HOLMAN, M. E. (1957). *J. Physiol., Lond.*, **136**, 569

HOLMAN, M. E. (1958). *J. Physiol., Lond.*, **141**, 464

HOTTA, Y. & TSUKUI, R. (1968). *Nature, Lond.*, **217**, 867

KURIYAMA, H., OSA, T. & TOIDA, N. (1966). *Br. J. Pharmac. Chemother.*, **27**, 366

KURIYAMA, H., OSA, T. & TOIDA, N. (1967). *J. Physiol., Lond.*, **191**, 225

LÜTTGAU, H. C. & NIEDERGERKE, R. (1958). *J. Physiol., Lond.*, **143**, 486

NIEDERGERKE, R. & ORKAND, R. K. (1966). *J. Physiol., Lond.*, **184**, 312

NONOMURA, Y., HOTTA, Y. & OHASHI, H. (1966). *Science, N.Y.*, **152**, 97

OZEKI, M., FREEMAN, A. R. & GRUNDFEST, H. (1966). *J. gen. Physiol.*, **49**, 1319

PORTZEHL, H., CALDWELL, P. C. & RÜEGG, J. C. (1964). *Biochim. biophys. Acta*, **79**, 581

REUBEN, J. P., BRANDT, P. W., GIRARDIER, L. & GRUNDFEST, H. (1967). *Science, N.Y.*, **155**, 1263

WASHIZU, Y. (1966). *Comp. Biochem. Physiol.*, **19**, 713

KINETIC ASPECTS OF CALCIUM CURRENT IN VENTRICULAR MYOCARDIAL FIBRES

H. REUTER

Department of Pharmacology, University of Berne, Switzerland

The application of the voltage clamp technique to cardiac Purkinje fibres (Deck, Kern & Trautwein, 1964) suggests that, in the region of the plateau of the action potential, there are permeability changes which are dependent on voltage and time and which involve most of the ions in the extracellular fluid (Noble & Tsien, 1969). Thus, beside sodium ions, calcium ions contribute to the total depolarizing membrane current during rather strong depolarization (Reuter, 1967). Recent experiments by Niedergerke & Orkand (1966) and by Rougier, Vassort, Garnier, Gargouil & Coraboeuf (1968) have shown that calcium ions may also contribute a charge during the action potential plateau of frog myocardium. In addition to the significance for the action potential, the flow of calcium ions across the membrane might be important for activation of the contractile proteins. Therefore it seemed worthwhile to investigate the kinetics of calcium current in cardiac preparations in greater detail.

We devised a voltage clamp technique applicable to mammalian cardiac preparations in order to obtain more information about this subject (Reuter & Beeler, 1969a,b). With this technique we examined dog ventricular myocardium preparations such as very thin (0·7 mm or less) trabeculae or papillary muscles. These preparations were pulled through tightly fitting holes in rubber membranes which divided a chamber into three compartments. The middle compartment was continuously perfused with isotonic sucrose solution while the outer compartments were perfused with Tyrode solution or a test solution. The high resistivity of the sucrose solution meant that current had to flow through intracellular pathways of the fibre bundle when a voltage was applied between the two outer compartments. Thus it was possible to change the membrane potential homogeneously in about 1 mm of the fibre bundle exposed to solution in one of the outer compartments. The membrane potential was recorded by means of intracellular electrodes and controlled through a negative feedback circuit. Tension was measured using a force displacement transducer. Tension, membrane potential, and membrane current were simultaneously recorded on an oscilloscope.

The inward sodium current

Figure 1 shows a series of voltage clamp steps typically recorded from a dog ventricular fibre bundle in Tyrode solution. The holding potential (here resting potential) was −77 mV. Net membrane current is zero at this potential. Stepwise changes of the membrane potential first produced surges of capacitative current. For small clamp steps up to −64 mV this was followed by constant outward current. Beyond this potential, however, a large inward current occurred, which turned into outward current within 20–50 msec. The amplitude of this current reached a maximum at about −60 mV and decreased during further depolarization. The inward current became zero at about +55 mV (equilibrium potential) and changed its sign, becoming outward current, during stronger depolarization. It disappeared

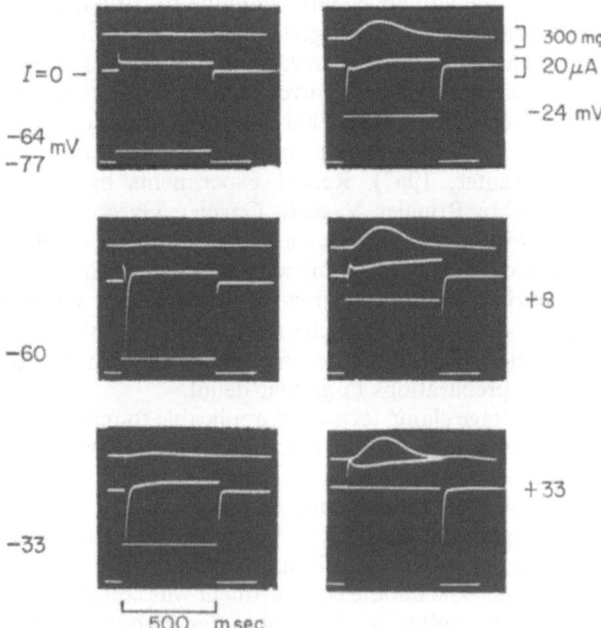

FIG. 1. Membrane currents (middle traces) and isometric contractions (upper traces) during depolarizing voltage clamp steps (lower traces) recorded from a ventricular trabecula (dog heart) in Tyrode solution. Upward deflection from $I = 0$ indicates outward current; downward deflection, inward current. Holding potential (= resting potential) −77 mV; the figures beside the voltage records indicate the membrane potentials during displacement from the holding potential. Records show steady state contraction after five or six depolarizations to each potential level. (From Reuter & Beeler, 1969a.)

in the absence of external sodium ions and became extremely small in the presence of tetrodotoxin (10^{-5} g/ml.). Moreover, the equilibrium potential of this inward current could be shifted more negative in a predictable manner (according to the Nernst equation $E_{Na} = (RT/F) \times \ln [Na]_0/[Na]_i$) by 29 ± 2.3 mV (mean $\pm$ S.E. of five preparations) when the external sodium concentration was reduced to 31% of that in Tyrode solution. This result suggests strongly that the large rapidly decaying inward current is solely carried by sodium ions. This sodium current (I_{Na}) is responsible for the rapid depolarization phase of the action potential.

The inward calcium current

In addition to I_{Na}, Fig. 1 shows a second much smaller and slower inward current which could be activated at -24 mV. At the same potential the steady state contraction became very strong, while it was weak and not much dependent on the membrane potential in the range -60 to -35 mV. The kinetics of the slow inward current could be investigated accurately only if I_{Na} did not interfere. Inactivation of I_{Na}, however, could be

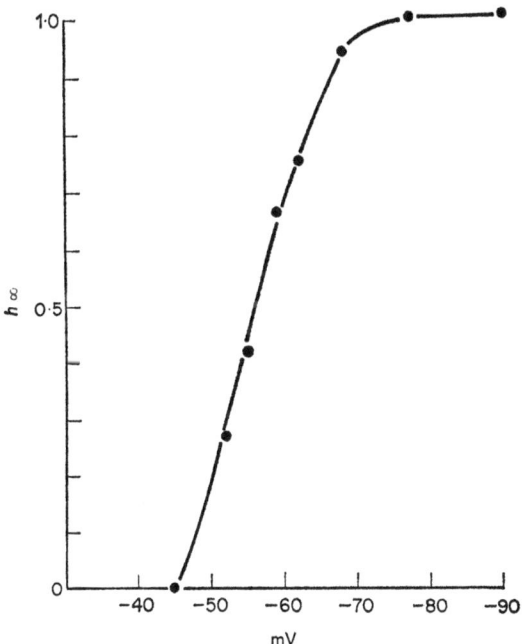

FIG. 2. Influence of membrane potential (abscissa) on peak sodium current associated with a potential step to -38 mV in dog ventricular trabecula. Ordinate, h_∞, is the fraction of the sodium system which is not inactivated and therefore available for activation in the steady state relative to maximum sodium current.

achieved by changing the membrane potential in two steps (Hodgkin & Huxley, 1952). The first step (V_1) was held for 300–400 msec and varied in amplitude while the second step (V_2; 200–400 msec) was kept constant at a potential where I_{Na} was still large (around -40 mV). The result of such an experiment is plotted in Fig. 2; h_∞ is the steady state fraction of maximum I_{Na} which can be recorded at V_2 as a function of the preceding membrane potential V_1. I_{Na} at V_2 was typically completely abolished if V_1 was as low as -45 mV.

When the membrane potential was depolarized in a first step to about -40 mV or when the holding potential was set to this level, the slow inward current could be investigated without interference from I_{Na}. Figure 3 shows double step voltage clamps. The first step, V_1, was always to -38 mV for 420 msec, the second step (V_2; 210 msec) was varied in amplitude. V_1 was sufficiently large and long to inactivate I_{Na} at V_2 completely. Thus, depolarization at V_2 to -33 mV produced only a small increase in outward current. Further depolarization elicited inward current reaching its

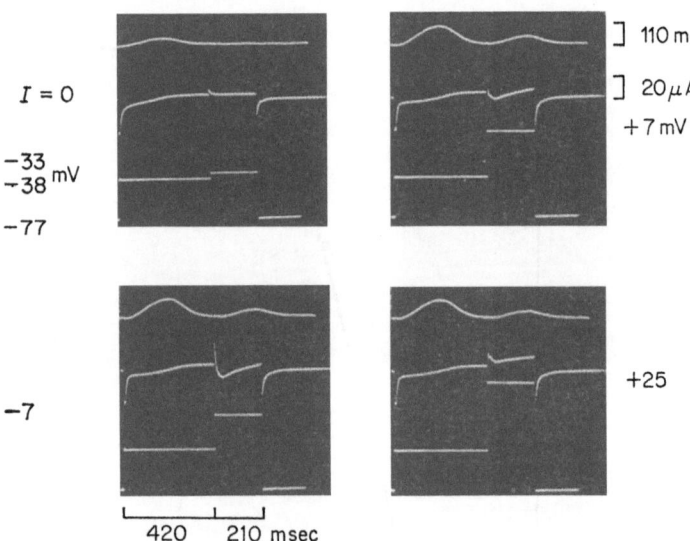

FIG. 3. Membrane currents (middle traces) and isometric contractions (upper traces) during double step voltage clamps (lower traces) recorded from a ventricular trabecula (dog heart) in Tyrode solution. Upward deflection from $I = 0$ indicates outward current, downward deflection, inward current; holding potential (= resting potential) was -77 mV; first clamp step (V_1) was always to -38 mV and second step (V_2) was variable in amplitude; figures beside voltage records indicate membrane potential during V_2. Note that contraction during V_1 increased after calcium inward current at V_2 had been activated during preceding depolarizations. (From Reuter & Beeler, 1969b.)

maximum at about −20 mV. The amplitude of this inward current decreased during stronger depolarizations, but net inward current could still be measured at +7 mV. When depolarizations were more positive than +25 mV it was not possible to separate the slow inward current accurately from outward current which increased steeply in this potential range. For this reason the equilibrium potential for the slow inward current could not be precisely determined. But from the data presented it is clear that the equilibrium potential of this current is in the positive potential range. Rough estimations revealed an equilibrium potential around +60 mV. Ionic requirement, amplitude, and time course of this current, however, are very different from I_{Na}.

The slow inward current was not much affected by complete replacement of sodium ions in the extracellular fluid by choline, Tris-(hydroxymethyl)-aminomethane, or sucrose. Also tetrodotoxin (10^{-5} g/ml.) did not affect the current. But the current was very sensitive to alterations in external

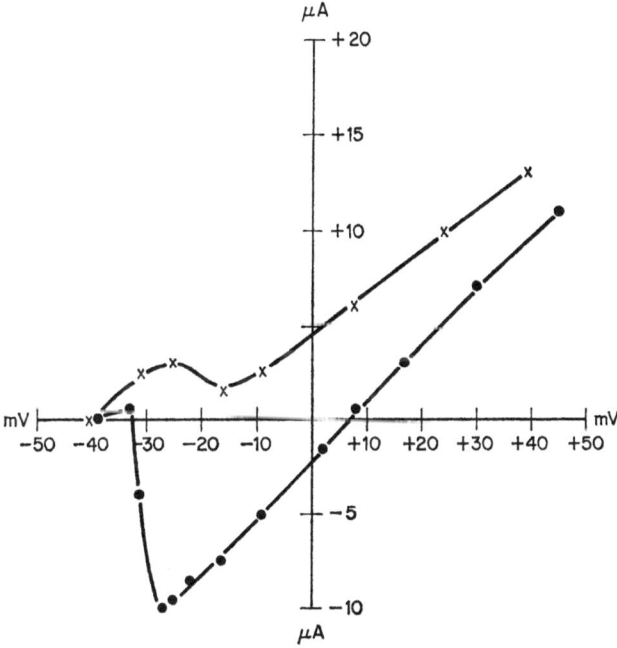

FIG. 4. Current voltage relations for Ca-dependent inward currents (I_{Ca}) measured in a dog ventricular fibre in Tyrode solutions with two different calcium concentrations ([Ca]$_0$ 0·1 mM crosses and 1·8 mM filled circles). Holding potential was −40 mV in order to inactivate the sodium system. Ordinate is maximum inward current (negative) or minimum outward current (positive) in microamperes; abscissa is membrane potential in millivolts. Note that I_{Ca} is not subtracted from outward current.

calcium ion concentration ($[Ca]_0$). The strongest evidence against the assumption that this inward current might be carried by sodium ions is obtained from the results which show the occurrence and persistence of the inward current beyond the sodium equilibrium potential (E_{Na}) soon after the switch from Tyrode solution to a sodium-free bathing fluid. Also the current cannot be generated by sodium ions in unstirred extracellular clefts which do not equilibrate readily with the external fluid since it is immediately affected by changes in $[Ca]_0$. It is therefore concluded that the slow inward current which carries appreciable charge during the plateau of the cardiac action potential is a calcium current (I_{Ca}).

Figure 4 shows the dependence of I_{Ca} on $[Ca]_0$ (0·1 and 1·8 mM) plotted as current voltage relations. The outward current has not been subtracted from the inward current to give the true values of the amplitude of I_{Ca} at each potential level because the accuracy of such a separation may be doubtful, especially in the positive potential range. Maximum inward or minimum outward currents are therefore plotted as functions of the membrane potential. Clearly the slow inward current became very small with 0·1 mM $[Ca]_0$. With 1·8 mM $[Ca]_0$ the maximum amplitude was attained at -27 mV. Further increase in $[Ca]_0$ shifted the threshold for I_{Ca} and the current maximum slightly along the voltage axis to more negative potentials. In addition, peak inward current became even larger under such conditions. A comparable shift of the threshold of I_{Ca} was also observed in the absence of $[Na]_0$.

Time course of I_{Ca}

In order to obtain information about the time course of I_{Ca}, experiments were performed with holding potentials at about -40 mV and varying duration of the depolarizing clamp steps (10 to 1000 msec). After repolarization to the holding potential, inward current tails were recorded which decreased in amplitude if the pulse duration was increased. The maxima of these inward current tails (after correction for the capacitative spikes) were plotted on a semilog scale as functions of the duration of the preceding depolarization. The envelope of the tails of current on return to the holding potential could be fitted by an exponential. The time constants of the exponentials, however, were strongly dependent on the membrane potential during the depolarizing clamp step. Because the inward current tails are a measure of calcium conductance activated during the preceding depolarization, the time constants (τ) of the exponentials give the values for the decay of I_{Ca} with time at each potential level. The values are plotted in Fig. 5 as rate constants ($1/\tau$) obtained from a single experiment. There is a steep voltage dependence of the rate of decay of I_{Ca}, especially in the potential range -30 to $+10$ mV. This means that I_{Ca} is inactivated rapidly at negative potential levels and slowly during stronger depolarization.

Calcium ions provide the charge for inward current during voltage

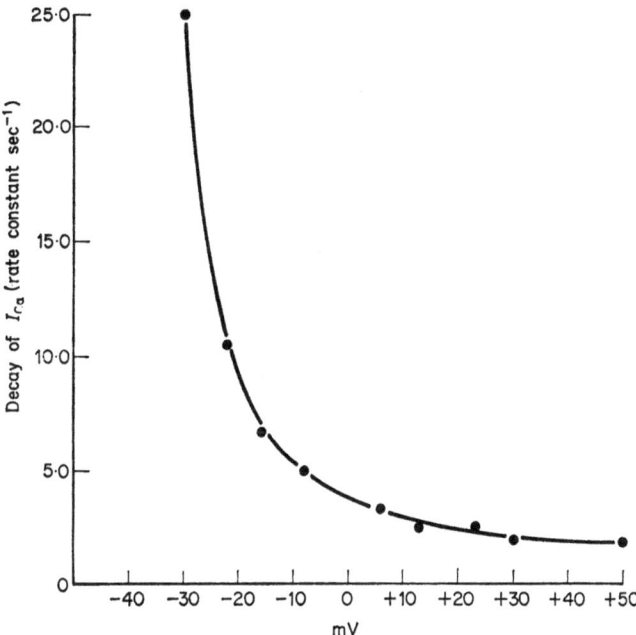

FIG. 5. Variation of rate of decay of calcium current (I_{Ca}) measured as reciprocal time constants (rate constants sec^{-1}; ordinate) with membrane potential (abscissa). Filled circles are experimentally determined values from a dog ventricular trabecula in Tyrode solution. The smooth curve has been fitted by eye.

oscillations in sodium-free solution (Fig. 6). In this experiment constant depolarizing current pulses of increasing strength were applied for 4 sec to a ventricular trabecula excised from a sheep heart. From a certain threshold (about −60 mV) one or more regenerative depolarizations occurred, depending on the current strength. Each membrane oscillation elicited a contraction. Oscillations and contractions were quickly abolished in calcium-free solution.

Conclusions
The results provide evidence for two separable inward currents in mammalian cardiac muscle. The first inward current is large and rapidly inactivated. It is carried solely by sodium ions and is responsible for the rapid depolarization phase of the action potential. The second inward current could be accurately resolved only if the sodium system was inactivated. It is much smaller and slower than I_{Na} and is carried by calcium ions. I_{Ca} is slowly inactivated in the (inside) positive potential range and rapidly at negative potentials. Calcium ions contribute an

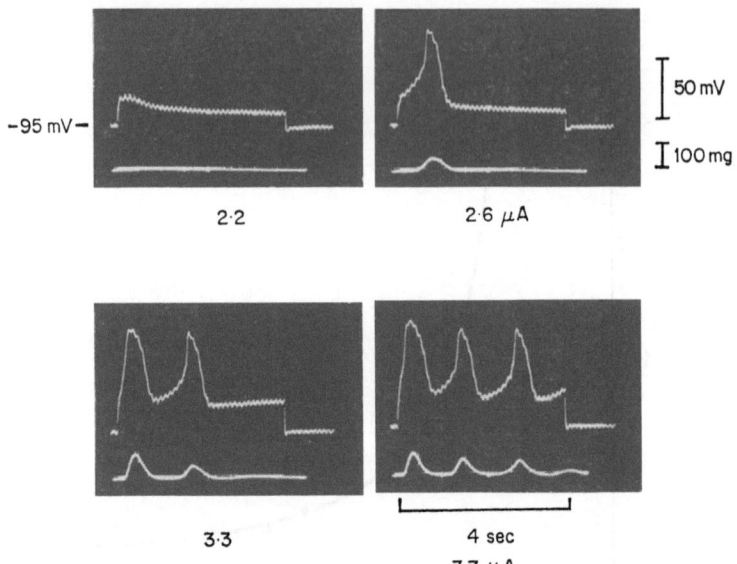

FIG. 6. Membrane potential oscillations (upper traces) and isometric contractions (lower traces) evoked by constant depolarizing current pulses (4 sec)
of different strengths (2·2–3·7 μA). Ventricular trabecula (sheep heart) in
sodium-free solution of following composition (in mmoles/l.): sucrose 274;
KCl 2·7; CaCl$_2$ 1·8; MgCl$_2$ 1·05; glucose 5·5; Tris-buffer + HCl to pH 7·4.
Membrane potentials are superimposed by small anodal current pulses (50
msec; 0·4 μA) in order to measure slope conductance. (From Reuter &
Scholz, 1968.)

appreciable amount of charge transfer during the plateau of the cardiac
action potential. In the absence of sodium ions they provide the only
positive charge for inward current during regenerative voltage oscillations.

The flow of calcium ions into the myocardium fibre during depolarization
is intimately related to activation of contraction. There are large differences, however, between activation of contraction by the influx of calcium
ions into myocardium fibres kept in sodium-free or sodium-containing
solutions. In sodium-free solution calcium ions which flow into the fibre
activate contraction directly (Reuter & Scholz, 1968; Reuter & Beeler,
1969b). In sodium-containing solution, however, the evidence suggests
that I_{Ca} serves primarily to fill some intracellular stores from which
calcium can be released by moderate depolarization (Reuter & Beeler,
1969b; Wood, Heppner & Weidmann, 1969).

Most of the results were obtained during the tenureship of an award as
Career Investigator Visiting Scientist of the American Heart Association
(sponsored by Dr. E. H. Wood) and supported by Research Grant AHA
67-053.

REFERENCES

DECK, K. A., KERN, R. & TRAUTWEIN, W. (1964). *Pflügers Arch. ges. Physiol.*, **280**, 50

HODGKIN, A. L. & HUXLEY, A. F. (1952). *J. Physiol., Lond.*, **116**, 497

NIEDERGERKE, R. & ORKAND, R. W. (1966). *J. Physiol., Lond.*, **184**, 291

NOBLE, D. & TSIEN, R. W. (1969). *J. Physiol., Lond.*, **200**, 205

REUTER, H. (1967). *J. Physiol., Lond.*, **192**, 479

REUTER, H. & SCHOLZ, H. (1968). *Pflügers Arch. ges. Physiol.*, **300**, 87

REUTER, H. & BEELER, Jr., G. W. (1969a). *Science, N.Y.*, **163**, 397

REUTER, H. & BEELER, Jr., G. W. (1969b). *Science, N.Y.*, **163**, 399

ROUGIER, O., VASSORT, G., GARNIER, D., GARGOUIL, Y. M. & CORABOEUF, E. (1968). *C.R. Acad. Sci. Paris*, **266**, 802

WOOD, E. H., HEPPNER, R. L. & WEIDMANN, S. (1969). *Circulation Res.*, **24**, 409

CARDIOACTIVE STEROIDS WITH SPECIAL REFERENCE TO CALCIUM

M. REITER
Institut für Pharmakologie und Toxikologie der Technischen Hochschule, München

The similarity between the positive inotropic effects of cardioactive steroids and calcium ions on the heart muscle led Loewi (1918) to suggest that glycosides sensitize the heart to calcium. There is evidence that calcium ions play an important role in activating myofibril-ATPase, so such a sensitization could mean that the cardioactive steroids act directly on the contractile proteins and make them sensitive to calcium ions. Many investigations, however, show that the contractile proteins are not directly affected by the glycosides in a manner which could explain their inotropic action.

There are numerous reports which indicate that the cardioactive steroids consistently produce effects on the membrane of the cardiac cell. It therefore seems reasonable to consider the possibility that the inotropic effect of these steroids is secondary to their interaction with the membrane. In particular, one might look for a membrane effect that could lead to an increase in the intracellular concentration of free calcium ions which is greater than that resulting normally from the influx of calcium across the membrane during stimulation and depolarization of the muscle.

Effects of cardioactive steroids and of calcium ions
It seems appropriate to start by comparing more closely the inotropic action of the cardioactive steroids with that of calcium ions. The super-imposed isometric tension curves from a guinea-pig papillary muscle (Fig. 1) show that the effects of increasing the calcium concentration in the medium are very similar to the effects of increasing the concentration of dihydro-ouabain. This glycoside is less active than ouabain but it offers the experimental advantage of a faster development of its inotropic action. The increase in tension height, or force of contraction (F), derives in both cases from an increase in steepness (S) of the isometric contraction curve which reflects an increase in the velocity of the shortening of the contractile proteins.

Not only the rate of tension development but also the relaxation velocity increases, so it may be assumed that the uptake of calcium ions by the sarcoplasmic reticulum which is, at least partly, responsible for the removal of intracellular calcium ions is not inhibited by dihydro-ouabain.

There is of course the possibility that more calcium ions are made available inside the cell in the early phase of depolarization. In the case of the inotropic action due to added calcium the situation seems relatively

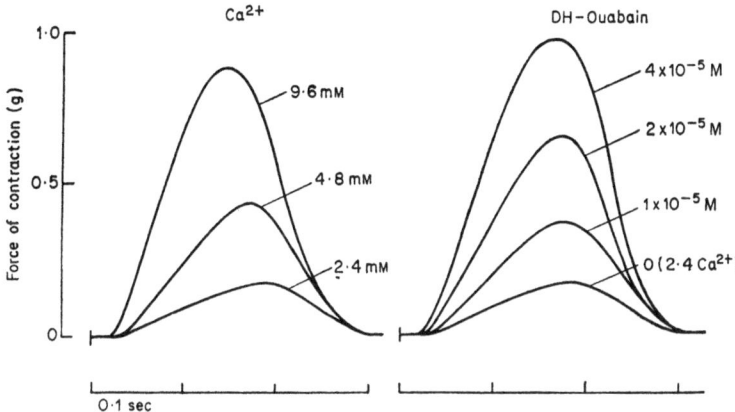

FIG. 1. Isometric tension curves of a guinea-pig papillary muscle under the influence of increasing concentrations of Ca^{2+} (left) and dihydro-ouabain (right). 35°C. Resting tension 0·4 g, stimulation frequency 1/sec.

simple: more calcium will enter the cell during depolarization as a result of the increased concentration gradient. It was shown by Wilbrandt & Koller (1948) and later by Lüttgau & Niedergerke (1958) that sodium competes with calcium in its inotropic action, and Niedergerke (1963) found an increased uptake of calcium by cardiac cells when the external sodium concentration was reduced. It appears, therefore, that Ca^{2+} and Na^+ compete for a common membrane carrier. A lowering of the external sodium concentration will shift the curve which relates force of contraction to $[Ca]_0$ to the left, as shown in the lower part of Fig. 2 by the calcium-response curves for the two sodium concentrations 140 mM and 70 mM. In the latter condition NaCl was replaced by sucrose to maintain the osmolarity.

Sodium ions and the inotropic effect of cardioactive steroids

With regard to the possible mechanism by which cardioactive steroids produce their inotropic effect, it is of interest to know whether this effect is influenced by changes in the extracellular concentration of either sodium or calcium. If the inotropic effect of a glycoside is tested on a papillary muscle in solutions differing in their sodium concentrations (Fig. 3), the

M. Reiter

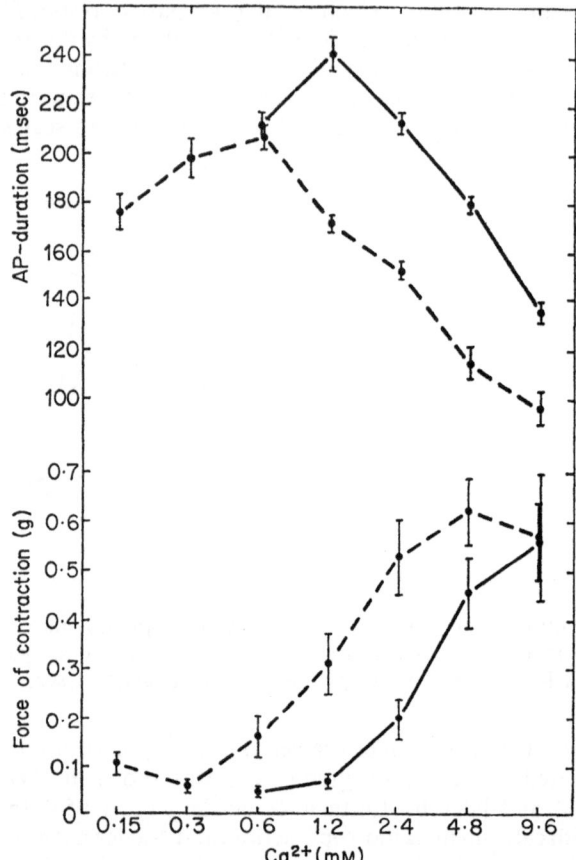

FIG. 2. Effect of extracellular [Na$^+$] on the dependence on extracellular [Ca^{2+}] of force of contraction and duration of action potential (AP), ———, 140 mM Na$^+$; – – –, 70 mM Na$^+$. Guinea-pig papillary muscle, same conditions as in Fig. 1. (From Stanley & Reiter, 1965.)

inotropic effect is more than doubled by increasing the sodium concentration from 70 to 140 mM. A similar observation was made by Farah & Witt (1963).

The sodium dependence cannot be explained by assuming that with low extracellular sodium the carrier is saturated by calcium ions because external sodium and calcium concentrations were reduced simultaneously. Hence the force of contraction did not change until the steroid was added. It is evident therefore that the role of sodium ions during the inotropic actions of the cardioactive steroids is different from the role they play when apparently competing with external calcium.

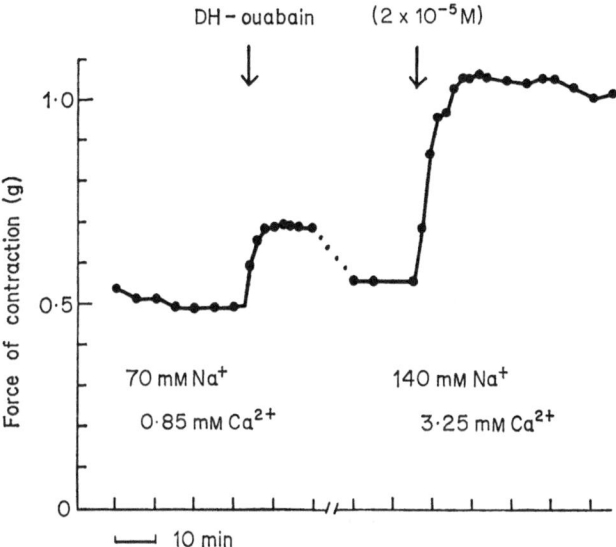

FIG. 3. Dependence on extracellular [Na⁺] of the inotropic effect of a cardio-active steroid. (From Reiter, 1963.)

The importance of sodium for the inotropic steroid action is further demonstrated (Fig. 4) when the effects on the calcium response curves of increasing concentrations of the glycoside are examined at two different sodium concentrations. The ordinate indicates the steepness of the isometric contraction curve rather than the force of contraction because

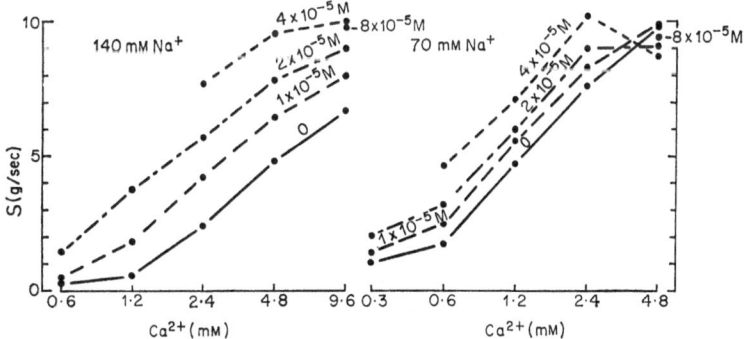

FIG. 4. Calcium-response curves of guinea-pig papillary muscle as influenced by increasing concentrations of dihydro-ouabain at two different sodium concentrations (140 mM and 70 mM + sucrose). Ordinate: mean steepness of tension curve (rate of development of tension); abscissa: [Ca²⁺] on a loga-rithmic scale. (From Reiter 1963.)

this is more directly related to the inotropic effect; overall force may be altered by changes in the time during which maximum contraction is attained. The glycoside shifts the calcium response curves upwards, which means that the inotropic action is virtually independent of the calcium concentration and of the ratio $[Ca]/[Na]^2$. The inotropic effect is diminished only at the beginning and at the end of the calcium response curves, and this indicates that an inotropic effect cannot be observed in these two conditions: when there is no calcium in the extracellular fluid, and at a very high calcium concentration which already causes maximal force of contraction. It is evident that the glycoside dependent shift of the calcium response curve is less pronounced at the lower sodium concentration.

Cardioactive steroids and the Na/Ca exchange mechanism

The sodium dependence of the inotropic effect is related neither to the external calcium concentration nor to the height of the contraction before the steroid is added. In its sodium dependence, the inotropic effect of the steroid differs not only from that of a high external calcium concentration but also from the inotropic effect of adrenaline which is completely independent of the sodium concentration in the bathing solution (Reiter & Schöber, 1965).

Does this sodium-dependence throw any light on the mechanism underlying the inotropic action of the glycosides? As pointed out earlier, the mechanism should be one by which the cardioactive steroids can produce an increase in the intracellular calcium concentration during depolarization. Measurements on calcium fluxes through the membranes of erythrocytes (Schatzmann, 1966) and of squid axons (Baker, Blaustein, Hodgkin & Steinhardt, 1967) did not reveal any direct effect of glycosides. The cardioactive steroids, however, inhibit the outward fluxes of sodium (that is, the 'sodium pump'). The question, then, is whether by an inhibition of the sodium pump the glycosides can indirectly lead to an increase in the net influx of calcium. This idea is supported by the evidence of a calcium–sodium exchange through different membranes. The investigations of Baker *et al.* (1967, 1969) led to a clear distinction in squid axon between two different outward fluxes of sodium, one ouabain-insensitive, which depends on external calcium, and a Na–K coupled pump which is inhibited by ouabain. The calcium influx via the sodium–calcium exchange mechanism can be increased by raising the intracellular sodium concentration. Evidence for the existence of a sodium–calcium exchange in guinea-pig atria was obtained by Reuter & Seitz (1968). Glitsch, Reuter & Scholz (1969) observed an increase in calcium uptake by this tissue when its sodium content was raised by different procedures. The inotropic effect may then be explained by assuming that inhibition of the Na-pump leads to an increase in internal Na which in turn increases the influx of calcium. Considering the extracellular sodium–calcium antagonism for the carrier system, it would be tempting to speculate that a similar competition between Na and Ca also exists inside the membrane and that a rise in internal sodium concentration would not only lead to an increase in Ca^{2+} influx but also to a reduction of its efflux, which would further increase the net uptake of calcium.

The sodium dependence of the inotropic glycoside effect can be explained on the basis of a sodium influx which is proportional to the extracellular sodium concentration. When the sodium influx is reduced as a

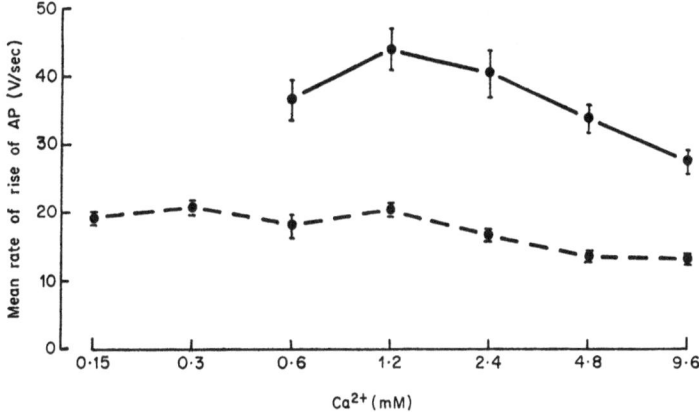

FIG. 5. Mean rate of rise of the action potential (rate of depolarization) in guinea-pig papillary muscle at different concentrations of sodium: ●—●, 140 mM Na^+; ●---● 70 mM Na^+. (From Stanley & Reiter, 1965.)

result of lowering the external Na concentration, an inhibition of the Na pump would lead to a smaller rise in $[Na]_i$ and, consequently, to a smaller increment in calcium influx. Evidence for a reduction in sodium influx after lowering the extracellular sodium is provided by the fact that the mean rate of depolarization in various tissues, including guinea-pig papillary muscle incubated in 70 mM Na^+, is only 50% of that in 140 mM Na^+ (Fig. 5).

The hypothesis that inhibition of the Na pump may increase the net influx of Ca by raising $[Na]_i$ is not invalidated by the fact that many investigators failed to find an increase in the sodium content of cardiac muscle during the action of a 'therapeutic' concentration of a glycoside. For the mechanism postulated here it should be sufficient if the slowing of the pump produces an increase of the sodium concentration near the membrane during the contraction period, simply by not pumping away fast enough the sodium that has entered during depolarization.

Antagonism of cardioactive steroids by potassium ions

There still remains an important question: if an increase in $[Na]_i$ caused by an inhibition of the sodium pump during the contraction period cannot be

10—C.C.F

measured directly, what evidence is there for the assumption that an interference with the sodium pump is really taking place at steroid concentrations which could only cause a low inotropic effect? Although the investigations by Glynn (1964) of the inhibition of the Na–K pump in erythrocytes and studies by Repke (1965) on a Mg^{2+} (and $Na^+ + K^+$) stimulated ATPase from membrane particles indicate that the inhibitory

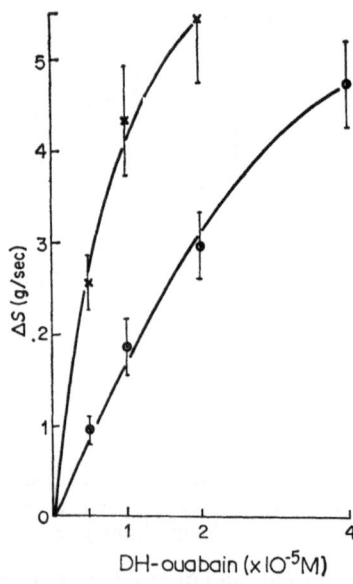

FIG. 6. Effect of $[K^+]_0$ on the inotropic action of DH-ouabain: $\times — \times$, 2·4 mM K^+; $\odot — \odot$, 9·6 mM K^+. Guinea-pig papillary muscle (mean values of twelve experiments). Ordinate: inotropic action ΔS (increase in mean contraction steepness).

activity of various steroids has the same structural requirements as their inotropic activity, this does not prove that there is a causal relationship between the two effects in cardiac muscle. Such a relationship is, however, suggested by the fact that the dependence of the two effects on external potassium is rather similar. The transport studies in erythrocytes by Glynn (1957) have shown that the inhibition of the sodium pump by glycosides can be antagonized by an increase in external potassium concentration. This observation was later confirmed by Kahn (1963). Although the reports on a potassium–glycoside antagonism in regard to the inotropic action are controversial, this antagonism can regularly be observed under appropriate experimental conditions (Caviezel & Wilbrandt, 1958; Reiter, Stickel & Weber, 1966). As shown in Fig. 6, the glycoside dependent increase in contraction steepness is significantly less with 9·6 mM K^+ than with

2·4 mM K$^+$ in the extracellular fluid. In its susceptibility to potassium the inotropic action of steroids again differs from that of adrenaline which is not influenced at all by changes in [K$^+$]$_o$ between 2·4 and 9·6 mM (Reiter *et al.*, 1966).

The antagonistic effect of potassium raises another question with regard to the proposed mechanism by which extracellular sodium influences

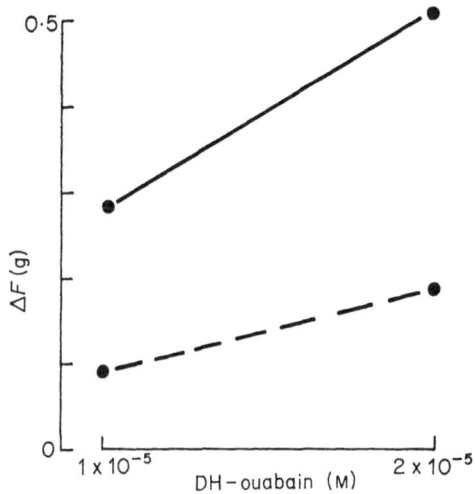

FIG. 7. Effect of reducing [Na$^+$]$_o$ and [K$^+$]$_o$ on the inotropic effect of DH-ouabain: ●—●, 140 mM Na$^+$, 5·9 mM K$^+$; ●--●, 70 mM Na$^+$, 2·95 mM K$^+$. Ordinate: increase in force of contraction (ΔF).

the inotropic steroid action. The work of Whittam & Ager (1962) and of Schatzmann (1965) suggests an antagonism between Na$^+$ and K$^+$ at the external side with regard to the activation of membrane ATPase. The question is, therefore, whether or not the sodium dependence of the inotropic steroid effect is caused by a displacement of potassium from its activating site: this possibility has been tested by simultaneously reducing external sodium and potassium (Fig. 7). Because the inotropic effects of the glycoside are greater in solutions with high sodium, even though the Na/K ratio is kept constant, it would seem that the influence of the external [Na$^+$] on the concentration of sodium inside the cell, during the contraction period, is a plausible explanation for the sodium dependence of the inotropic action of digitalis.

If the theory suggested is correct, the positive inotropic effect resulting from an extensive reduction in the extracellular potassium concentration should also be sodium-dependent because this reduction in [K]$_o$ inhibits the sodium pump. This is actually so (Table 1). The inotropic effect ΔF (increase in force of concentration) associated with lowering [K]$_o$ is

TABLE 1. *Positive inotropic effect of a reduction in* $[K^+]_o$ *and the influence of* $[Na^+]_o$ *upon it*

	140 mM Na$^+$ 3·2 mM Ca^{2+}		70 mM Na$^+$ 0·8 mM Ca^{2+}	
	5·9 mM K$^+$	1·2 mM K$^+$	5·9 mM K$^+$	1·2 mM K$^+$
MP (mV)	69 ± 1	89 ± 2	68 ± 1	89 ± 3
F (mg)	584 ± 35	838 ±	408 ± 35	366 ± 32
ΔF (%)		+ 44		− 10

Guinea-pig papillary muscle, five experiments with 140 mM Na and eight with 70 mM Na$^+$ + sucrose. 35°C, stimulation frequency 1/sec, resting tension 0·4 g (Stickel & Reiter, 1965). *MP*, membrane potential; *F*, force of contraction; Δ*F*, inotropic effect.

abolished by a reduction in the external sodium concentration from 140 to 70 mM (Stickel & Reiter, 1965). In this connection it is noteworthy that the inotropic effect resulting from an increase in contraction frequency is also sodium-dependent (Reiter, 1966).

The discovery of the existence in a cellular membrane of two different carrier systems for the extrusion of sodium ions, only one of which is susceptible to cardioactive steroids while the other is dependent on external calcium, provides, in its application to the heart muscle cell, a key for an understanding of the inotropic action of the steroids. It is important that the theory suggested does not postulate that the inotropic effect is associated with a marked change in the total Na$^+$ and K$^+$ content of the heart muscle. On the contrary, such a change, which is often observed in congestive heart failure, may be normalized with the improvement of contractility after digitalization (Clarke & Mosher, 1952). Moreover, neither the membrane potential nor the overshoot of the action potential are necessarily changing during the inotropic action of the steroids (Dudel & Trautwein, 1958).

An intracellular increase in sodium and a decrease in potassium concentration can, however, be demonstrated even with conventional methods after high steroid concentrations and probably also after an increase in stimulation frequency at low concentrations. In these situations, which are usually referred to as toxic, an increment in the resting tension of the muscle and sometimes oscillatory contractions (aftercontractions, Reiter, 1962) can be observed. For these particular mechanical phenomena a change in the calcium binding capacity of some intracellular structures (for example, mitochondria) which has been shown to occur as a result of a decrease in the intracellular K/Na ratio (Dransfeld, Greef, Hess & Schorn, 1967) may be responsible.

Summary

The inotropic effect of cardioactive steroids on the heart is dependent on the external sodium concentration. The sodium dependence of the inotropic effect cannot be explained by a sodium–potassium antagonism on an external activating site of a potassium–sodium carrier. It is assumed to be a consequence of the increase in internal sodium concentration during the depolarization caused by an inhibition of the sodium pump. The possibility is discussed that the increase in sodium leads to an increase in net uptake of calcium by a sodium–calcium exchange mechanism.

REFERENCES

BAKER, P. F., BLAUSTEIN, M. P., HODGKIN, A. L. & STEINHARDT, R. A. (1967). *J. Physiol., Lond.*, **192**, 43P

BAKER, P. F., BLAUSTEIN, M. P., HODGKIN, A. L. & STEINHARDT, R. A. (1969). *J. Physiol., Lond.*, **200**, 431

CAVIEZEL, R. & WILBRANDT, W. (1958). *Helv. physiol. Acta*, **16**, 12

CLARKE, N. E. & MOSHER, R. E. (1952). *Circulation*, **5**, 907

DUDEL, J. & TRAUTWEIN, W. (1958). *Arch. exp. Pharmak. Path.*, **232**, 393

DRANSFELD, H., GREEFF, K., HESS, D. & SCHORN, A. (1967). *Experientia*, **23**, 375

FARAH, A. & WITT, P. N. (1963). In *Proc. 1st Int. Pharmac. Meeting*, ed. WILBRANDT, W. & LINDGREN, P., vol. 3, p. 137. Oxford: Pergamon Press

GLITSCH, H. G., REUTER, H. & SCHOLZ, H. (1969). *Arch. exp. Pharmak. Path.*, **264**, 236

GLYNN, I. M. (1957). *J. Physiol., Lond.*, **136**, 148

GLYNN, I. M. (1964). *Pharmac. Rev.*, **16**, 381

KAHN, Jr., J. B. (1963). In *Proc. 1st Int. Pharmac. Meeting*, ed. WILBRANDT, W. & LINDGREN, P., vol. 3, p. 111. Oxford: Pergamon Press

LOEWI, O. (1918). *Arch. exp. Path. Pharmak.*, **83**, 366

LÜTTGAU, H. C. & NIEDERGERKE, R. (1958). *J. Physiol., Lond.*, **143**, 486.

NIEDERGERKE, R. (1963). *J. Physiol., Lond.*, **167**, 551

REITER, M. (1962). *Arch. exp. Path. Pharmak.*, **242**, 497

REITER, M. (1963). *Arch. exp. Path. Pharmak.*, **245**, 487

REITER, M. (1966). *Arch. exp. Path. Pharmak.*, **254**, 261

REITER, M. & SCHÖBER, H. G. (1965). *Arch. exp. Path. Pharmak.*, **250**, 9

REITER, M., STICKEL, F. J. & WEBER, S. (1966). *Experientia*, **22**, 665

REPKE, K. (1965). In *Proc. 2nd Int. Pharmac. Meeting*, ed. BRODIE, B. B. & GILETTE, J. R., vol. 4, p. 65. London and Prague: Pergamon Press and Czechoslovak Medical Press

REUTER, H. & SEITZ, N. (1968). *J. Physiol., Lond.*, **195**, 451

SCHATZMANN, H. J. (1965). *Biochim. biophys. Acta*, **94**, 89

SCHATZMANN, J. H. (1966). *Experientia*, **22**, 364

STANLEY, E. J. & REITER, M. (1965). *Arch. exp. Path. Pharmak.*, **252**, 159

STICKEL, F. J. & REITER, M. (1965). *Arch. exp. Path. Pharmak.*, **251**, 150

WHITTAM, R. & AGER, M. E. (1962). *Biochim. Biophys. Acta*, **65**, 383

WILBRANDT, W. & KOLLER, H. (1948). *Helv. physiol. pharmac. Acta*, **6**, 208

DISCUSSION TO SESSION IV

Mongar (*London*)

I want to discuss briefly yet another system in which calcium is required for the secretion of intracellular granules. Mast cells and basophil leucocytes, sensitized to a specific antigen, require calcium for the anaphylactic release of histamine from their basophylic granules (Mongar & Schild, 1958). Furthermore, the calcium can be replaced by strontium and antagonized by magnesium.

Figure 1 (Greaves & Mongar, 1968) shows the calcium dose-response curves for four examples of histamine-releasing systems: the mast cells of chopped guinea-pig lung, isolated peritoneal mast cells of the rat, basophil leucocytes of the rabbit and human leucocytes. They all exhibit a marked increase of release with increasing concentrations of calcium, although release from the leucocyte systems is inhibited by high calcium concentrations. Guinea-pig lung is peculiar in that the release of histamine is not completely abolished by simply removing the calcium from the incubating medium: a chelating agent, for example, 1 mM EGTA, must be used to achieve complete inhibition.

I have been looking recently at the effect of other ions on this system. Figure 2 summarizes some of the results. Release of histamine increases with increasing extracellular calcium concentration over a wide range of concentrations above as well as below the physiological range. No other ions apart from calcium are required; the system functions fairly well in isotonic sucrose.

Strontium is the only ion found to replace calcium. It is less effective than calcium; higher concentrations are needed for a similar release. Similar findings have been reported by Lichtenstein & Osler (1964) who were working with human leucocytes.

Barium will not replace calcium as it appears to do in some other systems. A concentration of 0·1 mM depresses the small residual release that is obtained when antigen is added to the lung tissue suspended in calcium-free saline solution (i.e., the Ba curve is below the hatched band).

Although magnesium cannot replace calcium it appears to augment submaximal releases which occur in low calcium concentrations. This is shown by the left-hand part of the Mg curve in Fig. 2. Much greater augmentation has been reported with human leucocytes (Lichtenstein & Osler, 1964). Concentrations of magnesium above the physiological range depresses release by antigen.

The antagonism of calcium by barium and magnesium has been studied quantitatively. In each case the log dose-response curves of calcium plus

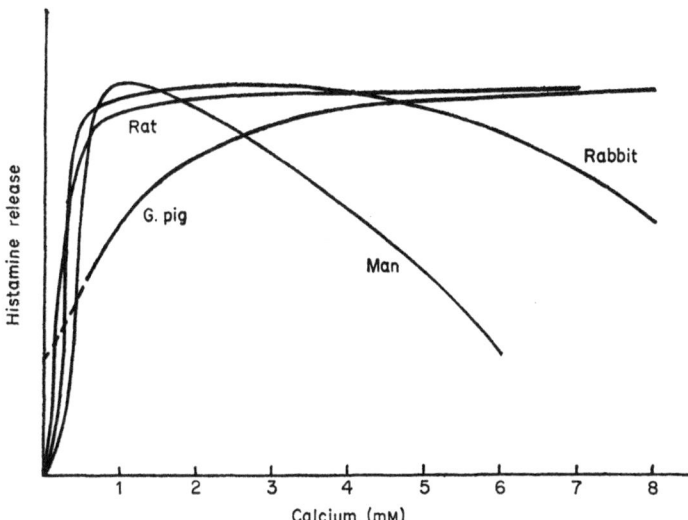

FIG. 1. Relationship between calcium concentration in medium and anaphylactic histamine release for four systems: rat = rat peritoneal mast cells; guinea-pig = chopped guinea-pig lung; man = human leucocytes; rabbit = rabbit basophil leucocytes.

antagonist are shifted to the right of the calcium curve but they remain parallel to it. If this is interpreted as a competitive antagonism at the 'calcium receptor' a dissociation constant can be determined for each antagonist. A value of about 3 mM is obtained for the Mg ion and 0·1 mM for the Ba ion. The slope of the plot 'dose-ratio minus one' against the concentration of antagonist (both on a log scale) is about unity for Mg as would be expected from a simple competitive mode of action (Arunlakshana & Schild, 1959). Barium gives a much smaller slope on this plot and it must be assumed that it has a more complex mode of action.

On the whole these effects of the alkaline earth metals on the anaphylactic release of histamine are rather similar to their effects on other systems. Strontium is able to replace calcium in all systems in which this ion has been studied but it is much less effective in some situations, for example, at motor nerve terminals (Dodge, Miledi & Rahamimoff, 1969) and on guinea-pig mast cells. On the chromaffin cells strontium behaves in all respects like calcium (Douglas, 1968).

Magnesium is well known to be an antagonist of calcium. At the motor nerve terminal this antagonism has been studied quantitatively by Jenkinson (1957) and by Dodge & Rahamimoff (1967). The latter authors found the magnesium dissociation constant to be 3 mM. This is the same as the value obtained in the present experiments on histamine release from guinea-pig lung.

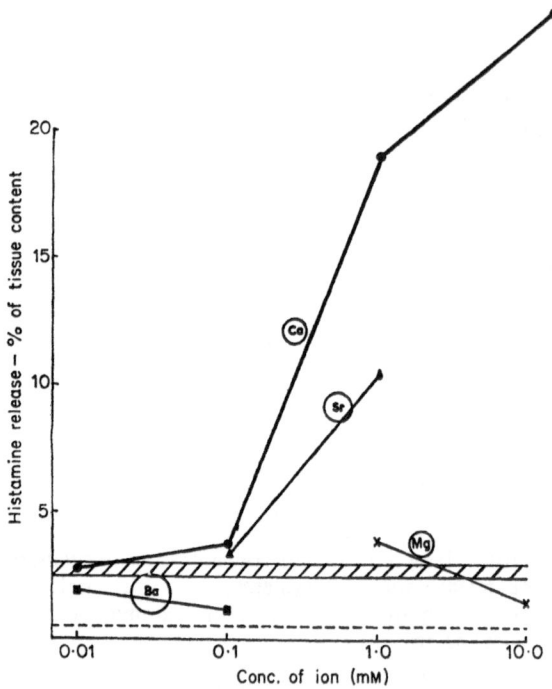

FIG. 2. Relationship between alkaline earth concentration and anaphylactic histamine release from chopped guinea-pig lung. The hatched band represents the approximate level of release in solutions free of any divalent ion and the dotted line the release when EGTA is added to the incubating solutions. The circle around each atomic symbol represents the approximate ionic radius of that element.

Magnesium has a dual role on secretion from the perfused adrenal medullary gland. It protects the chromaffin cells from the changes that would otherwise occur in them on calcium deprivation and which cause them to secrete when calcium is re-introduced. A somewhat similar secondary role can be inferred for physiological concentrations of magnesium in the anaphylatic system. The antagonistic action of magnesium on chromaffin cells can be seen when it is added to the normal perfusing solution when it inhibits the secretory response to acetylcholine.

The action of barium is not at all clear, probably because it has a direct stimulant action of its own on many tissues. This was certainly the case with the perfused adrenal gland (Douglas & Rubin, 1964) but when tested at low concentrations barium appeared to replace calcium. At motor nerve terminals barium could only occasionally be shown to replace calcium (Miledi, 1966). The anaphylactic release of histamine seems to be the only case where a powerful inhibitory action has been reported.

The ability of the alkaline earth metals to substitute for calcium seems to be related to the size of their ions. Strontium, the only effective substitute, differs only slightly from calcium (ionic radius 1·13 as compared with 0·99 Å). Magnesium with its antagonistic action is much smaller, only 0·69 Å. Barium on the other hand is much bigger than calcium—1·35 Å. Why this should result in its being a stimulant in some systems and an antagonist in others is a mystery; so little is known about the receptive substance on which calcium acts, despite its being so widely distributed in secretory systems.

Jones (Aberystwyth)

Dr. Rahamimoff, have you considered the possibility that the kinetics you describe may be those of Ca–ATP formation. I ask this because I am reminded of the fact that Kuperman, Okamoto & Gallin (1967) recently showed that in Ca^{2+} deficient nerve preparations, ATP inhibits firing, and these authors ascribed this to a direct effect of ATP on the membrane. Is it possible that here divalent cations are acting as very subtle ATP buffers, and thus, affecting ATP-membrane effects?

Rahamimoff (Jerusalem)

This is a very interesting suggestion but I do not have at the moment any experimental data to support it.

Jones (Aberystwyth)

I was interested in Dr. Woodin's statement that if leucocidin-treated leucocytes are cooled to 0°C, there is no longer any need for added ATP in order to accumulate Ca^{2+}. I would like to suggest an explanation. In a recent paper (Jones, 1969) I demonstrated that intracellular ATP levels rise on cooling. It may be, therefore, that the increase in intracellular ATP meets the Ca^{2+} binding requirement.

Woodin (Oxford)

A fall in temperature is not likely to stimulate ATP synthesis. In the experiments of Jones (1969) perhaps the higher ATP levels in the cold cells results from reduced rates of hydrolysis compared with the controls. The calcium accumulation I have described is considered to result from the accumulation of orthophosphate before the cooling of the leucocidin-treated cells.

Baker (Cambridge)

I should like to ask two rather general questions. First, is there any clear evidence that Ca ions must enter a cell to effect release of nervous transmitters or hormones? And second, what effects do metabolic poisons have on secretory processes? Slowing metabolism might be expected to increase the intracellular Ca concentration both by a reduced rate of Na extrusion favouring Ca entry into the cells and also by causing release of

Ca from intracellular stores. Both processes should lead to an increased rate of secretion provided the secretory mechanism itself is not very dependent on metabolism. In asking this question I have in mind the observations of Comline & Silver (1966) that anoxia increases catecholamine secretion from the foetal, but not the adult, gland. It seems possible that the foetal medulla obtains its energy primarily from oxidative metabolism whereas the adult may use alternative sources of energy.

Banks (Sheffield)

There is no doubt that Ca^{2+} ions are required for the release of catecholamines but it is not possible to say whether they enter the cell sap or whether they remain closely associated with the cell membrane. The relationship between Ca^{2+} and Na^+ in secretion is described more fully in my paper (see Banks, p. 148).

Vrbova (Birmingham)

Dr. Lüttgau, do you know whether caffeine enters muscle fibres at high concentrations of external calcium?

Lüttgau (Bochum)

Your question has—as far as I know—not yet been answered by direct measurements. Our experiments, which show that external Ca ions only modify the action of caffeine, suggest, however, that the drug reaches the site of action even at a high concentration of external calcium.

Caldwell (Bristol)

I would like to add a comment on Professor Lüttgau's paper.

The results of experiments in which caffeine is injected in crab muscle fibres also lead one to suspect that caffeine acts by displacement of calcium from a binding site in muscle rather than by inhibition of calcium uptake by the sarcoplasmic reticulum. This injection of various quantities of caffeine produces immediate contractions which last about the same length of time as those produced by the injection of an equivalent quantity of calcium (Caldwell & Walster, 1963). Further, the contraction produced by the injection of caffeine is suppressed if an amount of EGTA equivalent to the caffeine is injected with it (Ashley, quoted by Caldwell, 1964).

Rüdel (London)

May I ask Dr. Reuter whether the contraction associated with the Ca-current in low Na is different in its time course from the contraction in normal sodium. Is it sustained as long as Ca-current flows?

Reuter (Bern)

Compared with Na-containing solution the time course of contraction is slightly prolonged in fibre bundles soaked in Na-free solution. However, relaxation does occur even though there is still a small calcium inward current left after several hundred milliseconds depolarization. Therefore, one must assume that the relaxation mechanism is not much affected in the absence of external Na.

Hodgkin (Cambridge)
Dr. Reuter, does driving the internal potential more positive than the calcium equilibrium potential reduce or abolish the contraction?

Also do metabolic inhibitors lower the apparent calcium equilibrium potential—as one might expect if they release calcium from reticulum or vesicles?

Reuter (Bern)
Yes, contraction is markedly reduced at internal potentials in the neighbourhood of the calcium equilibrium potential or beyond it. Moreover, upon repolarization a large tail of calcium inward current flows which activates a secondary contraction even at this negative potential level. We did not use metabolic inhibitors.

Baker (Cambridge)
I should like to ask Dr. Tomita about the model he presented. He showed a site at the outside of the membrane at which competition between external Na and Ca ions determines whether or not a 'Ca-channel' is open in the membrane. As the conditions favouring the appearance of Ca-channels appeared to be those which also favour passive Ca entry into cells, I wondered whether a change in internal Ca might play some part in controlling the appearance of Ca-channels.

Tomita (Kyushu)
We need more experiments before we can speculate on the mechanisms controlling the Ca channels.

Caldwell (Bristol)
Concerning Dr. Reiter's paper, Dr. Marianna Talbot (1968) has found a similar sensitivity of glycoside action to sodium concentration with isolated frog ventricle strips. On lowering the external sodium concentration (but keeping Na/Ca constant) progressively longer latent periods were observed before a given concentration of ouabain produced an inotropic effect. In other work she has also obtained some evidence that the onset of the inotropic effect under these conditions may coincide with a reduction in potassium influx and hence with the onset of inhibition of sodium transport.

Baker (Cambridge)
There is another possible explanation of the reduced effectiveness of glycosides in low Na media, namely that the rate of interaction between

glycosides and the sodium pump may depend on the sodium concentration in the medium. In squid axons, the rate of inhibition of the sodium pump is dependent on the sodium content of the bathing fluid, being fast in NaCl and slow in choline or dextrose (Baker & Manil, 1968).

Reiter (Munich)
In our experiments with guinea-pig papillary muscle we did not find a difference in the rate of inotropic action with the fast acting glycoside DH-ouabain when the sodium concentration was reduced to 50% (see Fig. 3). It clearly was the absolute amount of increase in force of contraction which proved to be sodium dependent. I, therefore, doubt that in cardiac muscle a change in rate of interaction between glycosides and the sodium pump is decisive for the reduced inotropic effectiveness of cardioactive steroids in low Na media.

Bouman (Amsterdam)
Dr. Reiter has spoken about competition between sodium and calcium. Where does he think this competition is localized? Is there a competition to enter the cell, or is there also a competition for exit from the cell?

Reiter (Munich)
The antagonism between Na^+ and Ca^{2+} in the extracellular fluid in regard to the contractile force of the heart muscle indicates a competition between both ions in entering the cell. Assuming a sodium–calcium exchange mechanism in the membrane one should expect the existence of a similar competition between both ions inside the membrane which should effect the efflux of these ions. The calcium storage by fragmented sarcoplasmic reticulum, however, is apparently not affected by Na^+ (Martonosi & Feretos, 1963).

REFERENCES

ASHLEY, C., cited by CALDWELL, P. C. (1964). *Proc. R. Soc. B*, **160**, 512
ARUNLAKSHANA, O. & SCHILD, H. O. (1959). *Br. J. Pharmac., Chemother.* **14**, 48
BAKER, P. F. & MANIL, J. (1968). *Biochim. biophys. Acta*, **150**, 328
CALDWELL, P. C. & WALSTER, G. (1963). *J. Physiol., Lond.*, **169**, 353
COMLINE, R. S. & SILVER, M. (1966). *J. Physiol., Lond.*, **183**, 305
DODGE, F. A. & RAHAMIMOFF, R. (1967). *J. Physiol., Lond.*, **193**, 419
DODGE, F. A., MILEDI, R. & RAHAMINOFF, R. (1969). *J. Physiol., Lond.*, **200**, 267
DOUGLAS, W. W. (1968). *Br. J. Pharmac.* **34**, 451
DOUGLAS, W. W. & RUBIN, R. P. (1964). *J. Physiol., Lond.*, **175**, 231
GREAVES, M. W. & MONGAR, J. L. (1968). In *Immunopharmacology*, p. 50. London: Pergamon Press

JENKINSON, D. H. (1957). *J. Physiol., Lond.*, **138**, 434
JONES, P. C. T. (1969). *J. cell Physiol.*, **73**, 37
KUPERMAN, A. S., OKAMOTO, M. & GALLIN, E. (1967). *J. cell Physiol.*, **70**, 257
LICHTENSTEIN, L. M. & OSLER, A. G. (1964). *J. exp. Med.*, **120**, 507
MARTONOSI, A. & FERRETOS, R. (1963). *Fedn Proc.*, **22**, 352
MILEDI, R. (1966). *Nature, Lond.*, **212**, 1233
MONGAR, J. L. & SCHILD, H. O. (1958). *J. Physiol., Lond.*, **140**, 272
TALBOT, M. S. (1968). *Nature, Lond.*, **219**, 1053

INDEX

Acetates, alkaline, earths, 8
Acetylcholine,
  action on guinea-pigs uterine smooth
    muscle, 69, 70
  calcium-dependent release, dependence
    upon intraneuronal sodium con-
    centration, 155
  contractile action, magnesium effect
    on, 215
  effect on adrenal medulla calcium
    exchange, 199
  effect on calcium-dependent secretion
    of catecholamines from the adre-
    nal medulla, 149, 150, 151
  effect on freezing and thawing of
    guinea-pig uterine smooth muscle,
    65, 66
  effect on pancreas calcium exchange,
    199
  effect on salivary gland calcium ex-
    change, 199
  ouput, quantitative relation between
    calcium and, 134, 140
  release, and neuromuscular junction,
    factors affecting, 144
    cooperation of calcium and strontium
      in, 146
  response of adrenal medulla, to, 148
  storage in cholinergic neurone, 172
  synaptic release, calcium dependence,
    163
Acidophil, storage of growth hormone
  by, 172
Actin, similarity to major microtubules
  protein, 210
Actinomysin system, effect of caffeine on,
  241
Action potential,
  and calcium, 249
  effect of differing sodium concentra-
    tions on, papillary muscle, guinea-
    pig, 275
  effect of sodium concentration, taenia
    coli, 250, 251
Acyl phosphatase,
  leucocyte, 184
  on isolated cell membrane, 187
Adenohypophyseal cells, calcium effect on
  secretion by, 166

Adenohypophysis,
  calcium-dependent release of thyroid-
    stimulating hormone from, 163
  calcium exchange, effect of LHRF on,
    199
  effect of TRF on, 199
  storage of hormones, in, 172
Adenyl cyclase,
  effects of hormones on, 201, 210
  relationship between calcium and, 202,
    209, 211
  requirement in relationship between
    cyclic AMP and calcium, 200
ADP, effect on calcium transport in
  sarcoplasmic vesicles, 81
Adrenal glomerulosa, calcium exchange,
  effect of angiotensin on, 199
Adrenal medulla,
  calcium-dependent release of adrenaline
    and noradrenaline from, 163
  calcium exchange, effect of acetyl-
    choline on, 199
  calcium role as coupling agent for
    stimulus-secretion in, 148–160
  effect of cardiac glycosides on, 105
  model for events associated with
    catecholamine release from, 180
  secretions, calcium influx, sodium and,
    148
Adrenaline,
  and the heart, 203
  calcium-dependent release from the
    adrenal medulla, 163
  effect on heart calcium exchange, 199,
    203
  inotropic effect, 274
  lack of susceptibility to potassium
    compared to steroids, 277
Adrenergic neurone, storage of noradrenaline in, 172
Adrenocorticotrophic hormone,
  effect on cellular calcium exchange, 199
  storage, 170, 172
Aequorin,
  calcium luminescence, 42
  kinetics of reaction with calcium, 42–
    43
  molecular weight, 42
  source, *Aequorea forskalea*, 42

Age, effect on calcium transport in intestine, 114

Aldolase, distribution of, 188

Algal limestones, calcium occurrence as, 4

Aliphatic amines, displacement of calcium ions from phosphotidylinositol by, 38

Alkaline earth metals, effect on histamine release, 281, 282

Alkaline earths, stabilities of complexes, 8

AMP, release with amines, 159

AMP, Cyclic, 198–213

Amphetamine,
effect on calcium-dependent catecholamine secretion from the adrenal medulla, 149
induction of catecholamine secretion from adrenal glands by, 159

Amylase, relationship between calcium, cyclic AMP and, 200

Anaphylaxis,
histamine release, relationship between calcium concentration in medium and, 281
magnesium role in, 282
requirement of calcium for, 280

Angiotensin,
effect on adrenal glomerulosa calcium exchange, 199
effect on calcium-dependent catecholamine secretion from the adrenal medulla, 149

Anoxia, effect on catecholamine secretion, 284

Arginine-vasopressin, potentiation by magnesium, 215

ATP,
and calcium, 189
as chelator of calcium, 202
calcium uptake restoration by, 192
dependence of calcium extrusion from red cells on, 91
effect of sympathomimetic compounds on release from chromaffin granules, 159
fall in contents after catecholamine secretion by adrenal medulla, 158
hydrolysis enabling protein to pass out, 193
in calcium ion uptake, 242
in cell membrane, 180
level rise on cooling, leucocidin-treated cells, 283
release from chromaffin granules, 157, 160

release of hormone from isolated chromaffin and neurohypophyseal granules, by, 179
role in secretory process of acetylcholine, 159

ATPase,
activity, effect of phospholipases on, in sarcoplasmic vesicles, 82
in chromaffin granules, 158
in sarcoplasmic vesicles, restoration by unsaturated fatty acids and lysolethicin, 83
adenyl cylase, as, 202, 203
calcium-magnesium-activated, 92
effect of EDTA on, during tissue freezing and thawing, 73
effect of ouabain on, 184
in cell membrane, 180
leucocyte, 184
Mg^{2+} stimulated, effect of steroids on, 276
membrane, activation, 277
inhibition, 85

Ca–ATP, formation kinetics, 283

Calcium–ATPase, sensitivity to replacement of potassium by sodium, 125

Ca–Mg activated ATPase,
effects of mersalyl, 92, 93
inhibition by ethacrynic acid, 92
sensitivity of cardiac glycosides to, 92
stoichiometry, 93

Mg^{2+}-ATPase,
activation by low concentrations of calcium, 160
hydrolysis of membrane-bound ATP by, 159

Barium,
effect on cytoplasmic viscosity, 175
effect on smooth muscle, 257
effect on spike generation, 252, 259
inability to replace calcium, for secretion of histamine, 280, 282, 283

Barnacle muscle fibres, effect of EGTA on, 13

Basophil,
requirement of calcium for intracellular granule secretion, 280
storage of follicle-stimulating hormone and luteinizing hormone by, 172

Benzyl alcohol,
analysis of distribution between lipid and protein in intact erythrocyte, 238

Benzyl alcohol—*contd.*
and calcium, effect on erythrocyte membrane, 221
aromatic resonance of, 225, 226
nuclear magnetic relaxation, 226
calcium effect on action on haemolysis of erythrocyte membrane, 230
effect on membrane preparations containing in steroid nitroxide, spin labels, 236, 237
effects resembling xylocaine, 229
fluidizing effect on environment of the spin label in the ememmbrane, 233
fraction bound to the erythrocyle membrane, 225, 226
interaction with membrane, 238
effect of calcium on, 237
irreversible changes in the membrane induced at lytic concentrations, 238
irreversible perturbation of the membrane by, 228
line widths of aromatic protons of, in various membrane preparations, 227, 228, 229
partition coefficients, 228
relaxation changes caused by different membranes, 239
Beryllium, effect on transmitter liberation, 144
Bladder, toad, calcium exchange, effect of vasopressin on, 199
Bradykinin,
effect on calcium-dependent catecholamine secretion from the adrenal medulla, 149
polypeptides, magnesium effect on, 215
Brownian motion, 174, 175, 176, 185
Brucine, effect on Ca ions in phospholipid monolayers, 39

C-fibre, mammalian, calcium-dependent sodium efflux, 99
Caffeine, 241
action on activation of phasic muscle fibres, 241
and muscle contraction, 13, 15
effect on calcium transport system, 94
effect on coupling mechanism between excitation and contraction, 241
effect of external calcium on entrance into muscle fibres, 284
inhibition of phosphodiesterase by, 202
mode of action, 244
in crab muscle, 284
site of action, 243

Calcium,
and local anaesthetics, interaction with membrane, 229
and neuromuscular transmission, 131, 146
activity electrode, 108, 109
-binding protein, 120, 126, 127
binding to phosopholipid monolayers, 27, 28
bond strengths, 6
complexes, rate constants for formation, 54
co-ordination number, 7
current, 263, 265, 266
dependent sodium efflux, frog heart, 99
fluxes in red cells, 90
hydration of ions, 54
in chromaffin granules, 160
ionic potential, 5
kinetics of aequorin reaction with, 42, 43
occurrence as; Weddellite, 3; Whewellite, 3; algal limestones, 4
oxygen ligands, 8
polarizability, 7
relationship between cyclic AMP and, in physiological systems, 200
storage in sarcoplasmic vesicles, dependence on concentration, 79
transport, 75, 85–95, 108, 110–119, 114–116, 118, 120
water exchange, 8
Carbamyl choline,
competition with calcium in smooth muscle freezing and thawing, 71
effect on calcium-dependent catecholamine secretion from the adrenal medulla, 149, 152, 153, 155, 156
Cardiac glycosides,
actions explanation, 105
effect on adrenal medulla, 105
effect on calcium uptake of cerebral cortex slices, 105
effect on cardiac muscle calcium uptake, 106
effect on neuromuscular junction, 105
effect on pancreas, 105
effect on quantal content, 143
sensitivity of Ca-Mg-activated ATPase to, 92
Cardiac muscle,
calcium current in, 261–269
calcium current, time course of, 266
effect of cardiac glycosides on calcium uptake, 106

Cardiac muscle—*contd.*
  effect of tetrodotoxin on the inward
    sodium current, 262
  inward calcium current, 263
  inward sodium current, 262
Cardioactive steroids, 270–279
  and the Na/Ca exchange mechanism,
    274
  antagonism, by potassium ions, 275
  inotropic action, 270
  sodium ions and the inotropic effect of,
    271
  cardiolipin, effect of calcium on, 32
Catecholamine,
  effect of anoxia on secretion of, 284
  model for release from adrenal medulla,
    180
  release, 146–160
    by sympathomimetic compounds,
      159
  storage in chromaffin cell, 172
Cell, effect of EDTA on surface potential,
    60
Cell adhesion,
  and divalent cations, 59
  and ionic bridging, 60
Cell surface charge, and hormone release,
    177
Cerebral cortex slices,
  effect of cardiac glycosides on calcium
    uptake, 105
  effect of ouabain on calcium uptake by,
    105
Cerium, displacement of spin label by,
    233
Cetyltrimethylammonium, displacement
    of Ca ions from phosphatidylino-
    sitol, by, 36, 37
Chlorpromazine,
  displacement of calcium ions from
    monolayers, by, 35
  displacement of spin label by, 233
  effect on calcium transport in sarco-
    plasmic vesicles, 81
  effect on calcium transport system, 93
Cholesterol, effect of calcium, on 32, 33
  in hormone secretion, 173
  release from chromaffin granules, 158
  stabilization of lysosomes by, 190
Choline,
  effect on inward calcium current, cardiac
    muscle, 265
  effect on sodium–calcium exchange, 97
Choline plasmalogen, calcium binding, 18,
    19

Cholinergic neurone, storage of acetyl-
    choline in, 172
Chromaffin cell, storage of catecholamine
    in, 172
Chromaffin granules, 157
  ATPase in, 157, 158, 179
  calcium in, 160
  ionic relationships, 159
  membrane, phospholipid and magne-
    sium-activated ATPase in, 180
  release of chromogranins from, 157
  removal, of calcium from, 160
  surface charge and hormone release,
    177
Chromogranin A, release, during cate-
    cholamine secretion by adrenal
    medulla, 158
Chromogranins, release from chromaffin
    granules, 167
Chromophobe, storage of adrenocorti-
    cotrophic hormone in, 172
Cobalt, action in potentiating S-S poly-
    peptides, 215
Cobalt hexamine, action in potentiating,
    S-S polypeptides, 215
Colchicine,
  and the role of calcium and micro-
    tubules in mitosis, 210
  blocking action, 209, 210
Contraction activation, 241
Contracture, caffeine induced, 241
Cortisol, effect on transcellular calcium
    transport, 198
Cortisone, stabilization of lysosomes by,
    190
Coupling mechanism, between excitation
    and contraction, effect of caffeine
    on, 241
Crab axon, *Maia*, 99, 100, 101
Cyanide,
  effect on calcium efflux into squid and
    crab nerves, 100, 104
  effect on sodium–calcium exchange,
    97
  poisoning, responses of sodium and
    potassium effluxes from squid
    axons, 101
Cysternae,
  caffeine effect on calcium release from,
    243
  terminal, as part of sarcoplasmic
    reticulum, 243
Cytoplasm, effects on EDTA on, 62
Cytoplasmic viscosity in different cell
    types, 175

Debye–Hückel constant, 177
Deoxyribonuclease, specific activity, 188
Desaminooxytocin, effect of magnesium on the activity of, 214
Detergents, effect on slime mould amoebae, 63
DFP,
  effect on protein secretion, 190
  enhancement of β-glucuronidase release, by, 186, 187
  enhancement of leucocidin by, 187
Dicetylphosphate, calcium binding, 26
Dicetylphosphoric acid, calcium binding, 19
  effect of calcium on, 32, 33
  effect on monolayer surface potential, 21
  surface ionization, 29
Dibutryl 3′5′-AMP, 209
Dictyostelium discoideum, cells, effect of EDTA on surface properties of, 61
Digitalis, inotropic action sodium- dependence, 277
Dihydro-ouabain,
  and intracellular calcium ions, 271
  effect of sodium on inotropic action rate, 286
  effect on papillary muscle, guinea-pig, 270
  inotropic effect, 273, 276
  sodium concentration effect on action on calcium-response curves of guinea-pig papillary muscle, 273
Diisopropyl phosphorofluoridate, 183, 184
Dimethyl sulphoxide, effect on freezing experiments, 65, 66, 73
Dinitrophenol,
  effect on calcium transport system, 94
  effect on sodium–calcium exchange, 97
Dipalmitoyl lecithin, effect of calcium on, 32, 33
Disaggregation, of cells by EDTA, 59, 60, 62
Dodecyl trimethyl ammonium, displacement of spin label by 232, 233

EDTA,
  association constants, 10–12
  calcium removal from chromaffin granules by, 160
  effect on ATPase during tissue freezing and thawing, 73
  effect on cell mebrane permeability, 63
  effect on cell Zeta potential, 60

effects on cytoplasm, 62, 63
effect on leucocyte, 185
effect on oscillatory activity in calcium- free solution, 254
effects on pseudopodia, 62
effects on slime mould, *Dictyostelium discoideum*, 63
effect on surface properties of slime mould cells, 60
effect on toad bladder cell surface, 61, 62
Egg, sea urchin, relationship between calcium, cyclic AMP and, 200
EGTA,
  and muscular contraction, 12, 13
  association constants, 10–12
  effect on calcium efflux in squid and crab nerves, 100
  effect on oscillatory activity in calcium- free solution, 254
  effect on sodium–calcium exchange, 97
  effect on sodium efflux, 102
  inhibition of histamine release by, lung, 280
  suppression of caffeine action by, crab muscle, 284
  use in cell disaggregation, 59, 60
Eicosanyltrimethylammonium, calcium binding, 26
Electrical activity,
  effect of manganese on, mouse pancreas, 170
  in mouse pancreas islets, 169
Electron spin resonance, 219–241
End plate potentials, parameter for trans- mitter release, 133, 134, 135, 137, 142, 144
Epinephrine, *See* Adrenaline
Erythrocyte membranes and local anaes- thetics, 219–241
Ethacrynic acid, inhibition of Ca–Mg- activated ATPase by, 92
N-Ethyl maleimide,
  effect on granule movement in leuco- cidin-treated cell, 185
  inhibition of catecholamine secretion, 158
  penetration into cytoplasm, 186
  secretion inhibition by, 188
Exocytosis, 173, 189
  chromaffin cells, 157, 158
  role in neurohypophyseal hormone release, 174
  role of calcium in, 160
  secretion route by, 188

Fatigue, of skeletal muscle, 246
Ferritin,
　effect on secretion, 189
　role in T-system function, 243
Ferrous ion, action in potentiating S-S
　polypeptide, 215
Follicle-stimulating hormone, storage, 170,
　172
Freezing and thawing of cells, 65–71
Freezing experiments,
　effects of dimethyl sulphoxide on, 73
　effect of glycerol on, 73
Fructose, effect on calcium transport in
　intestine, 110

Glucagon,
　effect on liver calcium exchange, 199
　storage in α-cell, 172
Gluconeogenesis, 198
　relationship between calcium, cyclic
　　AMP and, 200
　renal, parathyroid hormone, calcium
　　and cyclic AMP relationship with,
　　205, 206
Glucose, concentration, relationship be-
　tween electrical activity of pancrea-
　tic islets and, 165
D-glucose, effect on action potential dis-
　charge on pancreatic islets, 164
Glucose,
　effect on calcium transport in intestine,
　　110
　stimulation of insulin secretion from
　　rabbit pancreas by, 168
β-glucuronidase,
　secretion, enhancement by DFP, 186,
　　187
　specific activity, 188
　vesicle content, 190
Glycerol,
　effect on freezing experiments, 73
　gluconeogenesis from, 206
Glycogenesis, effect of adrenaline, 203
Glycolysis, 198
Gouy equation, 31
Granules, fate of protein, 190
Growth hormone,
　release, relationship between calcium,
　　cyclic AMP and, 200
　storage, 170
　storage in acidophil, 172

Heart,
　calcium exchange, effect of adrenaline
　　on, 199, 203

frog, calcium-dependent sodium efflux,
　99
muscle, a calcium efflux from, 124
calcium influx, 124
Hela cells, parathyroid effect on calcium
　uptake by, 207
Hexadecyltrimethylammonium bromide,
　effect on slime mould amoebae, 63
Hexoses, effect on calcium transport by
　intestine, 110
Histamine,
　effect on calcium-dependent catechola-
　　mine secretion from the adrenal
　　medulla, 149
　effect in the freezing and thawing of
　　guinea-pig uterine smooth muscle,
　　65, 66
　release, effect on alkaline earth metals
　　on, 281, 282
　inhibition by EGTA, 280
　relationship between calcium con-
　　centration in medium and, 281
　requirement of calcium for, 280
　smooth muscle responses to, guinea-
　　pig uterus, 67
Hormone,
　release, and calcium, 163, 182
　in endocrine cells, calcium depen-
　　dence, 164
　storage and release, 172
Hormones,
　and calcium, 198–213
　interaction with adenyl cyclase, 201
5-Hydroxytryptamine, effect on calcium-
　dependent catecholamine secretion
　from the adrenal medulla, 149
Hypertensin, polypeptides, magnesium
　effect on, 215

Imipramine, effect on calcium transport
　in sarcoplasmic vesicles, 81
Insulin,
　effect of calcium omission on rabbit
　　pancreas, secretion of, 167
　effect of magnesium on secretion by
　　rabbit pancreas, 168, 169
　effect of ouabain on secretion by pan-
　　creas slices, 155
　relationship between calcium, cyclic
　　AMP and, 200
　release calcium-dependent, from pan-
　　creatic β-cell, 163
　from pancreas β-cell, calcium require-
　　ment, 209
　ionic requirements, 164

Insulin—*contd.*
  stimulation of secretion from rabbit pancreas by glucose, 168
  storage in β-cell, 172
Intestine, effect of vitamin D on the active transport of calcium by, 108, 111
Ionic bridging, and cell adhesion, 60
Isobutanol, effect on calcium efflux into squid and crab nerves, 100

Lactic dehydrogenase, non-release during catecholamine secretion from the adrenal medulla, 158
Langmuir adsorption isotherm, for calcium binding to monolayer phospholipids, 24
Lanthanum,
  effect on calcium of flux into squid and crab nerves, 100, 105
  effect on sodium–calcium exchange, 97
Leucine, effect on action potential discharge of pancreas islets, 164
Leucocidin,
  effect on leucocyte calcium exchange, 199
  inhibition and stimulation, 187
  leucocytes treated with, and calcium accumulation, 283
  primary action of, 185
  response of the cell to, 185
Leucocidin-treated cell, properties of, 183, 184
Leucocytes,
  calcium accumulation in, 190
  calcium exchange, effect of leucocidin on, 199
  human man, 183
    calcium dose response curve for, 280
  leucocidin-treated, properties of, 159, 184
  reactions mediated by calcium in, 187
  normal, properties, 184
    versus leucocidin-treated cell, 195
  polymorphonuclear treated with leucocidin, 183
  protein secretion in, 188
  rabbit, 183
LHRF, effect on adenohypophysis calcium exchange, 199
Liver, calcium, exchange, effect of glucagon on, 199
Lithium,
  effect on calcium efflux into squid and crab nerves, 100
  effect on sodium–calcium exchange, 97

importance in calcium-dependent sodium efflux, 98, 99
replacement of sodium chloride by, effect on catecholamine secretion from the adrenal medulla, 152, 155, 156
Local anaesthetics, 219–241
Lung, mast cells, calcium dose response for, guinea-pig, 280
Luteinizing hormone,
  release, from pituitary gland, rat, 171
  relationship between calcium, cyclic AMP and, 200
  storage, 170, 172
Luteotrophic hormone, storage, 170
Lysine-vasopressin, effect of magnesium on response to constant dose of, 215
Lysolethicin, and unsaturated fatty acids, restoration of ATPase activity in sarcoplasmic vesicles by, 83, 84
  calcium binding, 18, 19
  in hormone secretion, 173

Magnesium,
  and the secretory process, 180
  dependence of red cell calcium extrusion on, 88
  depresses histamine release, 280, 283
  effect of addition to calcium free solution, 255, 256, 258
  effect of concentration on electrical activity in mouse pancreas islet cells, 169
  effect of concentration on insulin secretion from rabbit pancreas, 169
  effect on calcium efflux into squid and crab nerves, 100
  effect on cytoplasmic viscosity, 175
  effect on glucose-induced action potentials of pancreatic isles, 165
  effect on insulin secretion by rabbit pancreas, 168
  effect on membrane, 259
  effect on membrane potentials of pancreatic islets, 165
  effect on transmitter release, 146
  effect on vasopressin response, 214, 215
  role in anaphylactic system, 282
  substitution by manganese to potentiate vasopressin effects, 214
Magnetic resonance spectroscopy, 219–241
Maia, muscle fibres, 12–14, 16

Manganese,
    action in potentiating S-S polypeptides, 215
    effect on electrical activity in mouse pancreas, 170
    effect on spike generation, 252
    inhibition of electrical activity of pancreatic islets, 166
    magnesium substitution by, to potentiate vasopressin effects, 214
Mannose,
    effect on action potential discharge of pancreas islets, 164
    effect on calcium transport in intestine, 110
Mast cells, calcium requirement for intracellular granules secretion, 280
Mechanical threshold, effect of caffeine on, 244
Melanin, calcium requirement for dispersal in melanocytes, frog, 200, 209
Melanocytes,
    calcium exchange, effect of MSH on, 199
    frog, requirement of calcium for dispersal of melanin in, 209
Melanocyte stimulating hormone,
    effect of melanocyte calcium exchange, 199
    of calcium requirement, 209
Mersalyl, effects on calcium–magnesium activated ATPase, 92, 93
Metal receptors, 214–218
'Metal-enzyme complexes', similarity with S-S polypeptide–magnesium system, 215
'Metallo-enzymes', lack of similarity to S-S polypeptide–magnesium system, 215
Methacholine, effect on calcium-dependent catecholamine secretion from the adrenal medulla, 149
3-O-methylglucose, effect on calcium transport in intestine, 110
Microtubular discharge, 173
Miniature endplate potentials, effect of ouabain on, 155
Mitochondria, electrical potential, effect on calcium distribution, 124
Mitosis, and microtubes, role of calcium in, 210
Monolayers,
    displacement of calcium from, by ions, 33–39

interfacial potential, 20, 23
phospholipid, calcium binding to, 24
    Langmuir adsorption isotherm for calcium binding to, 24
surface potential, 21, 29
    effects of diacetylphosphoric acid on, 21
surface radioactivity, 20, 21, 22
Muscarine, effect on calcium-dependent catecholamine secretion from the adrenal medulla, 149
Muscle,
    iliofibularis, frog, 242
    semitendinosus, frog, 242
    skeletal, fatigue of, 246, 247
muscle fibres,
    barnacle *Balanus*, sarcoplasmic calcium transport, 75
    effect of hypertonic salines, 50
    membrane potential responses, 43–50
    release of calcium during attraction, 43–50
    tension development, 43–50
    crab, *Maia*, sarcoplasmic calcium transport, 75
    tension development, 48
    frog, effect of hypertonic salines, 50
Myometrium, calcium exchange, effect of oxytocin on, 199

NEM, effect on calcium transport in sarcoplasmic vesicles, 81
Neostigmine, 141
Nernst equation, 109, 263
Neurohypophyseal granules, ATPase in, 179
Neurohypophysis,
    calcium-dependent release of vasopressin and oxytocin from, 163
    storage of hormones in, 172
Neuromuscular junction, effect of cardiac glycosides on, 105
Neuromuscular transmission, and calcium, 131–146
Neurone, storage of oxytocin and vasopressin in, 172, 174
Neurophysin, release of vasopressin and oxytocin from, by calcium ions, 174
Nickel, action in potentiating S-S polypeptides, 215
Nitroxide ESR spectra, 221, 223
Noradrenaline,
    release, calcium-dependent, from the adrenal medulla, 163

Noradrenaline, release—*contd.*
  from sympathetic nerves, 157
  mechanism, 157
  sodium deficiency enhancement of the release of, 157
  storage in adrenergic neurone, 172
  synaptic release calcium dependence, 163
Nuclear magnetic relaxation:
  results with benzyl alcohol, 226
  results with xylocaine, 226
Nuclear magnetic resonance,
  curves, relation to anaesthetic spectra, 231
  measurements, 225
Nucleoside diphosphatase, leucocyte, 184
Nucleoside triphosphatase in leucocyte, 184

Oligomycin, effect on calcium transport system, 94
Ornithine-vasopressin, potentiation by magnesium, 215
Ouabain, 208, 209
  effect on ATPase in leucocyt,e 184
  effect on calcium-dependent catecholamine secretion from the adrenal medulla, 151, 152, 154, 155
  effect on calcium efflux into squid and crab nerves, 100
  effect on miniature endplate potentials, 155
  effect on secretion of insulin by pancreatic slices, 155
  effect on two different outward fluxes of sodium, squid axon, 274
  inotropic effect, 285
Oxaloacetate, parathyroid normal effect on intracellular concentration changes, 208, 209
Oxytocin,
  action, effect of magnesium on, 215
  activity compared to vasopressin, 214
  calcium-dependent release from neurohypophysis, 163
  effect on myometrium calcium exchange, 199
  release from neurophysin by calcium ions, 174
  storage in neurone, 172, 174

Pack rat (*Neotoma* sp.), 121
Pancreas, 164, 165, 167
  calcium exchange, effect of acetylcholine on, 199

effect of cardiac glycosides on, 105
  electrical activity in, 169
  α-cell, storage of glucagon in, 172
  β-cell, calcium-dependent insulin release from, 163, 209
  storage of insulin in, 172
Pancreatic islets,
  action potential discharge, 164
  relationship between glucose concentration and electrical activity of, 165
  storage of hormone in, 172
Papillary muscle,
  guinea-pig, 272
  action potential rise at different sodium concentrations, 275
  comparison of effects of dihydro-ouabain and calcium ions on, 270
  effect of reduction in potassium, 278
  inotropic effect of dihydro-ouabain on, 276
Parathyroid hormone, and the renal tubule, 199, 205, 207, 211
Partition coefficient, 226, 229, 238
  membrane protein, 229
  of benzyl alcohol in intact membrane, 239
Peroxidase,
  distribution correlation with recovery of accumulated calcium, 191
  specific activity, 188
Phagocytin, specific activity, 188
Phagocytosis, 196
Phenethanolamine-N-methyltransferase, 173
Phenoxybenzamine, sodium deficiency enhancement of the release of, 157
Phenylethylamine,
  effect on calcium-dependent catecholamine secretion from the adrenal medulla, 149
  effect on catecholamine secretion by isolated chromaffin granules, 159
Phosphatase, acid and alkali, vesicle content, 190
Phosphatidic acid,
  calcium binding, 19
  phosphatidylserine, effect of calcium on, 32, 33
  surface ionization, 29, 30
Phosphatidylcholine,
  calcium binding, 18, 19, 26, 27
  surface ionization, 29
Phosphatidylethanolamine,
  calcium binding, 18, 19, 26
  surface ionization, 29

Phosphatidylinositol,
  calcium binding, 19, 24
  displacement of calcium from, by ali-
    phatic amines, 38
  by cetyltrimethylammonium, 36, 37
  by tetracaine, 36
  effect of surface pressure on calcium
    binding to, 27, 28
  surface ionization, 29
  surface potential, 31
Phosphatidylserine, calcium binding, 18,
    19, 24
  displacement of Ca ions from, 35
  effect of surface pressure on calcium
    binding to, 27
Phosphodiesterase,
  effect on caffeine on, 202
  effect of theophylline on, 202
  Phosphoenol pyruvate, carboxykinase,
    activation, 209
  carboxylase, control site in reaction
    series to glucose, 207
  parathyroid hormone effect on intra-
    cellular concentration changes, 208,
    209
Phospholipases, effect on ATPase activity
    in sarcoplasmic vesicles, 82
Phospholipids,
  calcium binding, 18, 19
  content of adrenal medulla after cate-
    cholamine secretion, 158
  in chromaffin granule membrane, 180
  in hormone secretion, 173
  release from chromaffin granules, 158
Phospholipid monolayers,
  calcium binding, 20–25
  effect of amphetamine on, 39
  effect of brucine on, 39
  effect of strychnine on, 39
Phosphorylase *b* kinase, calcium require-
    ment for, 204
Phosphorylase system, activation by
    cyclic AMB, 204
Pilocarpine, effect on calcium-dependent
    catecholamine secretion from the
    adrenal medulla, 149
S-S Polypetide receptor, 214–218
Posterior pituitary, calcium importance in
    the neurosecretory mechanisms of,
    166
Potassium ions, antagonism of cardio-
    active steroids by, 275
Potassium reduction, effect on guinea-pig
    papillary muscle, 278
Potentiation mechanism, 216

Pregnancy, effect on calcium transport in
    intestine, 115, 120
Prenylamine, effect on calcium transport
    in sarcoplasmic vesicles, 81, 82
Protein secretion, inhibition of, 189
Pseudopodia, effects of EDTA on, 62
Purkinje fibres, permeability changes, 261
Pyruvate, parathyroid hormone effect on
    intracellular concentration changes,
    208, 209
Pyruvate carboxylase, activation by cal-
    cium shift from mitochondria to
    cytoplasm, 198
Pyruvate kinase,
  control site in reaction series to glucose,
    207
  inhibition to calcium shift from mito-
    chondria to cytoplasm, 198

Quantal content,
  effect of cardiac glucosides on, 143
  parameter for transmitter release, 133,
    136, 137, 141

Red cell,
  calcium extrusion from, 87
    dependence on ATP, 91
    dependence on magnesium, 88
  calcium fluxes in, 90
  calcium uptake in, 86
  ghosts, human, 221
  intracellular calcium concentration,
    85
  membranes and local anaesthetics,
    219–241
  resealed, preparation of, 85
    transmembrane calcium movements
      in, 85–95
  strontium transport in, 899
Relaxation rate,
  membrane, 238
  proton nuclei of the molecule, 225
Renal tubule,
  and parathyroid hormone, 205, 217
  calcium exchange, effect of parathyroid
    hormone on, 199, 203
Reserpine, effect on calcium transport in
    sarcoplasmic vesicles, 81
Ribonuclease,
  specific activity, 188
  vesicle content, 190
R.P.M.l. No. 41 cells, calcium binding to,
    62

Saccharates, alkaline earths, 8
Salivary gland, calcium exchange, effect of acetylcholine on, 199
Sarcoplasmic membranes, rate of calcium transport, 75
Sarcoplasmic reticulum,
  and calcium distribution, 125
  calcium ions release from, 242
  calcium storage by, 286
  calcium transport, 75-84
  effect of steroids on calcium ion removal by, 271
  electrical potential effect on calcium distribution, 124
  inhibition of calcium uptake by, 284
  release of calcium from, 241
  terminal cysternae as part of, 243
Sarcoplasmic vesicles,
  calcium storage, dependence on concentration, 78
  calcium transport, effect of drugs, 81
  effect of phospholipases on ATPase activity, 82
  isolation of, 76
  precipitation of calcium oxalate in, 76
  restoration of ATPase activity with unsaturated fatty acids and lysolecithin, 83
  role of −SH groups, 79
Secretion,
  and calcium, 189
  of protein, 183
Semitendinosus muscle, frog, 242
Skeletal muscle,
  fibres, bullfrog, calcium binding, 34
  frog, calcium-dependent sodium efflux in, 99
  frog sartorius, transmitter release effect of calcium, 132
Slime mould,
  aggregation, relationship between calcium, cyclic AMP and, 200
  amoebae, effects of detergents on, 63
  effect of EDTA on, 63
    surface properties of, 61
  effect of hexadecyltrimethylammonium bromide on, 63
  effect of sodium dodecyl sulphate on, 63
Smooth muscle,
  action potential, 249
  effects of exposure to calcium-free solution on, 253
  effects of magnesium added to calcium-free solution, 255

effects of strontium and barium on, 257
guinea pig, uterus,
  freezing and thawing, dimethyl sulphoxide protective effect, 65
  effect of acetylcholine on, 65, 66
  effect of histamine on, 65, 66
  procedure, 66
  responses to acetylcholine, 69, 70
  responses to drugs at different calcium concentrations, 68
  responses to histamine, 67
  spontaneous activity, 66
Sodium–calcium exchange,
  effect of choline on, 97
  effect of cyanide on, 97
  effect of dinitrophenol on, 97
  effect of EGTA on, 97
  effect of lanthanum on, 97
  effect of lithium on, 97
  effect of tetrodotoxin on, 97
  mechanism, 274
Sodium
  chloride, effect of replacement on calcium transport by intestine, 116
  concentration effect on inotropic action rate of dihydro-ouabain, 286
  current, cardiac muscle, 262
  differing concentrations effect on action potential rise, guinea-pig papillary muscle, 275
  dodecyl sulphate, effect on slime mould amoebae, 63
  effect of concentration of taenia coli action potential, 250, 251
  efflux, from nerve, calcium dependence, 97
    potassium dependence, 97
  lithium importance in, 98, 99
  ions, effects on transmitter release, 139
Sphingomyelin, calcium binding, 18, 19, 26
Spike activity, in sodium-free solution in smooth muscle, 252
Spin labels, 219, 232
  displacement by cerium, 233
  response to perturbation by xylocaine, 233
Spin quantum number, of nitrogen-14, 221
Spin-spin relaxation time, 225
Squid axon, *loligo forbesi*, 96-106
Stern equation, 31
Steroid hormones, effect on transcellular calcium transport, 198

Steroid nitroxide,
  effect of benzoyl alcohol on membrane preparations containing, 234
  effect of xylocaine on membrane preparations containing, 234
  spin label, 235
Stochastic process, 174
Streptolysin O, lytic action, 190
Strontium,
  effect on calcium efflux into squid and crab nerves, 100
  effect on cytoplasmic viscosity, 175
  effects on smooth muscle, 257
  effect on spike production, 259
  effect on transmitter release, 144, 145
  influx into nerve, sodium dependent, 97
  replacement of calcium by, for secretion of histamine, 280, 281, 283
  transport in red cells, 89
Strychnine, effect on calcium ions in phospholipid monolayers, 39
Sucrose-gap method, use in investigating spike activity, 252
Surface potential of cells, effect of EDTA on, 60, 61
Sutherland's messenger concept, 201
Sympathetic nerves, noradrenaline release from, 157
Synaptic delay, 133, 134, 140

T-system,
  caffeine action on, 243
  ferritin role in function, 243
Taenia coli,
  action potential, effect of sodium concentration on. 250, 251
  effects of exposure to calcium-free solution, 253, 255
  effect of magnesium added to calcium-free solution, 255
  effects of strontium and barium on, 257, 258
  spike activity in, 252, 253
Tartrates, alkaline earths, 8
TEA, effect on protein secretion, 190
TEMPO, 232
  effect of benzyl alcohol on membrane preparations containing, 234
  effect of xylocaine on membrane preparations containing, 234
  properties with regard to lipid regions in membrane systems, 239
  spectra, 221
  spectrum in aqueous solution, 222, 223
  spin label, 220

Tetracaine,
  displacement of calcium ions from monolayers by, 36
  displacement of calcium ions from phosphatidylinositol by, 36
  effect on monolayer surface pressure, 36
tetraethylammonium, 183, 184
  action on leucocidin effects, 187
  inhibition of calcium accumulation, 193, 194, 195
2,2,6,6-tetramethylpiperidine-l-oxyl, spin label, 220
Tetrodotoxin,
  and membrane depolarization, 254, 255
  effect on calcium efflux into squid and crab nerves, 100
  effect on inward calcium current, cardiac muscle, 265
  effect on inward sodium current, cardiac muscle, 262
  effect on spike, 252
  effect on sodium–calcium exchange, 97
  use in timing calcium action in transmitter release, 133
Theophylline, inhibition of phosphodiesterase by, 202
Thyroid-stimulating hormone,
  calcium-dependent release from adenohypophysis, 163
  release, from pituitary gland, rat, 171
  relationship between calcium, cyclic AMP and, 200
  storage, 170
Thyroxine, release, relationship between calcium, cyclic AMP and, 200
Toad bladder, effect of EDTA on cell surface, 61, 62
Tolbutamide, effect on action potential discharge of pancreas islets, 164
Transmitter release,
  competition of calcium and magnesium, 136, 137, 138, 139
  effect of calcium, skeletal muscle, frog sartorius, 132
  effects of sodium ions on, 139
  effect of various different ions on, 144
  parameters, 133
  quantitative relation between calcium and acetylcholine output, 134
  timing of calcium action, 133
Transport carrier for calcium, 120
TRF, effect on adenohypophysis calcium exchange, 199
Triads, calcium accumulating part of, 243

Triphosphoinositide,
ATP effect on, 189
calcium binding, 18, 19, 27
effect on leucocidin, 185
possible physiological function for, 194, 198
(+)-tubocurarine, 145
post-synaptic sensitivity reduction with, 137
Tyramine,
effect on calcium-dependent catecholamine secretion from the adrenal medulla, 149
effect on catecholamine secretion by isolated chromaffin granules, 159
effect on chromaffin granules, 160

Unsaturated fatty acids and lysolethicin, restoration of ATPase activity in sarcoplasmic vesicles by, 83, 84
Ureter,
effects of magnesium added to calcium-free solution on, 257
effects of strontium and barium on, 257, 258
membrane depolarization, 254
spike activity in, guinea-pig, 252
Ussing flux ratio, 102
Uterus, potassium-depolarized, rat, 214

Vasopressin,
activity in solutions containing magnesium, 214
calcium-dependent release from neurohypophysis, 163
effect on toad bladder calcium-exchange, 199
effect, potentiation by manganese, 214
release from neurophysin by calcium ions, 174

response, effect of magnesium on, 214, 215
storage in the neurone, 172, 174
Ventricular myocardial fibres, calcium current in, 261-269
Vitamin A, lytic action, 190
Vitamin D,
effect on active calcium transport by intestine, 103, 111, 120, 126
effect on transcellular calcium transport, 198

Weddellite, calcium occurrence as, 3

Xylocaine,
and calcium, effect on erythrocyte membrane, 221
aromatic resonance of, 226
competititive nature of binding to membrane, 237
effect on membrane preparations containing steroid nitroxide spin label, 236
effect on membrane preparations containing TEMPO spin label, 235
effects resembling benzyl alcohol, 229
fraction bound to the erythrocyte membrane, 225, 226
N.M.R. curve, effect of calcium on, 230
nuclear magnetic relaxation, 229
response of spin label to perturbation by, 233
stabilization of erythrocyte membrane against hypotonic haemolysis, 229

Zeta potential, of cells in EDTA, 60
Zinc, action in potentiating S-S polypeptides, 215